# Topics in Applied Physics

Volume 152

Topics in Applied Physics is a well-established series of review books, each of which presents a comprehensive survey of a selected topic within the domain of applied physics. Since 1973 it has served a broad readership across academia and industry, providing both newcomers and seasoned scholars easy but comprehensive access to the state of the art of a number of diverse research topics.

Edited and written by leading international scientists, each volume contains high-quality review contributions, extending from an introduction to the subject right up to the frontiers of contemporary research.

Topics in Applied Physics strives to provide its readership with a diverse and interdisciplinary collection of some of the most current topics across the full spectrum of applied physics research, including but not limited to:

- Quantum computation and information
- Photonics, optoelectronics and device physics
- Nanoscale science and technology
- Ultrafast physics
- Microscopy and advanced imaging
- Biomaterials and biophysics
- Liquids and soft matter
- Materials for energy
- Geophysics
- Computational physics and numerical methods
- Interdisciplinary physics and engineering

We welcome any suggestions for topics coming from the community of applied physicists, no matter what the field, and encourage prospective book editors to approach us with ideas. Potential authors who wish to submit a book proposal should contact Zach Evenson, Publishing Editor:

zachary.evenson@springer.com

Topics in Applied Physics is included in Web of Science and indexed by Scopus.

Fumitaro Ishikawa · Hiroaki Ohfuji · Jun Kawano ·
Tetsuya Tohei
Editors

# Engineering Crystal Habit

## Applications of Polymorphism and Microstexture Learning from Nature

Springer

*Editors*
Fumitaro Ishikawa
Research Center for Integrated Quantum Electronics
Hokkaido University
Sapporo, Ehime, Japan

Jun Kawano
Department of Earth and Planetary Sciences
Faculty of Science
Hokkaido University
Sapporo, Hokkaido, Japan

Hiroaki Ohfuji
Department of Earth Science
Tohoku University
Sendai, Miyagi, Japan

Tetsuya Tohei
Graduate School of Engineering Science
Osaka University
Toyonaka, Osaka, Japan

ISSN 0303-4216 ISSN 1437-0859 (electronic)
Topics in Applied Physics
ISBN 978-981-96-0268-1 ISBN 978-981-96-0266-7 (eBook)
https://doi.org/10.1007/978-981-96-0266-7



This Springer imprint is published by the registered company Springer Nature Singapore Pte Ltd.
The registered company address is: 152 Beach Road, #21-01/04 Gateway East, Singapore 189721, Singapore

# Foreword

Crystals are everywhere in nature and the basis of our welfare society. Understanding the birth and growth of mineral crystals and how their morphology and microtextures are related to growth conditions is the gate to revealing bygone geochemical environments. It is also the rationale behind the endeavor of materials science and engineering to design and produce new advanced materials.

Rock-forming minerals are mostly crystalline materials. In very few cases, they are the large polyhedra that have fascinated humankind since remote times. Finding colossal crystals other than those dreamed up in literature and comic masterpieces is even more bizarre. Minerals most often occur in nature in polycrystalline masses that sometimes display morphologic patterns of enormous interest but uncertain origin. The study of these mineral patterns formed by crystalline aggregates has been neglected so far because of their inherent complexity. However, they represent an excellent opportunity to learn about the laws governing nature's symmetry. Deciphering the formation mechanism of these mineral growth patterns is a formidable challenge. However, the reward is immense.

As shown in Fig. 1, the diversity of mineral growth patterns is immense. They range from patterns controlled by the mineral crystal structure, such as those currently forming giant crystals of gypsum, to patterns with non-crystallographic symmetry, such as those controlled by the interaction of transport and chemical precipitation leading to mineral self-organization. We also may find mineral patterns fabricated by living organisms -called biominerals- with precise control of polymorphism and microtexture. Bizarre as it might seem, purely inorganic processes may also produce self-assembled complex crystalline materials—called biomorphs—that emulate biominerals' non-crystalline symmetry and textures. Decoding the information contained in these crystalline rock-forming minerals to reveal the physicochemical scenario from which they grew will help uncover our planet's geological history and learn new chemical engineering routes to materials innovation. This is a challenging task of formidable difficulty because, unlike for single crystals, the information on polycrystalline crystallization is scarce. Crystal growth studies will be much more helpful if there is an extensive exploration of polycrystalline pattern

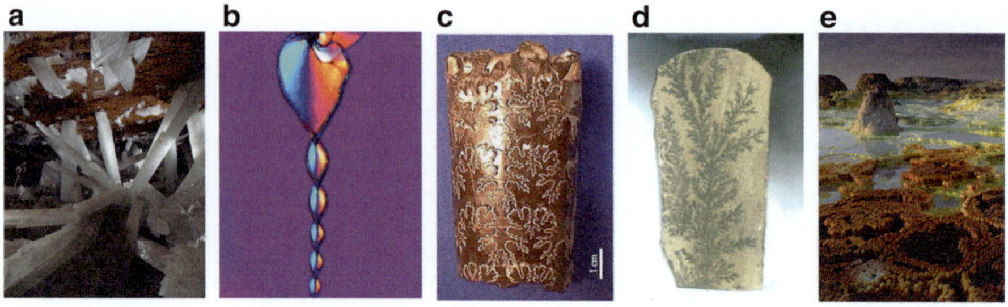


**Fig. 1** From left to right: **a** The giant crystals of gypsum in Naica, Mexico; **b** Silica-carbonate biomorphs under polarized light: **c** Suture lines in ammonite; **d** Fractal dendrites of manganese and iron oxides; **e** complex pattern formation in the Dallol salt volcano, Danakil depression, Ethiopia

formation. We need to answer questions like how minerals self-assemble into peculiar structures, how they do so when crystallizing massively from seas and lakes in magmatic of hydrothermal systems, and how they experience phase transitions during their spatiotemporal trip triggered by the slow but continuous dynamics of the planet

It is crucial to push the boundaries of crystal growth research beyond traditional studies. This requires using novel theoretical and experimental tools to decipher what mineral crystals can reveal about the dynamics of the Earth and other planets in the Solar System. Such an approach opens the possibility of conducting studies at the forefront of science and technology. And this is precisely what this book is about. It is a call to action, a challenge to explore new horizons in crystal growth research, and an opportunity to be part of groundbreaking scientific discoveries.

You are about to embark on a journey through a book that is the result of talent and unwavering determination. The editors, Prof. Fumitaro Ishikawa, Prof. Hiroaki Ohfuji, Prof. Jun Kawano, and Prof. Tetsuya Tohei, have meticulously curated an outstanding collection of contributions from experts in the subject. This book, while rooted in the classic scientific theme of crystal engineering, offers a fresh and innovative perspective. Its goal is to draw inspiration from the crystalline patterns found in nature, unravel their formation mechanisms, and apply this knowledge to the design and production of advanced materials. It is a book that will not only inform but also inspire and captivate you.

Crystal engineering has sought inspiration from ceramic materials that life has encountered over billions of years of an evolutionary process of natural selection. The manufacturing mechanisms of bones, teeth and shells still need to be fully understood but are being investigated in many laboratories worldwide. This book proposes that crystalline engineering should be inspired not only by the textural patterns found by living organisms but also by the morphological and textural assemblies in rocks. In other words, the book suggests that a geomimetic line of research should also be developed.

For this reason, the first part of the book explores the study of crystalline morphologies and textures found in rock. In this part, the reader will find excellent examples of mineral growth patterns, including the morphological richness of pyrite crystals, the mineral phases that form in volcanic settings and in the unknown but fascinating environments of extraterrestrial surfaces under vacuum. Then, in the second collection of articles of Part II, the reader will find a modern approach to the mechanisms controlling the birth and growth of minerals, exemplified by the case of aluminum and iron oxides, a beautiful demonstration of the flow of chemicals around a growing crystal and an innovative technology for the control of crystalline architectures by light-induced precipitation. Part III of the book contains several study cases reporting the successful control of polymorphism and growth patterns in fabricating advanced materials.

Juan Manuel García-Ruiz
Donostia International Physics Center
San Sebastián, Spain

# Preface

Welcome to the captivating realm of engineering crystal habit, where the inherent beauty of crystals converges with the vast potential of advanced materials. As we embark on this expedition through the intricate landscapes of crystallography, our journey aims to unveil the secrets encapsulated within the myriad forms of these natural marvels.

Within the domain of Crystal Habit Engineering for Advanced Materials, this book serves as a navigational tool through the complex terrain of crystalline structures, formation mechanisms, and the enthralling interplay between polymorphism and organization control. Our mission is to delve into the core essence of crystals, unlocking their latent potential to herald a new era of functional materials.

The odyssey commences with an introduction to the profound impact of crystallography on our comprehension of crystals. This preamble sets the stage for a thorough investigation into the formation mechanisms and shapes of crystals, unraveling their potential applications in the realm of cutting-edge functional materials.

Nature's artistic expression comes to life through crystals adopting diverse forms, ranging from needle-like structures to intricate dendrites to self-organized aasemblies. We will embark on a journey to explore the aesthetic allure of crystals, such as pyrite, diamond, garnet, corundum, and quartz, delving into their significance in advanced electronic applications, particularly within wide-gap materials like diamond and corundum.

Shifting our focus to the kaleidoscopic world of crystal colors, we delve into the mysteries of polymorphism exhibited by crystals like diamond and graphite. Advanced techniques, such as electron microscopy and photoluminescence mapping, will guide us through the correlation between crystal structure and color, providing insights reminiscent of the evaluation of optical devices in applied physics.

From the vacuum of space to the Earth's surface and deep underground, our exploration takes us through diverse environments where crystals manifest. The investigation delves into the creation of nanowhiskers and nanowires, uncovering their potential applications in semiconductor crystals and quantum structures. Needle crystals, exemplified by nanotrees, underscore the applicability of crystal engineering in creating functional semiconductor materials.

This segment leads us on a fascinating journey through crystal polymorphism, highlighting the discovery of functional semiconductors with metastable structures, such as oxides or other compound semiconductors. Controlled manipulation of metastable phase structures opens doors to inventing new materials with unique properties.

Our exploration culminates in envisioning practical applications arising from crystal growth technologies, observation techniques, and nanotechnology. By synthesizing crystals with intricate nanostructures, we anticipate the emergence of novel materials with enhanced functionalities, paving the way for innovations in light-emitting and light-absorbing devices. The collaborative spirit underlying our journey emphasizes the potential outcome of diverse knowledge and technological interactions, leading to the discovery of unexplored functional crystals.

The integration of modern knowledge and technology with classical crystallography holds the promise of discovering and improving new crystals. This collaborative effort, drawing inspiration from the diverse structures found in nature, propels us towards the creation of innovative functional crystals with applications across various fields. As you immerse yourself in the pages that follow, may the enchanting world of crystal habit engineering captivate your imagination and inspire new frontiers in advanced materials.

Sapporo, Japan — Fumitaro Ishikawa
Sendai, Japan — Hiroaki Ohfuji
Sapporo, Japan — Jun Kawano
Toyonaka, Japan — Tetsuya Tohei

# Contents

**Part I Polymorphism and Microtexture in Nature**

**1 Morphological and Textural Variety of Natural minerals—Pyrite as an Example** ..... 3
Hiroaki Ohfuji
1.1 Introduction ..... 3
1.2 Principle of Crystal Growth and Morphology ..... 4
1.3 Mineralogical Characteristics of Pyrite ..... 5
1.4 Morphology of Pyrite Crystals ..... 7
1.4.1 Single Crystal Morphology ..... 7
1.4.2 Spherulitic Aggregate ..... 8
1.4.3 Colloform Aggregates ..... 10
1.4.4 Framboidal Aggregate ..... 10
1.5 Concluding Remarks ..... 14
References ..... 14

**2 Extraordinary Crystals of Natural Minerals in Mining Sites and Volcanoes** ..... 17
Takuya Echigo
2.1 Utilization of Stable and Metastable Phases of Natural Minerals ..... 17
2.2 Natural Crystals of Organic Molecules in Hydrothermal Mercury Deposit ..... 19
2.3 Giant Plagioclase Crystals in Quenched Lava ..... 21
2.4 Conclusions ..... 25
References ..... 26

**3 Growth and Modifications of Crystals on Airless Surfaces** ......... 29
Toru Matsumoto
3.1 Extraterrestrial Materials Showing Space Weathering Features ........................................ 30
3.2 Ion Implantation in the Space Environment ................. 31
3.3 Crsytallographic Damages Formed by Solar Wind Implantation ............................................ 32
3.4 Vesiculation and Blistering by Accumulation of Solar Wind Ions ............................................ 34
3.5 Micrometeoroid Bombardment ............................ 35
3.6 Condensation Products ................................... 35
3.7 Formation of Nanoparticles of Iron Compounds .............. 37
3.8 Formation of Iron Whiskers ............................... 38
3.9 Conclusion ............................................ 39
References ................................................ 40

**Part II Nucleation and Growth of Minerals**

**4 The Chemistry of Phase Separation in Aqueous Hydrolyzing Metal Systems: Aluminium(III) and Iron(III)** ................... 45
Miodrag J. Lukić and Denis Gebauer
4.1 Nucleation of Al(III) and Fe(III) (Oxy)(Hydr)oxides from Homogeneous Aqueous Solutions .................... 47
4.2 The Role of Water in the Nucleation of Al(III) and Fe(III) (Oxy)(Hydr)oxides from Solution ......................... 62
4.3 Nucleation in the Presence of Organic Additives and Biomineralization .................................. 66
References ................................................ 71

**5 Visualization of pH and Ionic Concentration for Dissolving/Forming Crystals** ........................................ 79
Jun Kawano and Takashi Toyofuku
5.1 Visualization Using Fluorescent Probes ..................... 79
5.2 Selection of Fluorescent Probes ........................... 81
5.3 Imaging ............................................... 83
5.4 Application Examples .................................... 84
5.4.1 pH and Ionic Concentration Changes Around Dissolving Crystals ............................. 84
5.4.2 Visualization of pH Changes During Crystal Precipitation in Gels ............................ 85
5.5 Future Prospects ........................................ 88
References ................................................ 89

**6 Crystal Structure Control Technology for Organic and Biomaterials Using Light** ... 93
M. Maruyama, Y. Tsuri, H. Y. Yoshikawa, K. Takano, M. Yoshimura, and Y. Mori
6.1 Introduction ... 94
6.2 What is Crystal Polymorphism? ... 94
6.3 Indomethacin Crystal Polymorphism Control by Laser Irradiation ... 95
6.4 Attempts to Elucidate the Mechanism of Laser Nucleation Induction ... 98
6.5 Strategies for Most Difficult Polymorphism Control: Aspirin ... 102
6.6 Conclusions and Future Approaches ... 102
References ... 104

**Part III Applied Microtexture and Polymorphism**

**7 Microtexture and Polymorphism Observed During the Molecular-Beam Epitaxial Growth of Group III–V Semiconductor Nanostructures** ... 109
Fumitaro Ishikawa, Kazuki Nagashima, Takeshi Yanagida, Robert D. Richards, and Irina A. Buyanova
7.1 Molecular-Beam Epitaxy of Compound Semiconductors ... 110
7.2 III–V Compound Semiconductors ... 111
7.3 Growth of Nanostructures and Nanowires ... 112
7.4 Growth of Core–Shell Nanowires by Molecular-Beam Epitaxy Using Precisely Controlled Shell Structures ... 114
7.5 MBE Growth of Nanowires with Well-Ordered Microstructures: A Case Study of GaAs/GaNAs Core–Shell Nanowires ... 116
7.6 Microtextures Observed in the Epitaxial Nanostructure and Introduction of Bi During the Growth of GaAs Nanowires ... 117
7.7 Microstructure in GaAs/GaAsBi Core–Shell NWs ... 119
7.7.1 Using Composition and Structural Modulations of Nanowires to Form Quantum Structures ... 126
7.8 Polymorphism in Nanowires ... 127
7.8.1 Polytypism in GaAs/GaNAs Nanowires ... 127
7.8.2 Distributions of Stable ZB and Metastable WZ Phases ... 129

7.8.3 Polytypism Transfer Across the Nanowire Axis ....... 131
7.9 Functional Benefits of Polymorphism: Strong Second-Harmonic Generation in GaAs Nanowires Using Crystal-Structure Engineering ........ 133
7.10 Conclusions ........ 134
References ........ 136

**8 Crystal Phase Engineering in Self-Assisted GaAs Nanowires ...... 143**
T. Dursap, C. Botella, P. Regreny, N. Chauvin, M. Bugnet, A. Danescu, and J. Penuelas
8.1 RHEED Applied to the Nanowire Growth ........ 145
8.2 The Zinc Blende/Wurtzite Alternation ........ 148
8.3 Simulation of the Wetting Angle and Length Evolution ........ 151
8.4 Simulation of the Wetting Angle as a Function of the III and V Fluxes ........ 153
8.5 Evidences of the Pure-WZ Phase ........ 159
8.6 Conclusion ........ 165
References ........ 166

**9 Exploration of Nanowire Photonics Towards Advent of Super-Smart Societies ........ 171**
Jun Tatebayashi and Yasufumi Fujiwara
9.1 Introduction ........ 171
9.1.1 Growth and Optical Characteristics of Rare-Earth-Doped ZnO NWs ........ 173
9.2 Summary ........ 181
References ........ 181

**10 Phase-Selective Synthesis of Oxide Materials via Mist Chemical Vapor Deposition ........ 185**
Hitoshi Takane, Takeru Wakamatsu, Katsuhisa Tanaka, and Kentaro Kaneko
10.1 Introduction ........ 185
10.2 Rutile-Type Germanium Oxide ........ 186
10.3 Growth of r-$GeO_2$ Thin Film by Mist CVD ........ 188
10.4 Growth of r-$Ge_xSn_{1-x}O_2$ Alloy Thin Film by Mist CVD ........ 191
10.5 Characteristics of Crystallization of r-$GeO_2$ by the Mist CVD ........ 193
10.6 Conclusions ........ 195
References ........ 195

**11 Crystalline Microstructure and Versatile Resistive Switching Property in Rutile $TiO_{2-x}$ Four-Terminal Memristors ........ 199**
Tetsuya Tohei and Akira Sakai
11.1 Introduction ........ 199
11.2 Crystal Structure, Device Structure and Electrical Measurements ........ 201

11.2.1 Crystal Structure of Rutile $TiO_2$ and Device Geometries ........ 201
11.2.2 Basic Operation of Four-Terminals Memristive Devices ........ 202
11.2.3 Resistive Switching Reversibility in (100) and (001) Devices ........ 204
11.2.4 Resistive Switching in (001) $TiO_{2-x}$ Thin-Film Device ........ 207
11.3 Crystal Microstructure and Electronic States by STEM Observations ........ 209
11.4 Versatile Synaptic Functionality in Four-Terminals $TiO_2$ Memristors ........ 212
11.4.1 Heterosynaptic Implementation by Four-Terminals Memristor ........ 212
11.4.2 Multi-valued RS Property by Changing Sweep Rate ........ 212
11.4.3 Gate Tunable Heterosynaptic Plasticity by Pulse Drive Mode ........ 213
11.4.4 Tunable Spike-Timing-Dependent Plasticity (STDP) ........ 215
11.4.5 Pavlovian Conditioning ........ 218
11.5 Conclusions ........ 220
References ........ 221

**Index** ........ 223

# Contributors

**C. Botella** UMR 5270, Univ Lyon, Ecole Centrale de Lyon, CNRS, INSA Lyon, Université Claude Bernard Lyon 1, CPE Lyon, INL, Ecully, France

**M. Bugnet** CNRS, INSA Lyon, Université Claude Bernard Lyon 1, MATEIS, UMR 5510, Villeurbanne, France

**Irina A. Buyanova** Department of Physics, Chemistry and Biology, Linköping University, Linkoping, Sweden

**N. Chauvin** UMR 5270, Univ Lyon, Ecole Centrale de Lyon, CNRS, INSA Lyon, Université Claude Bernard Lyon 1, CPE Lyon, INL, Ecully, France

**A. Danescu** UMR 5270, Univ Lyon, Ecole Centrale de Lyon, CNRS, INSA Lyon, Université Claude Bernard Lyon 1, CPE Lyon, INL, Ecully, France

**T. Dursap** UMR 5270, Univ Lyon, Ecole Centrale de Lyon, CNRS, INSA Lyon, Université Claude Bernard Lyon 1, CPE Lyon, INL, Ecully, France

**Takuya Echigo** Graduate School of International Resource Sciences, Akita University, Akita, Japan

**Yasufumi Fujiwara** Graduate School of Engineering, Osaka University, Osaka, Japan

**Denis Gebauer** Institute of Inorganic Chemistry, Gottfried Wilhelm Leibniz University, Hannover, Hannover, Germany

**Fumitaro Ishikawa** Research Center for Integrated Quantum Electronics, Hokkaido University, Sapporo, Japan

**Kentaro Kaneko** Research Organization of Science and Technology, Ritsumeikan University, Shiga, Japan

**Jun Kawano** Department of Earth and Planetary Sciences, Faculty of Science, Hokkaido University, Sapporo, Japan

**Miodrag J. Lukić** Institute of Inorganic Chemistry, Gottfried Wilhelm Leibniz University, Hannover, Hannover, Germany;
Laboratory of Physics, "Vinča" Institute of Nuclear Sciences, National Institute of the Republic of Serbia, The University of Belgrade, Belgrade, Republic of Serbia

**M. Maruyama** Graduate School of Engineering, Osaka University, Yamadaoka, Suita City, Osaka, Japan;
Graduate School of Life and Environmental Sciences, Kyoto Prefectural University, Kyoto, Kyoto, Japan

**Toru Matsumoto** The Hakubi Center for Advanced Research/Division of Earth and Planetary Sciences, Kyoto University Kitashirakawa-Oiwakecho, Sakyo, Kyoto, Japan

**Y. Mori** Graduate School of Engineering, Osaka University, Yamadaoka, Suita City, Osaka, Japan

**Kazuki Nagashima** Research Institute for Electronic Science, Hokkaido University, Sapporo, Japan

**Hiroaki Ohfuji** Department of Earth Science, Graduate School of Science, Tohoku University, Sendai, Japan

**J. Penuelas** UMR 5270, Univ Lyon, Ecole Centrale de Lyon, CNRS, INSA Lyon, Université Claude Bernard Lyon 1, CPE Lyon, INL, Ecully, France

**P. Regreny** UMR 5270, Univ Lyon, Ecole Centrale de Lyon, CNRS, INSA Lyon, Université Claude Bernard Lyon 1, CPE Lyon, INL, Ecully, France

**Robert D. Richards** Department of Electronic and Electrical Engineering, University of Sheffield, Sheffield, UK

**Akira Sakai** Graduate School of Engineering Science, Osaka University, Toyonaka, Osaka, Japan

**Hitoshi Takane** Department of Material Chemistry, Kyoto University, Kyoto, Japan

**K. Takano** Graduate School of Life and Environmental Sciences, Kyoto Prefectural University, Kyoto, Kyoto, Japan

**Katsuhisa Tanaka** Department of Material Chemistry, Kyoto University, Kyoto, Japan

**Jun Tatebayashi** Graduate School of Engineering, Osaka University, Osaka, Japan

**Tetsuya Tohei** Graduate School of Engineering Science, Osaka University, Toyonaka, Osaka, Japan

**Takashi Toyofuku** Institute for Extra-Cutting-Edge Science and Technology Avant-Garde Research (X-Star), Japan Agency for Marine-Earth Science and Technology (JAMSTEC), Yokosuka, Japan

**Y. Tsuri** Graduate School of Engineering, Osaka University, Yamadaoka, Suita City, Osaka, Japan;
Division of Materials Science, Nara Institute of Science and Technology, Ikoma City, Nara, Japan

**Takeru Wakamatsu** Department of Material Chemistry, Kyoto University, Kyoto, Japan

**Takeshi Yanagida** Graduate School of Engineering, The University of Tokyo, Tokyo, Japan

**H. Y. Yoshikawa** Graduate School of Engineering, Osaka University, Yamadaoka, Suita City, Osaka, Japan

**M. Yoshimura** Institute of Laser Engineering, Osaka University, Yamadaoka, Suita City, Osaka, Japan

# Part I
# Polymorphism and Microtexture in Nature

# Chapter 1
# Morphological and Textural Variety of Natural minerals—Pyrite as an Example

**Hiroaki Ohfuji**

**Abstract** In this chapter, the variety of morphology and textures displayed by natural mineral crystals in macroscopic to microscopic scales is introduced in relation to the classical theory of crystal growth. In fluid-mediated mineral formations on the Earth's surfaces, crystal morphology and texture of minerals are influenced by a variety of environmental factors, including temperature, mass transport (supersaturation), and growth time, among which the degree of supersaturation of solution is particularly important. It gives the driving force behind crystallization and directly links to the "balance" between nucleation rate and crystal growth rate. Here, we focus on pyrite, the most ubiquitous sulfide mineral on the Earth, as an example to illustrate the causal relationship between the various crystal morphologies/textures from single (euhedral) crystals to spherulitic and framboidal aggregates, and their growth mechanisms and environments.

## 1.1 Introduction

In nature, minerals exhibit a wide variety of crystal morphologies and textures in addition to colors and brilliance which stimulate our interest and curiosity. The euhedral (polyhedral) morphology of mineral crystals is basically defined by the lattice type and symmetry of the constituent unit in which the atoms are arranged in a regular manner. However, even the same kind of minerals composed of the identical unit cells sometimes show various and unique "habits" in crystal morphology depending on the growth conditions and environments such as the degree of supersaturation, influence of impurities hydrodynamics, etc. Furthermore, minerals often show a wide variety of textures made up of an aggregate/assemblage of multiple crystals such as dendritic, colloform, spheroidal forms.

H. Ohfuji (✉)
Department of Earth Science, Graduate School of Science, Tohoku University, 6-3 Aramaki-Aza-Aoba, Aoba-Ku, Sendai 980-8578, Japan
e-mail: ohfuji@tohoku.ac.jp

 
F. Ishikawa et al. (eds.), *Engineering Crystal Habit*, Topics in Applied Physics 152,
https://doi.org/10.1007/978-981-96-0266-7_1

Such various morphologies and textures of natural minerals are produced as a result of the precipitation from the Earth's surface water and hydrothermal fluids, solidification from magma, chemical reaction and phase transition in metamorphic rocks and other geological environments, etc. Investigating the crystallization process of minerals, which do not have DNA for self-replication like living organisms, is important not only to understand the physical and chemical principles of the mineral formation, but also to lead to applications in the fields of materials science and engineering, such as structural and functional control of materials.

This article introduces the various shapes and textures formed by natural minerals, taking pyrite, iron sulfide ($FeS_2$) as an example, which is one of the most ubiquitous minerals on the Earth's surface and known to show various unique "habits" depending on the formation environments.

## 1.2 Principle of Crystal Growth and Morphology

Since the discovery of the "law of the constancy of interfacial angles" by Nicolaus Steno at the end of the seventeenth century through extensive observation of the crystal morphology of natural quartz, the study focusing on the geometrical properties of mineral crystal developed rapidly. After nineteenth century, the application of microscopic techniques enabled detailed observation of crystal surfaces at the micrometer scale, which led to the foundation of modern crystal growth theory such as layer growth theory (e.g. [1]) and spiral growth theory [2].

The morphology and texture of crystals depend on the driving force of the growth, which is simply given by the amount of deviation from the equilibrium state in the growth environment. It is defined by the degree of supersaturation (ratio of saturation concentration to solution concentration during growth) for the growth from solution, the degree of supercooling (difference between equilibrium temperature and growth temperature) for melt growth, and the ratio of saturated vapor pressure to vapor pressure during growth for vapor growth. The growth mechanism of crystals in those environments varies depending on the magnitude of the driving force, resulting in diversity in the morphology and texture. A comprehensive model that summarizes the relationship among the crystal morphology, growth mechanisms and driving forces was proposed by Sunagwa [3]. According to the model (Fig. 1.1), under low driving force conditions, it is difficult for crystals to overcome the energy barrier necessary for the formation of new two-dimensional nuclei, and the crystal growth occurred by "spiral growth mechanism" in which growth units are preferentially incorporated into the step around helical dislocations [4]. The crystal growth proceeds by the advancement of such steps and ends up with smooth surfaces. As the driving force increases, growth units are more likely to be captured on the crystal surface, providing new steps and kinks, to which new growth units are preferentially incorporated and the overall growth occurs by layer-by-layer growth (two-dimensional nucleation mechanism [1, 5]. With further increase in the driving force, the growth units can be attached directly to the crystal surface and the growth occurs by uniform and

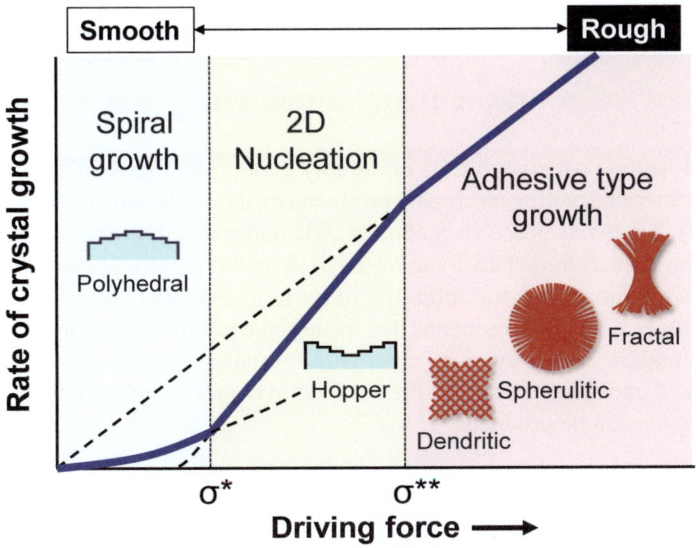


**Fig. 1.1** Schematic illustration of the changes of crystal morphology and growth rate depending on supersaturation (modified after Sunagawa [3])

continuous advancement of the interface. Crystal becomes rough-surfaced and often shows more unique forms such as skeletal, dendritic and spherulitic shapes. The Sunagawa's model was established through detailed and extensive morphological and microtextural observations of a variety of natural minerals, and it can indeed be applied to many cases of mineral crystal growth observed in nature and in the laboratory.

## 1.3 Mineralogical Characteristics of Pyrite

Pyrite is one of the ubiquitous minerals on the Earth and is known as "fool's gold" because it is often mistaken (by untrained people) for gold due to its shining brass-yellow color (Fig. 1.2). It has a chemical composition of $FeS_2$ and a rock-salt type structure, in which iron atoms occupy the Na sites and S–S dumbbell-shaped disulfide ions occupy the Cl positions (Fig. 1.3). Pyrite occurs in a variety of geological environments, including in almost all types of rocks including igneous, metamorphic and sedimentary rocks and in hydrothermal alteration and ore formation. The formation of pyrite is not as simple as that of NaCl, which is crystalized by the combination of dissolved $Na^+$ and $Cl^-$ ions in aqueous solution, since $S^{2-}$ ions can hardly exist in solution. It is considered to occur by sulfidation via reaction of the precursor FeS with polysulfide or $H_2S$ (e.g. [6–9]), as shown in the following reactions.

$$FeS + S_n^{2-}{}_{(aq)} \rightarrow FeS_2 + S_{n-1\,(aq)}^{2-} \tag{1.1}$$

$$FeS + H_2S_{(aq)} \rightarrow FeS_2 + H_{2(g)} \tag{1.2}$$

The polysulfide pathway occurs preferably under relatively oxidized conditions, in which zero-valent sulfur species are produced by the oxidation of dissolved sulfide, while the $H_2S$ pathway proceeds predominantly under strictly anoxic, reduced conditions by the oxidation of FeS by aqueous $H_2S$ without zero-valent sulfur sources [10]. It is well known that the solubility of pyrite is extremely low even at elevated temperatures (e.g. [11]). This means that the enrichment of key species such as $FeS^0$, $H_2S$, $HS^-$ and $S_2{}^-$ is essential for the pyrite formation from solution [12]. During this chemical enrichment process, the solution becomes supersaturated and various forms of pyrite can be produced.

**Fig. 1.2** Examples of various morphologies of natural pyrite crystals, **a** euhedral cubic crystals in talc, **b** aggregation of dodecahedral (pyritohedral) crystals, **c** spherulitic aggregate composed of radiating fibrous crystals, covered by a "crust" made up of many euhedral crystals, **d** disk-shaped aggregate composed of radiating fibrous crystals, so called "pyrite sun"

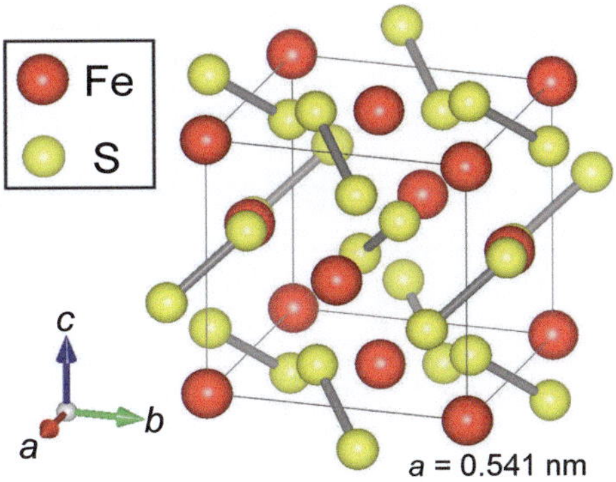


**Fig. 1.3** Schematic illustration of crystal structure of pyrite

## 1.4 Morphology of Pyrite Crystals

Pyrite shows a variety of crystal morphologies (Fig. 1.2) such as cubic, octahedral and pentagonal dodecahedral shapes and unique polycrystalline textures such as spherulitic, colloform and framboidal (raspberry-like) aggregates, depending on the formation environments.

### *1.4.1 Single Crystal Morphology*

Cubic and pentagonal dodecahedral (so-called "pyritohedral") habits (Fig. 1.2) are the most fundamental shapes of single crystals and are commonly shown by pyrite occurring in ores of hydrothermal metal deposits and hydrothermal alteration zones. In addition, small, millimeter-sized cubic (occasionally, octahedral and dodecahedral) crystals can be often found in almost all types of rocks including volcanic, metamorphic and sedimentary rocks. These morphological variations of pyrite are thought to be related to the temperature conditions and the degree of supersaturation during formation: with increase in temperature and supersaturation, crystal habit changes from cubic to octahedral and dodecahedral according to a series of hydrothermal experiments (250–500 °C) by Murowchick and Barnes (1987).

On the other hand, crystal habits can also vary depending on the size of the crystal, even when the formation conditions such as temperature and pressure are nearly the same. According to Sunagawa's observation of pyrite crystals in clays associated with hydrothermal ore deposits [13], the smaller the crystal, the higher the frequency of cubic habits surrounded by {100}, and as the crystal size increases, dodecahedral habits characterized by {210} increases. This is because the {210} planes appear as a result of repeated stacking of square step layers developed on the {100} plane (Fig. 1.4). Octahedral habit was observed in the intermediate size range between them, and Sunagawa thus concluded that pyrite crystals change their morphology from cubic to octahedral to dodecahedral as they grow.

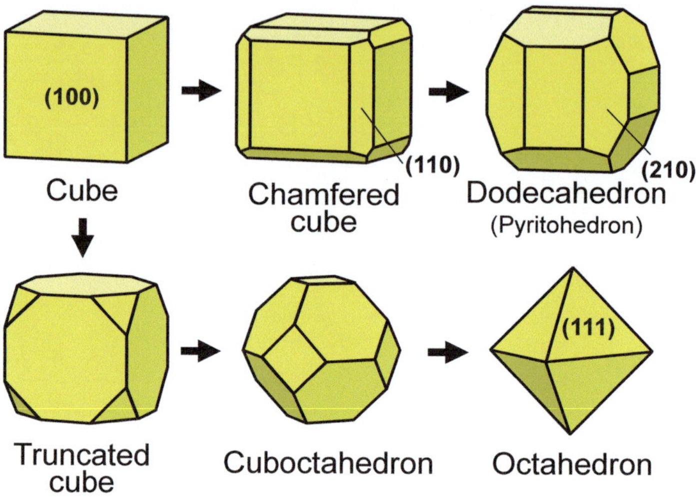


**Fig. 1.4** Variation in crystal habit from simple cube to dodecahedron and to octahedron

When the supply of materials from the surroundings is (partially) restricted or there are some growth-inhibiting factors, the crystal growth rate varies with directions, resulting in the formation of unusual habits such as skeletal, needle-like and plate-like. For example, Bonev et al. [14] found unique whiskers and thin platelets of microscopic pyrite in small cavities in chalcopyrite from a hydrothermal metal deposit in Bulgaria (Fig. 1.5a). They considered that pyrite whiskers and platelets are produced from unstirred highly-supersaturated solution, which provided a diffusion-controlled regime inside the cavities in chalcopyrite. Under such conditions, elongation growth of crystals is encouraged due to the unidirectional supply (diffusion) of materials through a sharp chemical gradient and the competition with surrounding crystals, resulting in the formation of anisotropic morphologies.

## *1.4.2 Spherulitic Aggregate*

In sedimentary rocks such as mudstone and fine-grained sandstone, pyrite is often found as nodules (small rounded masses or lumps of mineral aggregate) composed of fibrous to acicular crystals extending radially from the center (e.g. [15–17] (Figs. 1.2c, d and 1.5b). Unlike typical spherulites of silica and feldspar produced by rapid cooling of silicic magma, pyrite spherulites form rather slowly at much lower (near ambient) temperature in sedimentary rocks. As the sediment undergoes diagenesis, the gaps between detrital mineral particles decreases, and the supply of material is rate-limiting by diffusion. Pyrite is formed by the reaction between $H_2S$ produced through sulfate-reduction (often facilitated by sulfate-reducing bacteria) and Fe components (precursor FeS) in the sediment, in which a sharp chemical gradient is present around

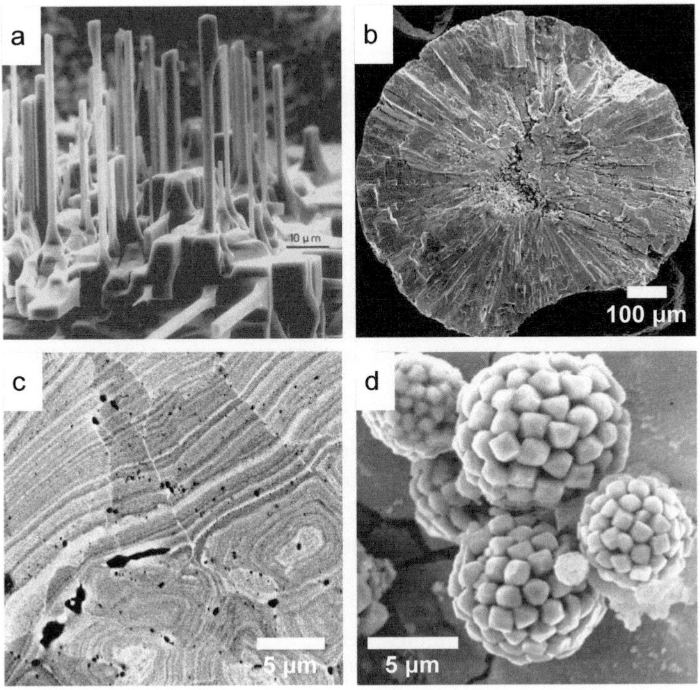


**Fig. 1.5** Unique habits and textures shown by natural pyrite crystals, **a** whisker crystals elongating along [100] direction [14], **b** fibrous crystals radiating from the center observed on a facture of pyrite spherulite, **c** colloform texture formed by fine acicular crystals shown by the channeling contrast imaging of a polished section, **d** framboidal aggregate composed of pyrite microcrystals with uniform size and shape

the initially formed pyrite seeds (nucleus) due to the diffusion-controlled environment. As a result, pyrite crystals grow and elongate outward from the surface of the seeds, forming spherulitic aggregates. Similar spherulite formation is also seen in calcite spherulites in sedimentary rocks (e.g. [18, 19]), indicating that it is one of the typical growth textures of mineral crystals in environments where the supply of materials is rate-limited by diffusion. The elongation direction of individual fibers/needles is usually [100] of pyrite [20]. In shales composed of densely packed fine clays with laminated, peelable features due to compaction, material supply across the lamination is strongly restricted and intra-layer (in-plane) diffusion becomes more dominant, resulting in the formation of radial aggregates in a disk shape called "pyrite sun" [21, 22] (Fig. 1.2d). The formation of spherical and disk-shaped aggregates by the radial growth of elongated crystals is interpreted as based on the rules of geometrical selection in the crystal growth process [13], as described in the next section.

### 1.4.3 Colloform Aggregates

Another assemblage of pyrite crystals found frequently in metal ore deposits and sometimes in sedimentary rocks is colloform texture showing rhythmic (wavy) bandings composed of acicular crystals (Fig. 1.5c). Pyrite spherulite is also a type of colloform texture, and indeed, the growth process of the individual crystals is basically the same. However, colloform texture is also formed in the form of veins and filling voids, in which the crystals grow vertically from the interface of the host rocks. The elongation direction of the acicular pyrite crystals is usually [100] and occasionally [110] and [111], and can vary even within a series of colloform aggregate, as reported by electron back scatter diffraction (EBSD) analysis of the crystallographic preferred orientation (CPO) [23–25]. The change in CPOs of pyrite crystals is likely caused by the change in the physico-chemical conditions of the source fluid such as temperature and the degree of supersaturation, because it correlates with abrupt changes in grain size, sulfur isotopic composition, trace elements [23–25]. EBSD analysis by Huang et al. [25] showed that the crystallographic orientation of pyrite is almost random near the substrate (at the bottom), but gradually becomes preferred oriented along [100] toward the rim of a colloform aggregate. This is thought to be the result of geometrical selection in crystal growth from the multiple nuclei. When multiple nucleation occurs freely on a flat substrate without any interfacial controls such as epitaxy, the orientation of the individual crystals is random. Among the crystals growing from these nuclei, those which are inclined with respect with the substrate surface stop growing when they come into contact with those growing perpendicularly. In contrast, the crystals growing perpendicularly are able to grow into a higher concentration environment (assuming isoconcentration parallel to the substrate), which further enhances their growth. As a result, among the many crystal nuclei generated on the substrate, only vertically oriented crystals survive and continue to grow, resulting in a decrease in the total number of crystals and the formation of a polycrystalline texture composed of elongated crystals aligned in a certain direction. Such colloform or layered texture produced through the geometrical selection is a typical polycrystalline texture observed in various natural minerals precipitated from highly-supersaturated solution, not limited to pyrite.

### 1.4.4 Framboidal Aggregate

The most common and unique morphology found in microscopic pyrite is the framboidal (raspberry-like) aggregate (Fig. 1.5d). The term "framboidal" is derived from the French word "*frabmoise*" meaning raspberry. Framboidal pyrite is normally 1 to 100 μm in diamter and consists of a large number of euhedral to sub-euhedral microcrystals (0.1 to 10 μm). It occurs commonly in sedimentary rocks, modern sediments and even in anoxic water column in ocean (as suspended particles) (e.g. [26]), but can also be found in a variety of geological environments such as in volcanic rocks [27], in metal deposits (e.g. [28]), near high-temperature hydrothermal vents (e.g. [29]). This suggests that the framboidal morphology is indeed a universal and fundamental product of the crystallization of microscopic pyrite in nature. On the other hand, laboratory synthesis of framboidal pyrite has been attempted for a long time, but only a few cases have produced framboidal-like aggregates similar to those found in nature [30]. This suggests that the conditions necessary for the formation of framboidal pyrite cannot readily be reproduced in laboratory for some reason.

The formation process of framboidal pyrite has been the subject of much debate, and various theories have been proposed to explain its unusual texture; mineralized replacement of bacteria or their colonies [31], replacement of organic particles originating from humic soil [32] and magnetic aggregation of ferromagnetic greigite ($Fe_3S_4$), a precursor mineral and further sulfidation [33]. However, none of them is yet convincing enough to explain its origin and the self-organization mechanism remains unclear. One thing is clear that the fact that a large number of microcrystals of completely uniform size and morphology make up a single framboid means that their formation is a result of a multiple and spontaneous nucleation event in a relatively short time period rather than continuous, attachment growth. Base on this property, the size distribution of framboidal pyrite in modern and ancient sediments is commonly used as a paleoenvironmental indicator of redox condition (e.g.[34], 35]).

The packing structure of microcrystals in framboidal pyrite is classified into two, regular packing and irregular packing, which can clearly be recognized in the cross sections. In the former microcrystals are arranged in regular manner (Fig. 1.6a), while in the latter they are assembled randomly with no obvious alignments (Fig. 1.6b). There are two types of regular microcrystal packing, (1) face-centered cubic packing (Fig. 1.6a) and (2) icosahedral packing (Fig. 1.6c). Face-centered cubic (fcc) packing consists of microcrystals arranged in ABCABC… stacking and shows uniform, regular patterns, represented by cubic (Fig. 1.6d) and hexagonal (Fig. 1.6e) patterns depending on the direction of the cross-section (Fig. 1.6f). Icosahedral packing is characterized by symmetrical patterns shown by aligned microcrystals with different orientations depending on each domain (sector) on the cross-section (Fig. 1.6c). Figure 1.7 shows examples of such microcrystal arrangements, in which domain structures with twofold, threefold and fivefold symmetry can be seen clearly. These patterns are interpreted as different sections of an icosahedral microcrystal packing, which consists of 20 tetrahedral subunits that share a vertex at the common center, and the microcrystals are arranged in the fcc packing in each unit [36]. Since an icosahedron has twofold, threefold, and fivefold axes at the edge, face and corner, respectively, the cross sections perpendicular to each of these axes show symmetric domain patterns created by the tetrahedral units. Note that framboidal pyrite is the only natural example of such an icosahedral structure composed of dense aggregation of multiple particles, suggesting that it may represent a new aspect of mineral crystallization observed at the microscopic level.

The two types of regular microcrystal packing found in framboidal pyrite share some similarities with the structure of atomic clusters and nanoparticles of various metals and metal compounds such as Au, Pt, Al-Cu-Fe, etc. at the nanoscale (e.g.

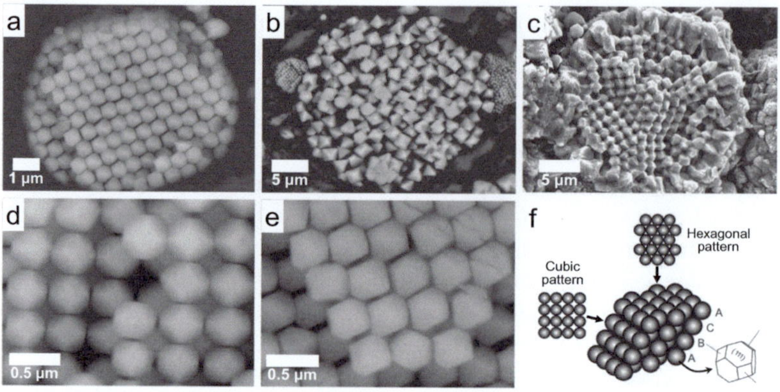


**Fig. 1.6** Microcrystal packing in framboidal pyrite, **a** regular (fcc) packing, **b** irregular packing, **c** regular (icosahedral) packing, **d** cubic array of microcrystals, **e** hexagonal array of microcrystals, **f** schematic illustration of the fcc packing of microcrystals

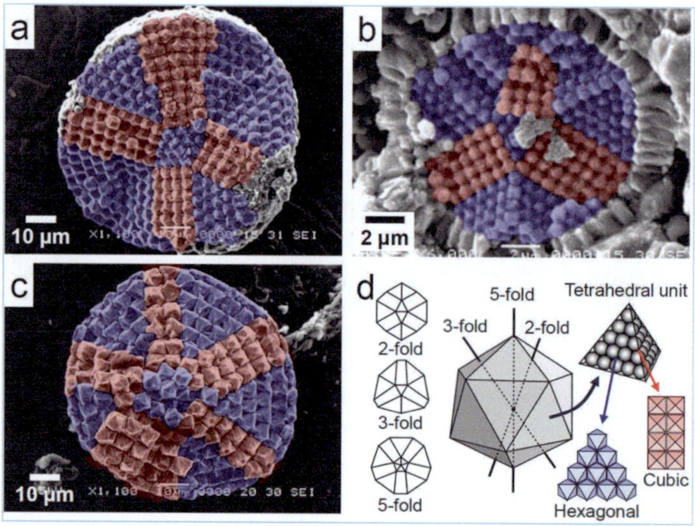


**Fig. 1.7** Symmetrical arrangements of microcrystals in **a** twofold, **b** threefold, and **c** fivefold, where the microcrystals colored red and blue in the images correspond the cubic (Fig. 1.6d) and hexagonal (Fig. 1.6e) 2D arrays of microcrystals, respectively. **d** Schematic illustration of icosahedral microcrystal packing composed of tetrahedral unit in which microcrystals are arranged in fcc. Cross-sections perpendicular to the 2-, 3- and fivefold axes yield a symmetrical domain structure. Sections parallel to the faces of the tetrahedral units show hexagonal microcrystal array, while those parallel to the edges show cubic array

[37]). The face-centered packing with the cuboctahedra shape composed of layer-packing around 12 nearest-neighbor spheres (Fig. 1.8a) is the most common and stable structure of atomic clusters and is inherited by bulk crystals as it grows. Icosahedral packing is also made up of a layered packing centered on 12 nearest-neighbor spheres (Fig. 1.8b), although the distance between each sphere is slightly longer than that of fcc packing [38]. It can be more stable structure in terms of surface energy, since it is terminated by triangular surfaces in which the spheres are arranged in 2D closest packing (Fig. 1.8c, d). However, icosahedral clusters become unstable and transform to bulk crystals when the particle size exceeds 100–200 nm [37]. This is due to the structural distortion intrinsic to icosahedral packing. Since the distance between the center and vertex of an icosahedron is about 5% shorter than the vertex-vertex distance, the misfit (gap) between the tetrahedral units (Fig. 1.8e) becomes larger with increasing cluster size by layer growth. It is therefore, surprising that framboidal pyrite can maintain the icosahedral packing up to micrometer scale (>100 μm), where the structural distortion is overcome by introducing stacking faults in microcrystal arrangements along the boundaries of tetrahedral units. A recent study demonstrated that the magic number rule, which places a number of particles in a completely closed (concentric) shells such as Mackay icosahedra [38] (Fig. 1.8d), is the solution for minimizing the total free energy of self-assembled particles not only for atomic clusters but also for colloidal aggregation [39]. Framboidal pyrite represents the natural example of such surface energy minimization involved in self-assembled colloidal crystals.

Although the detailed mechanism of the formation of monodisperse colloidal crystals and their self-assembly producing framboidal morphology are still unknown, they should occur under very high driving force (highly supersaturated condition),

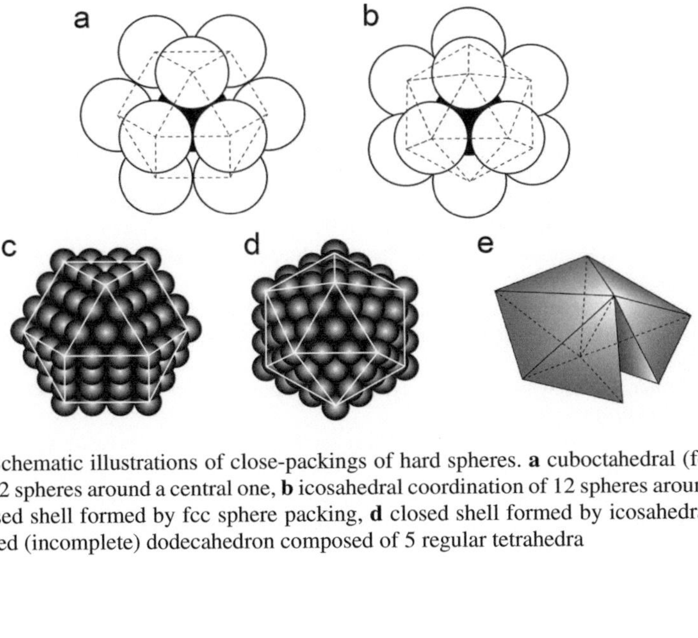


**Fig. 1.8** Schematic illustrations of close-packings of hard spheres. **a** cuboctahedral (fcc) coordination of 12 spheres around a central one, **b** icosahedral coordination of 12 spheres around a central one, **c** closed shell formed by fcc sphere packing, **d** closed shell formed by icosahedral packing, **e** non-closed (incomplete) dodecahedron composed of 5 regular tetrahedra

where the rate of nucleation significantly exceeds the rate of crystal growth. Such extremely high chemically enriched conditions are not readily achieved inorganically and may be provided through biological activity by sulfate-reducing bacteria [40], which would be one of the reasons why framboidal texture is unique to pyrite.

## 1.5 Concluding Remarks

As described above, pyrite crystals show a wide variety of morphologies and textures depending on the growth conditions in various surface environments of the Earth. The important factors that give diversity to "crystal habits" include: (1) competition and variation of crystal faces depending on the growth size and growth orientation of crystals, (2) anisotropic growth such as elongation and geometrical selection promoted by sharp chemical and/or thermal gradients, and (3) relative balance between nucleation rate and crystal growth rate which are determined by supersaturation and hydrodynamics of the solution. In the field of mineralogy, we have deepened our knowledge and understanding of the crystal growth mechanism and the physicochemical factors that give rise to the diversity and complexity of crystal habit through detailed observation of natural minerals and their reproduction in the laboratory. The theory of crystal growth established through these efforts has been shared in a variety of fields, such as material sciences, engineering, medicine and life sciences, and the principles and techniques are now utilizing in various applied research and manufacturing (See other chapters in this book). Learning from the complex phenomena and products of nature and applying them to our lives are essential for sustainable development.

## References

1. I.N. Stranski, Zur Theorie des Kristallwachstums. Phys. Chem. **136**, 259–278 (1928)
2. F.C. Frank, The influence of dislocations on crystal growth. Discuss. Faraday Soc. **5**, 48–54 (1949)
3. I. Sungawa, Characteristics of crystal growth in nature as seen from the morphology of mineral crystals. Bull. Minér. **104**, 81–87 (1981)
4. W.K. Burton, N. Cabrera, F.C. Frank, The growth of crystals and the equilibrium structure of their surfaces. Phil. Trans. R. Soc. A **234**, 299–358 (1951)
5. W. Kossel, Über Krystallowachstum. Die Naturwissenschaten **18**, 901–910 (1930)
6. R.A. Berner, Sedimentary pyrite formation: an update. Geochim. Cosmochim. Acta. **48**, 605–615 (1984)
7. G.W. Luthter, Pyrite synthesis via polysulfide compounds. Geochim. Cosmochim. Acta. **55**, 2839–2849 (1991)
8. M.A.A. Schoonen, H.L. Barnes, Reactions forming pyrite and marcasite from solution: I. Nucleation of FeS below 100°C. Geochim. Cosmochim. Acta. **55**, 1495–1504 (1991)
9. D. Rickard, Kinetics of pyrite formation by the H S oxidation of iron (II) monosulfide in aqueous solutions between 25 and 125°C: the rate equation. Geochim. Cosmochim. Acta. **61**, 115–134 (1997)

10. D. Rickard, G.W. Luther, Kinetics of pyrite formation by the H S oxidation of iron (II) monosulfide in aqueous solutions between 25 and 125°C: the mechanism. Geochim. Cosmochim. Acta. **61**, 135–147 (1997)
11. H. Ohmoto, K. Hayashi, Y. Kajisa, Experimental study of the solubilities of pyrite in NaCl-bearing aqueous solutions at 250–350 °C, GCA (1994)
12. D. Rickard, G.W. Luther, Chemistry of iron sulfides. Chem. Rev. **107**, 514–562 (2007)
13. I. Sunagawa, Variation in crystal habit of pyrite. Report No. 175, Geological Survey of Japan, pp. 1–47 (1957)
14. K. Bonev, M. Reiche, M. Marinov, Morphology, perfection and growth of natural pyrite whiskers and thin platelets. Phys. Chem. Miner. **12**, 223–232 (1985)
15. A. Agangi, A. Hofmann, B. Eickmann, F.O. Ossa, P. Tyler, B. Wing, A. Bekker, A multiple sulfur record of super-large volcanic eruptions in Archaean pyrite nodules. Earth Planet. Sci. Lett. **594**, 117737 (2022)
16. N.R. Ayupova, V.V. Maslennikov, D.A. Artem'ev, Mineralogical and geochemical features of pyrite nodules from sulfide turbiditesin the Talgan Cu-An massive sulfide deposit (Southern Urals). Lithol. Min. Resour. **54**, 447–464 (2019)
17. A. Bekker, H.D. Holland, P.-L. Wang, D. Rumble III., H.J. Stein, J.L. Hannah, L.L. Coetzee, N.J. Beukes, Dating the rise of atmospheric oxygen. Nature **427**, 117–120 (2004)
18. J. Aguillar, J. Schmatz, A. Busch, R. Swennen, Model for calcite spherulite formation in organic, clay-rich, lacustrine carbonate shales (Barbalha Formation, Aptian, Araripe Basin, NE Brazil). Mar. Pet. Geol. **128**, 104988 (2021)
19. H. Claes, T. Miranda, T.C. Falcão, J. Soete, Z. Mohammadi, L. Zieger, M.M. Erthal (2021)
20. E.A. Bannister, The distinction of pyrite from marcasite in nodular growths. Miner. Mag. **23**, 179–187 (1932)
21. A.R. Myers, C. Chenoweth, Directory of coal mines in Illinois 7.5 minute quadrangle series, Tilden quadrangle, Randolph, Washington and St Clair Counties, Illinois. Illinois State Geol. Surv. Urbana 44 (2010)
22. G.J. Retallack, Reassessment of the Silurian problematicum *Rutgersella* as another post-Ediacaran vendobiont. Alcheringa Australas. J. Palaeontol. **39**, 573–588 (2015)
23. C.D. Barrie, A.J. Boyce, A.P. Boyle, P.J. Williams, K. Blake, T. Ogawara, J. Akai, D.J. Prior, Growth controls in colloform pyrite. Am. Miner. **94**, 415–429 (2009)
24. K. Freitag, A.P. Boyle, E. Nelson, M. Hitzman, J. Chrchill, M. Lopez-Pedrosa, The use of electron back-scatter diffraction and orientation contrast imaging as tools for sulphide textural studies: example from the Greens Creek deposit (Alaska). Miner. Deposita **39**, 103–113 (2004)
25. F. Huang, S. Gao, L. Chen, L. Su, Y. Li, L. Meng, K. Liu, C. Chai, X. Qi, Micro-texture and in situ sulfur isotope of pyrite from the Baiyunpu Pb-Zn deposit in central Hunan, South China: implications for the growth mechanism of colloform pyrite aggregates. J. Asian Earth Sci. **193**, 104302 (2020)
26. J. Skei, Formation of framboidal iron sulfide in the water of a permanently anoxic Fjord-Rrmvaren, south Norway. Mar. Chem. **23**, 345–352 (1988)
27. L.G. Love, G.C. Amstutz, Framboidal pyrite in two andesites. News Jahrb. Mineral. Monatsh **3**, 97–108 (1969)
28. B.M. England, J. Ostwald, Framboidal-derived structures in some Tasman fold belt basemetal sulphide deposits, New South Wales. Ore Geol. Rev. **7**, 381–412 (1993)
29. K. Liu, F. Huang, S. Gao, Z. Zhang, Y. Ren, B. An, Morphology of framboidal pyrite and its textural evolution: evidence from the Lagotchev area, mid-atlantic ridge. Ore Geol. Rev. **141**, 104630 (2022)
30. H. Ohfuji, D. Rickard, Experimental syntheses of framboids—a review. Earth Sci. Rev. **71**, 147–170 (2005)
31. L.G. Love, Microorganisms and the presence of syngenetic pyrite. Q. J. Geol. Soc. Lond. **113**, 429–440 (1957)
32. H. Papunen, Framboidal textures of the pyritic layer found in a peat bog in S.E Finland. Bull. Comm. Geol. Finlande **222**, 117–125 (1966)

33. R.T. Wilkin, H.L. Barnes, Formation process of framboidal pyrite. Geochim. Cosmochim. Acta **61**, 323–339 (1997)
34. R.T. Wilkin, H.L. Barnes, S.L. Brantley, The size distribution of framboidal pyrite in modern sediments: an indicator of redox conditions. Geochim. Cosmochim. Acta. **60**, 3897–3912 (1996)
35. P.B. Wignall, R. Newton, M.E. Brookfield, Pyrite framboid evidence for oxygen-poor deposition during the Permian-Triassic crisis in Kashmir. Palaeogeogr. Palaeoclimatol. Palaeoecol. **216**, 183–188 (2005)
36. H. Ohfuji, J. Akai, Icosahedral domain structure of framboidal pyrite. Am. Miner. **87**, 176–180 (2002)
37. H. Hofmeister, Forty years study of fivefold twinned structures in small particles and thin films. Cryst. Res. Technol. **33**, 3–25 (1998)
38. A.L. Mackay, A dense non-crystallographic packing of equal spheres. Acta Crystallogr. A **15**, 916–918 (1962)
39. J. Wang, C.F. Mbah, T. Przybilla, B.A. Zubiri, E. Spiecker, M. Engel, N. Vogel, Magic number colloidal clusters as minimum free energy structures. Nat. Commun. **9**, 5259 (2018)
40. A. Duverger, J.S. Berg, V. Busigny, F. Guyot, S. Bernard, J. Miot, Mechanisms of pyrite formation promoted by sulfate-reducing bacteria in pure culture. Front. Earth Sci. **8**, 588310 (2020)
41. J.B. Burowchick, H.L. Barnes, Effects of temperature and degree of supersaturation on pyrite morphology. Am. Miner. **72**, 1241–1250 (1987)

# Chapter 2
# Extraordinary Crystals of Natural Minerals in Mining Sites and Volcanoes

**Takuya Echigo**

**Abstract** Many mineral crystals in natural environments exhibit unique morphology, size, color, microtexture, and other properties that have not yet been reproduced artificially, and the understanding of their formation mechanisms may provide insights for the development of new crystalline materials. This review introduces two specific examples of such natural minerals: karpatite, natural crystal of coronene in hydrothermal metal deposits and Al/Si disordered anorthite, metastable aluminosilicate mineral in volcanic rocks. The former mineral shows distinctive fluorescent property which is different from the artificial crystal of coronene and the latter mineral is characterized by very large size (>1 cm single crystal) despite its metastability. Their occurrences are "out of the ordinary" because they cannot be explained by simple crystal growth such as cooling or fractionation. However, careful observation and analysis of those natural minerals provide a reasonable explanation for the formation of extraordinary crystals in natural environments. It is expected that the knowledge obtained from such natural minerals will be applied to the crystal engineering for developing functional crystalline materials.

## 2.1 Utilization of Stable and Metastable Phases of Natural Minerals

The earliest example of intentional use of natural mineral crystals in the human's history is pigment [1]. Most early pigments are the mixture of glue and mineral powders with beautiful coloring properties and used as painting on rock walls (fresco). One of the most famous example is the cave paintings of Lascaux in which iron (hydr)oxides such as hematite ($\alpha$-$Fe_2O_3$) and goethite [$\alpha$-FeO(OH)] were used as brownish mineral pigments. Because these minerals are the most stable phases in the atmosphere, they record ancient human activity without decomposition for about twenty thousand years.

T. Echigo (✉)
Graduate School of International Resource Sciences, Akita University, Akita 010-8502, Japan
e-mail: echigo@gipc.akita-u.ac.jp

 
F. Ishikawa et al. (eds.), *Engineering Crystal Habit*, Topics in Applied Physics 152,
https://doi.org/10.1007/978-981-96-0266-7_2

While stable phases such as hematite and goethite do not degrade for long periods of time, metastable phases which transform into stable phases over time also have been used as pigment. The blue sky in the van Eycks' magnificent centre panel of the Ghent Altarpiece, the *Adoration of the Lamb*, from about 1425–32 relies on blue azurite [$Cu_3(CO_3)_2(OH)_2$] pigment, where it has now turned slightly green malachite [$Cu_2CO_3(OH)_2$] in parts [2]. Azurite is a bright blue crystal and was used to for painting of blue dresses and skies, however, it gradually transforms to malachite because malachite is more stable phase at near-surface condition ($p_{CO2} \sim 10^{-3.5}$ MPa ~ $\log p_{CO2} = -3.5$) [3]. Figure 2.1 shows $p_{O2} - p_{CO2}$ phase diagram, illustrating the stability of copper minerals in the presence of water at 25 °C and 0.1 MPa. The partial pressure of $CO_2$ at which azulite and malachite coexist in equilibrium is log $p_{CO2} = -0.13$ and azurite is a metastable phase in the air. Though this equilibrium condition was not known in the fifteenth century, this historical painting is one of the best examples that humans have been using metastable phases with favorable properties for a long time.

Crystals of natural minerals were also used in engineering fields, not only in cultural fields, e.g., large and transparent single crystals of natural minerals were used as lenses before the development of crystal growth technology. Quartz ($SiO_2$) and fluorite ($CaF_2$) are typical examples, and even in the twentieth century, natural crystals of those minerals were often used [4]. Since crystal growth technologies have been developed, the use of natural mineral crystals "as is" has become rare. However, many crystals of natural minerals exhibit some physical properties that cannot be reproduced by artificial crystals, e.g., color, size, morphology, micro-texture, and so on. Understanding the formation mechanism of such minerals in the natural environment may lead to the invention of new crystalline materials. This chapter introduces examples of research on natural mineral crystals that are "out of the ordinary" (=extraordinary) and cannot be explained by simple crystal fractionation or crystal growth mechanisms.

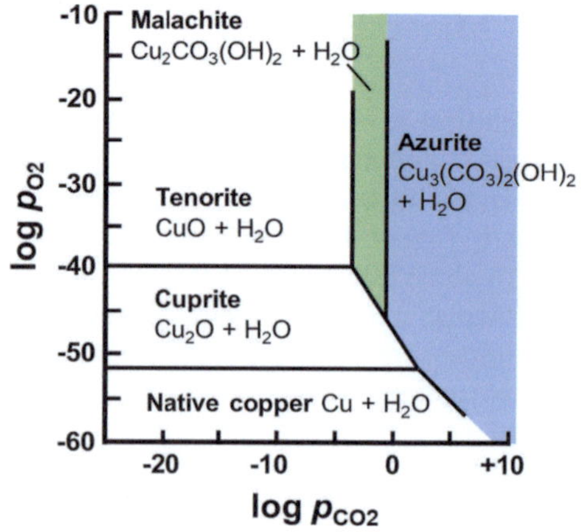


**Fig. 2.1** $p_{O2}-p_{CO2}$ phase diagram, illustrating the stability of copper minerals in the presence of water at 25 °C and 0.1 MPa (pressure is in bars, 1 bar = 0.1 MPa)

## 2.2 Natural Crystals of Organic Molecules in Hydrothermal Mercury Deposit

The first example of extraordinary mineral is natural crystals of organic compounds in epithermal mercury deposits, specifically karpatite ($C_{24}H_{12}$) which is a crystal of polycyclic aromatic hydrocarbon molecules called coronene [5]. In general, hydrothermal ore deposits are formed by high temperature magma-hydrothermal system around the igneous intrusion in which most organic compounds thermally decompose into $CO_2$ under oxidative condition or graphite under reductive condition [6]. Although its occurrence has been known for a long time [7, 8], the formation mechanism and the origin remained unknown. Figure 2.2 shows the occurrence of karpatite; yellow crystals of karpatite precipitated on the surface of the quartz vein. The red mineral is cinnabar (HgS), a mercury sulfide mineral, and the brown mineral is magnesite ($MgCO_3$), a magnesium carbonate mineral. This occurrence suggests that silica-rich hydrothermal fluid infiltrated into the magnesite to form the quartz vein. Then coronene molecules and mercury dissolved in the hydrothermal fluid precipitated after the quartz precipitation.

To elucidate the origin of carbon, the analysis of isotopic composition $\delta^{13}C$ (= $[(^{13}C/^{12}C)_{sample}/(^{13}C/^{12}C)_{standard}]-1] \times 1000$) is very effective, for example, diamond derived from mantle carbon shows relatively heavy $\delta^{13}C$ ($-15$ ~ $0$ ‰) and coal/petroleum derived from organic matters in sedimentary rocks shows relatively light $\delta^{13}C$ ($-30$ ~ $-20$ ‰) [9]. The carbon isotopic composition of the karpatite is $\delta^{13}C = -22.39$ ‰ [5], which suggests that the coronene molecules of karpatite are derived from organic matter in oceanic sediments. In addition, single-crystal XRD, micro-Raman, micro-infrared-absorption and thermogravimetric analyses revealed that the karpatite contains no impurity because of the high degree of aromaticity of aromaticity of coronene molecules [5]. Based on the analytical results, the formation mechanism of karpatite was summarized as follows (Fig. 2.3) [10]: Organic matter in sediments undergoes hydrothermal alteration into polycyclic aromatic hydrocarbon

**Fig. 2.2** Photograph of karpatite from the Picacho Peak Area, San Benito County, California, which is crystallized on quartz containing cinnabar grains (Ka = karpatite $C_{24}H_{12}$, Qtz = quartz $SiO_2$, Cin = cinnabar HgS, Mgs = magnesite $MgCO_3$)

compounds, then the less stable molecules decompose and the coronene molecules are concentrated in the hydrothermal fluids, resulting in crystallization as karpatite. In general, thermodynamic stability of hydrocarbons at high temperature increases with decreasing H/C ratio, for a given carbon number [11]. Hence the PAHs of lower molecular weight such as benzene ($C_6H_6$), naphthalene ($C_{10}H_8$), and phenanthrene ($C_{14}H_{10}$) are depleted from high temperature fluids because of their lower thermodynamic stability. It is notable that the organic mineral was formed by relatively short time reaction with hydrothermal fluids at high temperature and pressure, rather than in an environment where organic matter is transformed into coal or petroleum by relatively long time (diagenetic) reaction at low temperature.

Interestingly, one recent study [12] clarified why natural and synthetic crystals of coronene exhibit the different optical properties when they are irradiated with ultraviolet light, the former (karpatite) emits blue-white fluorescence and the latter does yellow-green UV fluorescence. This phenomenon itself has been known for a long time, but it was thought to be probably due to impurities. Potticary et al. [12] demonstrated that the difference of microtexture between them cause the difference of fluorescence. The natural crystal is made up of multiple thin layers (c.a., 100 nm thickness) stacked along the *c*-axis, while the synthetic crystal has no such structure. It is thought that the stacked structure of layered crystals with a thickness of approximately 100 nm in the c-axis direction reduces the energy difference between stacked free excitons and self-bound excitons, resulting in a shortening of the fluorescence wavelength. As above-mentioned, natural crystals are crystallized from hydrothermal fluids, so it may be possible to create unique nanostructures by hydrothermal reactions instead of evaporation of organic solvents.

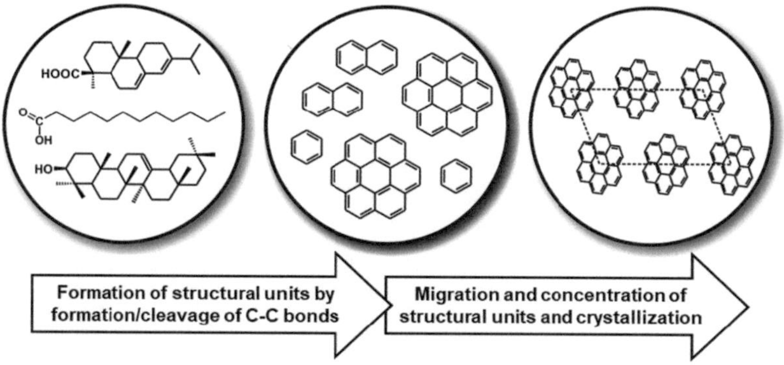


**Fig. 2.3** Schematic representation of the two stages in the formation of organic minerals. The first stage is the formation of structural units, e.g., coronene molecules for karpatite, from biomolecules with formation or cleavage of C–C bonds. The second stage is the migration and concentration of structural units and subsequent crystallization of the mineral on cooling

## 2.3 Giant Plagioclase Crystals in Quenched Lava

The second example is giant plagioclase in quenched lava. Lava is defined as magma that is erupted to the ground surface from the underground and not limited to liquid state (lava flow) but also in a solid state (eruptive rocks). Because lava is exposed to the cool air, it is solidified in short time period. Those rocks solidified in short time period and by rapid cooling are called volcanic rocks and generally consist of small crystals of phenocrysts, which is generally smaller than 1 mm, and matrix called groundmass containing microlite and/or volcanic glass [13]. On the other hand, the other rocks solidified from magma in long time period and by slow cooling at the deep underground are called plutonic rocks and generally consist of large crystals of similar size [14]. In plutonic rocks, giant crystals exceeding 1 m often occur and crystal in volcanic rocks seldom exceeds 1 cm. However, plagioclase (solid solution of Ca- and Na-feldspar) crystals larger than 1 cm occur in several locations and such plagioclase are called "giant plagioclase" [15]. Particularly Ca-rich giant plagioclase is called "anorthite megacryst (Fig. 2.4)" and uniquely occur in Japanese island arc [16]. Although its occurrence has been known for a long time [17], the formation mechanism and the origin remained unknown. Echigo et al. [18] revealed that, in a single crystal of anorthite megacryst, both stable phase (Al/Si ordered structure) and metastable phase (Al/Si disordered structure) coexist and discussed how anorthite megacrysts have grown in magma chamber. From these natural minerals, we may be able to gain idea for growing crystals of metastable phases with high functionality such as diamond [19] and $\alpha$-$Ga_2O_3$ [20] and this section introduces the crystallographic characteristics of the anorthite megacryst

Feldspar minerals are the most abundant minerals in the Earth crust [21] and their formation mechanism is well investigated for a long time [22]. They consist of the three end-members of K-feldspar ($KAlSi_3O_8$), Na-feldspar ($NaAlSi_3O_8$) and Ca-feldspar ($CaAl_2Si_2O_8$). The former two phases make the complete solid solution called alkali-feldspar series [$(K,Na)AlSi_3O_8$] and the latter two also make the complete solid solution called plagioclase series [$(Ca_{1-x},Na_x)Al_{2-2x}Si_{2+2x}O_8$], though

**Fig. 2.4** Photo showing the occurrence of anorthite magacryst (giant plagioclase), which is the large white crystal in the center of specimen, in lava from Hachijo-island

the K-feldspar and Ca-feldspar only make very limited solid solution due to the significant difference in ionic radii of $K^+$ (1.51 Å in coordination number 8) and $Ca^{2+}$ (1.12 Å in coordination number 8) [23]. All of the feldspar minerals consist of aluminosilicate framework with negative charge and interstitial cations; K- and Na-feldspars have $AlSi_3$-framework and Ca-feldspar has $Al_2Si_2$- framework, respectively. Because of the similarity of ionic radii of $Al^{3+}$ and $Si^{4+}$, those cations can be arranged in the framework in ordered manner and disordered manner. For alkaline feldspar series, both Al/Si ordered (orthoclase) and disordered (sanidine) K-feldspar are found in natural crystals (Table 2.1). For Na-feldspar as well, both Al/Si ordered (low-albite) and disordered (high-albite) phases occur in natural environments (Table 2.1). However, for Ca-feldspar, only Al/Si ordered phase is reported (Table 2.1) and disordered phase was not reported until 2014 [18].

There is a widely accepted empirical rule, which is called the Al-avoidance rule or Loewenstein rule [24] constraining the ordering of Al and Si in the aluminosilicate framework. In ordered $Al_2Si_2$-framework, the $Al^{3+}$ and $Si^{4+}$ cations with tetrahedrally coordinated $O^{2-}$ are arranged alternatively, i.e., $–^{IV}Al^{3+}–O^{2-}–^{IV}Si^{4+}–O^{2-}–^{IV}Al^{3+}–O^{2-}–^{IV}Si^{4+}–O^{2-}–^{IV}Al^{3+}–O^{2-}–^{IV}Si^{4+}–O^{2-}–^{IV}Al^{3+}–O^{2-}–^{IV}Si^{4+}–O^{2-}–$, and the total positive charge supplied to the bridging oxygen is +1.75 (=3/4 + 4/4). The negative charge of the bridging oxygen is -2 and the deficient charge is supplied by the interstitial cations such as $K^+$ and $Na^+$. On the other hand, in disordered $Al_2Si_2$-framework, the $Al^{3+}$ and $Si^{4+}$ cations are arranged randomly, e.g., $–^{IV}Al^{3+}–O^{2-}–^{IV}Al^{3+}–O^{2-}–^{IV}Si^{4+}–O^{2-}–^{IV}Al^{3+}–O^{2-}–^{IV}Si^{4+}–O^{2-}–^{IV}Si^{4+}–O^{2-}–$ and the chemical bonds excluded in ordered arrangement, such as $^{IV}Al^{3+}–O^{2-}–^{IV}Al^{3+}$ and $^{IV}Si^{4+}–O^{2-}–^{IV}Si^{4+}$, must be created. Though the latter bonds are stable, the former $^{IV}Al^{3+}–O–^{IV}Al^{3+}$ bonds are unstable because the total positive charge supplied to the bridging oxygen is +1.5 (=3/4 + 3/4) and is significantly deficient for the negative charge of oxygen (−2). Thus it has been considered reasonable that Al/Si disordered Ca-feldspar is absent in the phase diagram of Na–Ca feldspar join (plagioclase series).

Based on the above-mentioned constrain, Al/Si disordered Ca-feldspar is considered as unstable phase, however some experimental studies suggested that it is metastable phase. The extreme heating of Ca-feldspar close to the melting point (1555 °C) and the subsequent quenching were carried out to produce partially Al/Si disordered anorthite [25, 26]. Although only the limited changes in Al/Si ordering

**Table 2.1** Chemical composition, structural state and lattice parameters of selected feldspar minerals

| Mineral name | Chemical composition | Tetrahedral Al/Si order | Lattice parameters | | | | | |
|---|---|---|---|---|---|---|---|---|
| | | | *a* (Å) | *b* (Å) | *c* (Å) | *α* (°) | *γ* (°) | *β* (°) |
| Microcline | $KAlSi_3O_8$ | Ordered | 8.59 | 12.97 | 7.22 | 90.6 | 115.9 | 87.7 |
| Sanidine | $KAlSi_3O_8$ | Disordered | 8.6 | 13.03 | 7.18 | 90.0 | 116.0 | 90.0 |
| Low-albite | $NaAlSi_3O_8$ | Ordered | 8.14 | 12.79 | 7.16 | 94.3 | 116.6 | 87.7 |
| High-albite | $NaAlSi_3O_8$ | Disordered | 8.16 | 12.87 | 7.11 | 93.5 | 116.4 | 90.3 |
| Anorthite | $CaAl_2Si_2O_8$ | Ordered | 8.18 | 12.88 | 14.17 | 93.2 | 115.8 | 91.2 |

can be achieved by heat treatment, the first crystals to form on heating anorthite glass at temperatures 1100∼1400 °C have substantial Al/Si disorder [27]. Investigating the Al/Si disordered anorthite synthesized from anorthite glass at 1070 °C using TEM, John and Müller [28] concluded that this phase occurs only as a metastable phase and microscopic scale. The $^{29}Si$ MAS NMR study on the similar anorthite glass [29] also suggested that the Al/Si disordered arrangement may be locked up upon crystallization of Ca-feldspar from a glass. Afterwards, Phillips et al. [30] examined the similar synthetic Ca-feldspar crystallized from glass using $^{29}Si$ MAS NMR and determined the average number of $^{IV}Al^{3+}$–O–$^{IV}Al^{3+}$ linkages per formula unit as 0.4. The determination indicated that anorthite does not always obey the Al-avoidance rule.

Echigo et al. [18] found completely Al/Si disordered Ca-feldspar from the basaltic lava of the Miyake-jima island, Japan, one of the volcanic islands belonging to the Izu-Ogasawara arc located along the boundary between the Pacific and Philippine Seat Plates [31, 32]. Anorthite from Miyake-jima were often used in several past studies, e.g., in-situ high temperature XRD [33], NMR analysis [30], melting/quenching experiments [34], high resolution TEM study [35], while their eruption age and location were not described in the papers. Kimata and coworkers [16] published a review about mineralogical and geochemical characteristics of anorthite megacrysts from Japanese Island arc basalts, dealing with a description of the Miyake-jima anorthite. The authors reported that the anorthites from the Japanese Island arc basalts commonly contain defects in A-sites (non-tetrahedral sites) and $\square Si_4O_8$ end-member (silica solid solution) up to 1.4 mol%.

In Al/Si disordered Ca-feldspar from Miyake-jima, defect in A-sites was detected by both chemical analysis with electron microprobe and structural analysis with single crystal XRD. Echigo et al. [18] discussed the effect of such chemical non-stoichiometry on the stability of metastable Al/Si phase. Since early 1970s, crystal chemistry of non-stoichiometric feldspar has been studied by preparing synthetic feldspars as summarized in Table 2.2. Most of the synthetic feldspars with $Al_2Si_2$-framework have similar lattice parameters *c* ~7Å, except for synthetic Ca-feldspar [36]. As shown in the comparison of crystal structures of Al/Si ordered [37] and disordered [18] anorthite (Fig. 2.5), the lattice parameter *c* ~7Å suggests that their Al/Si configuration in the tetrahedral framework is disordered. Besides that, Ito [36] reported that his synthetic Ca-feldspar has body-centered structure (I-anorthite) with partially disordered Al/Si configuration even though it was synthesized by slow cooling (2 °C/hour). For natural Ca-feldspar, Laves and Goldsmith [34] conducted melting/quenching experiments using Ca-feldspar from Miyake-jima [$(Ca_{0.97},Na_{0.03})Al_{1.93}Si_{2.06}O_8$] and revealed that rapid crystallization of anorthite melt that contains small amount of Na produced an metastable Al/Si disordered crystal. These crystal chemical investigations suggest that the slight vacancy in extra-framework sites and the Al-deficiency in tetrahedral sites due to incorporation of $\square Si_4O_8$ end-member [16] are significant characteristics common to Al/Si disordered $Al_2Si_2$-feldspars. One possible interpretation on this observation is that $\square Si_4O_8$ component stabilizes the Al/Si disordered $Al_2Si_2$-feldspars by introducing $^{IV}Si^{4+}$–$O^{2-}$–$^{IV}Si^{4+}$ bonds into the tetrahedral framework.

**Table 2.2** Chemical composition and lattice parameters of c-axis of synthetic $Al_2Si_2$-feldspars with partly or completely Al/Si disordered configuration

| Cations in A-site | Chemical composition | Lattice parameter $c$ | References |
|---|---|---|---|
| $Ca^{2+}$ | $(Ca_{0.884}Na_{0.006}\square_{0.11})Al_{1.80}Si_{2.20}O_8$ | $c = 14.186$ Å | [36] |
| $Sr^{2+}$ | $(Sr_{0.84}Na_{0.03}\square_{0.13})Al_{1.69}Si_{2.29}O_8$ | $c = 7.1358$ Å | [38] |
| $Eu^{2+}$ | $(Eu_{0.96}\square_{0.04})Al_{1.92}Si_{2.08}O_8$ | $c = 7.124$ Å | [39] |
| $Mn^{2+}$ | $(Ca_{0.72}Mn_{0.20}Na_{0.04}\square_{0.04})Al_{1.867}Si_{2.133}O_8$ | $c = 7.069$ Å | [40] |
| $Pb^{2+}$ | $Pb_{1.02}Al_{1.98}Si_{2.00}O_8$ | $c = 7.174$ Å | [41] |

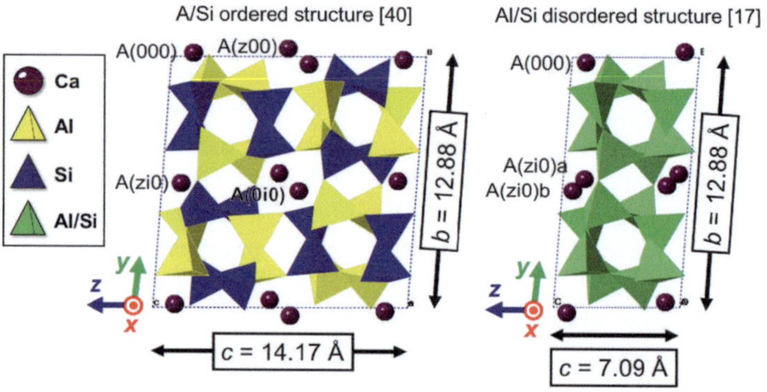


**Fig. 2.5** Crystal structures of Al/Si ordered (left) [37] and disordered (right) [18] anorthite along *a*-axis

Although the formation mechanism of anorthite megacryst remains unclear, crystal growth under temperature cycling would be one possible mechanism that produces anorthite megacryst with disordered Al/Si configuration. Mills and Glazner [42] demonstrated that coarsening of plagioclase crystals is greatly accentuated by temperature cycling (1150 °C at 1 atm, temperature amplitude up to ± 10 °C and oscillation period up to 60 min). On the basis of their experiments, it is suggested that sufficiently small crystals dissolve completely at the high-temperature point whereas larger crystals survive and nucleation is suppressed due to the low degree of undercooling during the temperature cycle and the presence of existent crystals. The decrease in the number of crystals may cause the inverse relationship between crystal size and surface energy (Ostwald ripening), which leads to slow destruction of small crystals and increasing of average crystal size. Mills and Glazner [42] also discussed the growth rates of plagioclase crystals in terms of "size-proportional crystal growth" as follows: during cooling low-energy (large) crystals grow at a faster linear rate than the high-energy (small) crystals. This size-proportional crystal growth [43, 44] upon cooling leads to a decrease in the number of crystals after multiple cycles and to growth of megacrysts which are much larger than other phenocrysts in pre-eruptive magma.

It is well known that variations in the activity of $H_2O$ in magma have a profound effect on crystallization/melting temperature [14], hence repeated injection of hydrous magmas into magma chamber may produce similar fluctuations for growth of larger crystals in magma by repeated release of H2O during successive crystallization events [45]. Island arc magma from which anorthite megacrysts precipitated contains significant amount of water (c.a., 3–6%) [46]. Słaby and co-workers [47] described that the incorporation of hydrogen $H^+$ into feldspar structure can increase Al-$O^-$-Al defects density by compensation of electric charge and water in magma may have contributed to the formation of the Al/Si disordered feldspar. In addtion, the melting experiments of natural basaltic lava showed that nearly $H_2O$-saturated conditions of 2–3 kbar is optimal for the crystallization of the most Ca-rich plagioclase [48].

## 2.4 Conclusions

This chapter reviewed extraordinary crystals of natural minerals found in mining sites and active volcanoes. These involve organic minerals occurring in hydrothermal mercury deposits, with a perspective on the fact that such organic compounds intuitively seem to be decomposed easily under high temperature and pressure environments. In addition, Because of its microtexture, natural karpatite exhibits fluorescent properties that synthetic crystals do not have. If the microtexture of organic crystals can be artificially controlled, it may be possible to produce crystals with the desired optical properties. Research on natural organic crystals would help us to develop such technology.

The second example is large plagioclase crystals, which is called anorthite megacrysts, in quenched lava. Though plagioclase is one of the most common mineral groups in the crust of Earth, it is remarkable that anorthite megacrysts partly contain single crystals of Al/Si disordered phase which is metastable and not crystallized under equilibrium condition. This metastable phase did not transform into more stable phase with ordered Al/Si configuration due to the incorporation of small amounts of $Na^+$ and defect in the extra-framework sites. In addition, dynamic processes such as repeated magma injection and convection in a huge open-system reaction vessel (magma chamber) promoted the growth of metastable phase. In natural environments, we can find many examples of metastable phases such as aragonite (high-pressure form of $CaCO_3$) in living bodies, adularia (high-temperature form of $KAlSi_3O_8$) and wurtzite (high-temperature form of ZnS) in low-temperature hydrothermal ore deposits. Vaterite and marcasite are also metastable phases which are often found in living bodies and hydrothermal ore deposits, respectively, while both of them do not have fields of stability. The metastable phases often show superior or preferable properties to the stable phase, and understanding the formation and stabilization mechanisms of natural metastable phases may be useful for synthesis of large crystals of high-performance device. Close collaboration between researchers of natural minerals and electronic materials could create new crystalline materials that contribute to developing a sustainable society through the studies of extraordinary mineral crystals.

## References


1. R.M. Cornell, U. Schwertmann, *The Iron Oxides: Structure, Properties, Reactions, Occurrences and Uses* (Wiley, 2006)
2. J. Sanyova, G. van der Snickt, F. Mederos-Henry, The challenges of a complex stratigraphy from a chemical point of view, in *The Ghent Altarpiece* (Brepols Pub, 2021), pp. 111–114
3. H.R. Wenk, A. Bulakh, *Minerals: Their Constitution and Origin* (Cambridge University Press, 2016)
4. A.C.S. van Heel, A luminous quartz fluorite achromatic lens for magnification—1. Physica **1**(1–6), 35–40 (1934). https://doi.org/10.1016/S0031-8914(34)90005-7
5. T. Echigo, M. Kimata, T. Maruoka, Crystal-chemical and carbon-isotopic characteristics of karpatite ($C_{24}H_{12}$) from the Picacho Peak Area, San Benito County, California: evidences for the hydrothermal formation karpatite. Am. Miner. **92**(8–9), 1262–1269 (2007). https://doi.org/10.2138/am.2007.2509
6. J. Hedenquist, J. Lowenstern, The role of magmas in the formation of hydrothermal ore deposits. Nature **370**, 519–527 (1994). https://doi.org/10.1038/370519a0
7. F.E. Wright, E.T. Allen, Curtisite a new organic mineral from Skaggs Springs, Sonoma County, California. Am. Mineral. **15**(5), 160–173 (1930)
8. T.A. Geissman, K.Y. Sim, J. Murdoch, Organic minerals picene and chrysene as constituents of the mineral curtisite (idrialite). Experientia **23**(10), 793–794 (1967). https://doi.org/10.1007/BF02146842
9. J. Hoefs, *Stable Isotope Geochemistry* (Springer, 2004)
10. T. Echigo, M. Kimata, Crystal chemistry and genesis of organic minerals—a review on oxalate and polycyclic aromatic hydrocarbon minerals. Can. Mineral. **48**(6), 1329–1358 (2010). https://doi.org/10.3749/canmin.48.5.1329
11. S. E. Stein, On the high temperature chemical equilibria of polycyclic aromatic hydrocarbons. Jour. Phys. Chem. **82**(5), 566–571 (1978). https://doi.org/10.1021/j100494a600
12. J. Potticary, T.T. Jensen, S.R. Hall, Nanostructural origin of blue fluorescence in the mineral karpatite. Sci. Rep. **7**, 9867 (2017). https://doi.org/10.1038/s41598-017-10261-w
13. H.U. Schmincke, *Volcanism* (Springer, Berlin Heidelberg, 2012)
14. Gill, R., *Igneous Rocks and Processes: A Practical Guide* (Wiley, 2011)
15. M.D. Higgins, D. Chandrasekharam, Nature of sub-volcanic magma chambers, Deccan Province, India: evidence from quantitative textural analysis of plagioclase megacrysts in the giant plagioclase basalts. J. Petrol. **48**(5), 885–900 (2007). https://doi.org/10.1093/petrology/egm005
16. M. Kimata, N. Nishida, M. Shimizu, S. Saito, T. Matsui, Y. Arakawa, Anorthite megacrysts from island arc basalts. Mineral. Mag. **59**(394), 1–14 (1995). https://doi.org/10.1180/minmag.1995.59.394.01
17. T. Ishikawa, Petrological significance of large anorthite crystals included in some pyroxene andesites and basalts in Japan. J. Fac. Sci., Hokkaido Univ. Ser. 4, Geol. Mineral. **7**(4), 339–354 (1951). https://eprints.lib.hokudai.ac.jp/dspace/handle/2115/35845
18. T. Echigo, M. Hoshino, M. Kimata, M. Shimizu, T. Matsui, N. Nishida, Single crystal X-ray and electron microprobe study of Al/Si-disordered anorthite with low content of albite. Zeitschrift für Kristallographie—Cryst. Mater. **229**(6), 435–449 (2014). https://doi.org/10.1515/zkri-2013-1713
19. S. Ohmagari, H. Yamada, H. Umezawa, N. Tsubouchi, A. Chayahara, Y. Mokuno, Growth and characterization of freestanding p+ diamond (100) substrates prepared by hot-filament chemical vapor deposition. Diam. Relat. Mater. **81**, 33–37 (2018). https://doi.org/10.1016/j.diamond.2017.11.003

20. K. Kaneko, S. Fujita, T. Shinohe, K. Tanaka, Progress in α-$Ga_2O_3$ for practical device applications. Japan. J. Appl. Phys. **62,** SF0803 (2023). https://doi.org/10.35848/1347-4065/acd125
21. W.D. Nesse, *Introduction to Mineralogy,* 3rd edn (Oxford University Press, 2016)
22. I. Parsons, *Feldspars and Their Reactions* (Springer, Netherlands, 2012)
23. R.D. Shannon, Revised effective ionic radii and systematic studies of interatomic distances in halides and chalcogenides. Acta Crystallogr. A **32**(5), 751–767 (1976). https://doi.org/10.1107/S0567739476001551
24. W. Loewenstein, The distribution of aluminum in the tetrahedra of silicates and aluminates. Am. Miner. **39**(1–2), 92–96 (1954)
25. E. Bruno, G. Chiari, A. Facchinelli, Anorthite quenched from 1530 °C I structure refinement. Acta Crystallogr. B **32**, 3270–3280 (1976). https://doi.org/10.1107/S056774087601008X
26. G. Chiari, A. Facchinelli, E. Bruno, Anorthite quenched from 1530 °C II discussion. Acta Crystallograph. Sect. B., **1757–1764**(1978). https://doi.org/10.1107/S0567740878006652
27. M.A. Carpenter, Mechanisms and kinetics of Al-Si ordering in anorthite: I. Incommensurate structure and domain coarsening. Am. Mineral. **76**(7–8), 1110–1119 (1991). https://pubs.geoscienceworld.org/msa/ammin/article-abstract/76/7-8/1110/42554/
28. R.J. John, W.F. Müller, Experimental studies on the kinetics of order-disorder processes in anorthite, $CaAl_2Si_2O_8$. Neues Jb. Mineral. Abh. **159**(3), 283–295 (1988)
29. B.H.W.S. De Jong, C.M. Schramm, V.E. Parziale, Polymerization of silicate and aluminate tetrahedra in glasses, melts, and aqueous solutions—V the polymeric structure of silica in albite and anorthite composition glass and the devitrification of amorphous anorthite. Geochim. Cosmochim. Acta. **48**(7), 2619–2629 (1984). https://doi.org/10.1016/0016-7037(84)90310-7
30. B.L. Phillips, R.J. Kirkpatrick, M.A. Carpenter, Investigation of short-range Al, Si order in synthetic anorthite by $^{29}Si$ MAS NMR spectroscopy. Am. Miner. **77**(5–6), 484–494 (1992)
31. S. Maruyama, Y. Isozaki, G. Kimura, M. Terabayashi, Paleogeographic maps of the Japanese Islands: plate tectonic synthesis from 750 Ma to the present. Island Arc. **6**(1), 121–141 (1997). https://doi.org/10.1111/j.1440-1738.1997.tb00043.x
32. R.J. Stern, M.J. Fouch, S.L. Klemperer, An overview of the Izu-Bonin-Mariana subduction factory, in *Inside the Subduction Factory,* ed. by J. Eiler (American Geophysical Union, 2003), pp. 175–222. https://doi.org/10.1029/138GM10
33. F.F. Foit, J.R. Peacor, The anorthite crystal structure at 410 and 830 °C. Am. Miner. **58**(7–8), 665–675 (1973)
34. F. Laves, J.R. Goldsmith, The effect of temperature and composition on the Al-Si distribution in anorthite. Zeitschrift für Kristallographie—Cryst. Mater. **106**(1–6), 227–235 (1955). https://doi.org/10.1524/zkri.1954.106.16.227
35. A.C. McLaren, The domain structure of a transitional anorthite; a study by direct lattice-resolution electron. Contrib. Mineral. Petrol. **41**, 47–52 (1973). https://doi.org/10.1007/BF00377651
36. J. Ito, High temperature solvent growth of anorthite on the join $CaAl_2Si_2O_8$–$SiO_2$. Contrib. Mineral. Petrol. **59**, 187–194 (1976). https://doi.org/10.1007/BF00371307
37. J.E. Wainwright, J. Starkey, A refinement of the structure of anorthite. Zeitschrift für Kristallographie—Cryst. Mater. **133**(1–6), 75–84 (1971). https://doi.org/10.1524/zkri.1971.133.16.75
38. H.D. Grundy, J. Ito, The refinement of the crystal structure of a synthetic non-stoichiometric Sr feldspar. Am. Miner. **59**(11–12), 1319–1326 (1974)
39. M. Kimata, The crystal structure of non-stoichiometric Eu-anorthite: an explanation of the Eu-positive anomaly. Mineral. Mag. **52**(365), 257–265 (1988). https://doi.org/10.1180/minmag.1988.052.365.13
40. T. Matsui, M. Kimata, Crystal chemistry of synthetic Mn-bearing anorthite; incorporation of $MnAl_2Si_2O_8$ end-member into feldspar. Eur. J. Mineral. **9**(2), 333–344 (1997). https://doi.org/10.1127/ejm/9/2/0333
41. P. Benna, M. Tribaudino, E. Bruno, The structure of ordered and disordered lead feldspar ($PbAl_2Si_2O_8$). Am. Miner. **81**(11–12), 1337–1343 (1996). https://doi.org/10.2138/am-1996-11-1205

42. R.D. Mills, J.J. Ratner, A.F. Glazner, Experimental evidence for crystal coarsening and fabric development during temperature cycling. Geology **39**(12), 1139–1142 (2011). https://doi.org/10.1130/G32394.1
43. T.F. Canning, A.D. Randolph, Some aspects of crystallization theory: systems that violate McCabe's delta *L* law. AIChE J. **13**(1), 5–10 (1967). https://doi.org/10.1002/aic.690130104
44. D.D. Eberl, D.E. Kile, V.A. Drits, On geological interpretations of crystal size distributions: constant vs. proportionate growth. Am. Mineral. **87**(8–9), 1235–1241 (2002). https://doi.org/10.2138/am-2002-8-923
45. A.F. Glazner, The ascent of water-rich magma and decompression heating: a thermodynamic analysis. Am. Miner. **104**(6), 890–896 (2019). https://doi.org/10.2138/am-2019-6925
46. M. Hamada, T. Kawamoto, E. Takahashi, T. Fujii, Polybaric degassing of island arc low-K tholeiitic basalt magma recorded by OH concentrations in Ca-rich plagioclase. Earth Planet. Sci. Lett. **308**(1–2), 259–266 (2011). https://doi.org/10.1016/j.epsl.2011.06.005
47. E. Słaby, H. Martin, M. Hamada, M. Śmigielski, A. Domonik, J. Götze, J. Hoefs, S. Hałas, K. Simon, J.L. Devidal, J. Moyen, l. Jayananda, Evidence in Archaean alkali feldspar megacrysts for high-temperature interaction with mantle fluids. J. Petrol. **53**(1), 67–98 (2012). https://doi.org/10.1093/petrology/
48. D. Takagi, H. Sato, M. Nakagawa, Experimental study of a low-alkali tholeiite at 1–5 kbar: optimal condition for the crystallization of high-An plagioclase in hydrous arc tholeiite. Contrib. Mineral. Petrol. **149**(5), 527–540 (2005). https://doi.org/10.1007/s00410-005-0666-7

# Chapter 3
# Growth and Modifications of Crystals on Airless Surfaces

## The Effects of Space Weathering on Extraterrestrial Materials

**Toru Matsumoto**

**Abstract** Understanding the modification and synthesis of crystals in the space environment, which is very different from the terrestrial laboratory, can be useful in exploring new possibilities for synthesis methods of nano-engineered materials. This chapter reviews the analysis of extraterrestrial samples from asteroids and the Moon, focusing on crystal growth and modifications induced by ion irradiation and micrometeoroid impacts that occur on airless surfaces in our solar system. Solid materials exposed to the space environment are irradiated by high-energy ions originating from the Sun, known as solar wind and solar flares. Airless surfaces are also impacted by nanometer- to micrometer-sized dust particles (called micrometeoroids) ejected from asteroids and comets. The solar wind irradiation and micrometeoroid impacts are ongoing on many small bodies in the solar system, such as the Moon and asteroids, because these small bodies, unlike Earth, cannot maintain a dense atmosphere by their own gravity and are directly exposed to the space environment. Several types of crystal growth and nanostructure modifications of space-exposed materials have been reported, such as amorphization, blistering, nanoparticle synthesis, the vapor–liquid-solid (VLS) growth and whisker formation. Evidence of interaction between solid materials and the space environment was first recognized by the analysis of lunar samples brought back by NASA's Apollo program. In addition, the asteroid sample return missions of the Japanese spacecraft Hayabusa and Hayabusa-2 have provided opportunities for the analysis of space-exposed samples from asteroids Itokawa and Ryugu. In the field of planetary materials science, modifications in the space environments on airless bodies are referred to as space weathering. Space weathering has attracted attention in planetary science and space exploration because it alters optical properties of the surface reflection spectrum, which is important for the interpretation of remote-sensing data, and may promote re-distribution of surface materials. The structure of space weathering is often less than a micrometer in size and has been

T. Matsumoto (✉)
The Hakubi Center for Advanced Research/Division of Earth and Planetary Sciences, Kyoto University Kitashirakawa-Oiwakecho, Sakyo 606-8502, Kyoto, Japan
e-mail: matsumoto.toru.2z@kyoto-u.ac.jp



F. Ishikawa et al. (eds.), *Engineering Crystal Habit*, Topics in Applied Physics 152,
https://doi.org/10.1007/978-981-96-0266-7_3

observed mainly using scanning electron microscopy (SEM) and transmission electron microscopy (TEM). Experimental study to understand modification processes in crystals due to space weathering are performed by ion irradiation experiments simulating solar wind irradiation and femtosecond laser pulse irradiation simulating energy injection during micrometeoroid impacts.

## 3.1 Extraterrestrial Materials Showing Space Weathering Features

Our solar system contains many small bodies, such as asteroids, comets, and satellites, which have no atmosphere and are directly exposed to the space environment. The term "space weathering" refers to as the unique modifications of solid materials on airless surfaces as a result of their interaction with the space environment [1]. The major causes of the modifications are solar wind irradiation and bombardments of micrometeorites (Fig. 3.1).

The airless small bodies are covered by unconsolidated layers of rock and soil, called regolith, which is formed through fragmentation of large rocks by meteorite impacts and by thermal fatigue [2]. Space weathering has been studied extensively through the analysis of regolith grains brought back to Earth from the lunar surface and asteroids. Lunar regolith grains have been recovered by the Apollo and the Luna mission and more recently by the China's Chang'e mission. Asteroid regolith samples have been collected and delivered to Earth from the asteroids Itokawa and Ryugu by the Japanese spacecrafts, Hayabusa and Hayabusa2, respectively (Fig. 3.2). Lunar

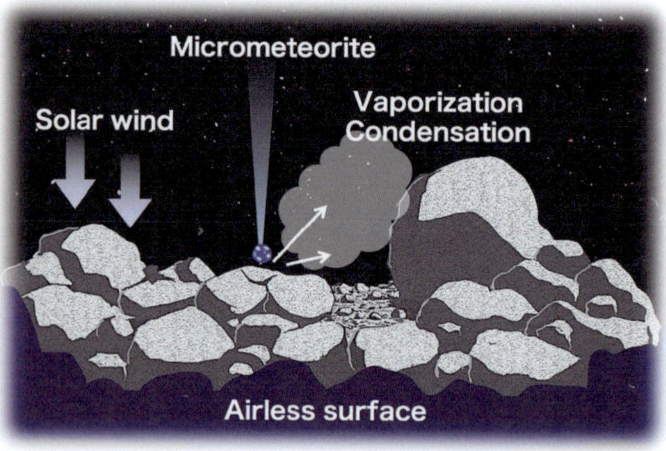


**Fig. 3.1** Schematic image of space weathering effects on airless surface, including solar wind irradiation, micrometeoroid bombardment and subsequent vaporization and condensation

**Fig. 3.2** Asteroid Itokawa **a**, asteroid Ryugu **b**, and fine particles from asteroid Itokawa (**c**). Credit for **a** and **b**: © ISAS/JAXA, Univ. of Tokyo, Kochi Univ. Rikkyo Univ., Nagoya Univ., Meiji Univ., CIT., Aizu Univ., and AIST

minerals consist mainly of silicate minerals including plagioclase, pyroxene and olivine, as well as titanium oxides (ilmenite) and minor mineral phases such as iron sulfides [3]. Itokawa regolith grains are composed of silicate minerals, iron sulfides and iron-nickel metals and correspond to ordinary chondrite meteorites (Fig. 3.2) [4]. In contrast to these anhydrous constituents, Ryugu samples have experienced aqueous alteration and contain hydroxyl ions and water molecules in their layered silicate minerals [5]. Space weathering features have been identified in the major minerals of these retuned samples, demonstrating that mineral modifications due to solar wind irradiation and micrometeorite bombardments are common phenomena on airless bodies in the solar system. In addition to the regolith grains obtained by the direct sampling by the space missions, meteorites (pieces of rock that have fallen to Earth, originated from other celestial bodies), micrometeorites (submillimeter-sized meteorites), and interplanetary dust particles (submillimeter-sized dust particles that have entered the Earth and were collected in the stratosphere at an altitude of 20–25 km [6]) are extraterrestrial materials available for laboratory analysis of space weathering. Meteorites classified as "regolith breccia" are composed of compacted rock fragments and fine grains compacted near the surface of asteroids and contain relatively high concentrations of noble gases derived from solar wind implantation [7]. Silicate grains in the regolith breccia have small craters formed by micro-impacts [8], redeposition layers formed by micrometeoroid impacts [9], and impact-induced glasses [10]. Silicate minerals in micrometeorites and interplanetary dust particles show damaged features caused by solar wind and solar flare irradiation during their journey through interplanetary space [11].

## 3.2 Ion Implantation in the Space Environment

Solar wind is a quasi-stationary outflow of ionized gas from the outer atmosphere called the solar corona. It is composed of electrons and ions with 95% of protons, 5% of helium, and minor other elements. Ions from the solar corona are accelerated along the magnetic field surrounding the Sun and delivered to interplanetary space as solar wind with a typical velocity of ~300 km/s to ~800 km/s (~1 keV/amu:

atomic mass unit) [12]. The flux of the solar wind is the order of $10^8$ $cm^2/s$ at the orbit of Earth, 1 astronomical unit (AU) from the Sun [13], and is decreases with increasing heliocentric distance from the Sun. Solar wind with the flux of this magnitude affect the lunar surface. The asteroid Itokawa and Ryugu have orbits of about 1AU in heliocentric distance, therefore, their surfaces also have experienced solar wind irradiation. The outflow of solar wind extends to ~100 AU in radial extent, defining the region surrounding the Sun, called the heliosphere. Solar wind is a steady flow of ions but is often derived from transient events, known as coronal mass ejections, which are originated above magnetically active regions on the Sun's surface and cause large streams of solar wind [14]. The solar corona also contains a large-scale magmatic energy release that results in eruptive phenomena called solar flares. It produces energetic ions with energies ranging from ~1MeV/amu to ~ 100 MeV/amu. The penetration of solar wind and solar flare to the Earth's surface is disrupted by their interaction with Earth's atmosphere and magnetic field. On the contrary, solar wind and solar flares are directly implanted into surface minerals of small celestial bodies that lack a thick atmosphere. The upper surface of the regolith layers of these bodies are exposed to the Sun for a certain period of time. Solar wind with the energy of 1 keV/amu is implanted into the minerals within the depth of 100 nm from the surface, causing amorphization of the crystal structures (Fig. 3.3), accumulation of solar gases, and formation of new crystal phases. The following sections provide detailed descriptions of the modifications due to solar wind irradiation. Energetic ions with MeV energy included in the solar flares can penetrate a few millimeters into the regolith layers and produce a trail of atomic displacement in minerals (Fig. 3.3). The tracks of displacement damage are called solar flare tracks. Their density provides the time interval that mineral grains stayed within a few millimeters from airless surface (e.g., [11]). Ion irradiation is also expected to occur in the interstellar environment (interstellar medium: ISM) and modify dust grains floating in the ISM. Ionized atoms accelerated by shock waves from supernovae in the ISM have been proposed to cause amorphization of crystalline silicates that were produced around evolved stars and transported to the ISM (e.g., [15]).

## 3.3 Crsytallographic Damages Formed by Solar Wind Implantation

When ions with the order of keV energy are implanted in solid materials, collision cascades of atoms are generated along their path through the crystals before they lose their kinetic energy, resulting in the formation of large numbers of vacancies and interstitial atoms. The weakest modification is observed as distortions of the crystal lattice within the depth of a few tens of nanometers, where solar wind ions are implanted and trapped (Fig. 3.3) [16]. Moderate modification by solar wind is partial or complete amorphization caused by the atomic displacement over a region

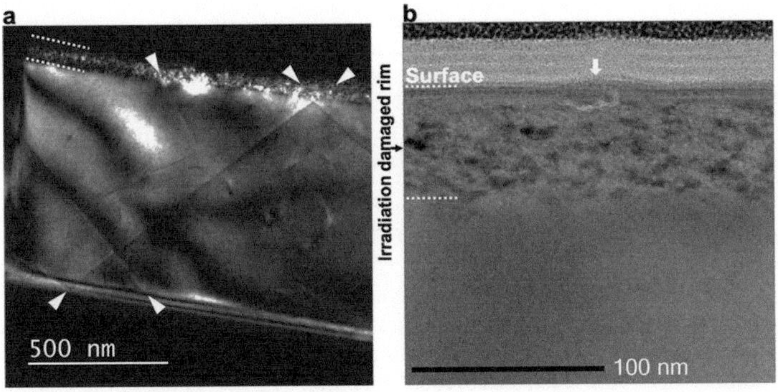


**Fig. 3.3** Transmission electron microscopy image (TEM) of olivine of Itokawa particles. **a** Solar flare tracks are indicated by triangles. Dotted lines indicate surface amorphous rim formed mainly by solar wind implantation. **b** Dotted lines indicate amorphous rim formed by crystallographic damages during solar wind irradiation. The damaged rim shows patchy contrast due to crystallographic misorientation and amorphization in the rim, whereas, unmodified substrate beneath the rim has uniform contrast. Blistering structure with vesicle is marked with an arrow

where the cascade occurs [17–19]. Silicate minerals (olivine, pyroxene, and plagioclase) in Itokawa particles show partially amorphous rims (damaged region mixed with amorphous and crystalline regions) probably formed by the implantation of hydrogen ions and helium ions in the solar wind (Fig. 3.3) [18]. Completely amorphous rims have been identified in silicate minerals of lunar regolith. Amorphous rims have also been found in phyllosilicate minerals in Ryugu samples [19]. The damage to mineral surfaces by ion irradiation in space is similar to ion damage in nuclear reactor materials and has been better understood by comparison with ion irradiation experiments. Irradiation experiments have been carried out with samples set up in an ultra-vacuum chamber and reproduced amorphization of silicate minerals by irradiation of ions with 1 keV/amu (e.g., [20]).

Unlike silicate minerals, complete amorphization does not appear to occur in space-exposed iron sulfides in Itokawa grains and lunar grains [21], and experimentally ion-irradiated sulfides [22]. Instead, the disappearance of their superstructures has been observed in electron diffraction patterns obtained from the damaged region within the depth of ~100 nm below the surface (Fig. 3.4). This indicates the disordering of atomic arrangements longer than the basic crystal lattice called "1C" in iron sulfides as a result of ion implantation. The reason for the lack of amorphization in iron sulfides is the rapid recovery from the atomic displacement due to their high thermal conductivity and their simple atomic arrangement compared to silicate minerals [22].

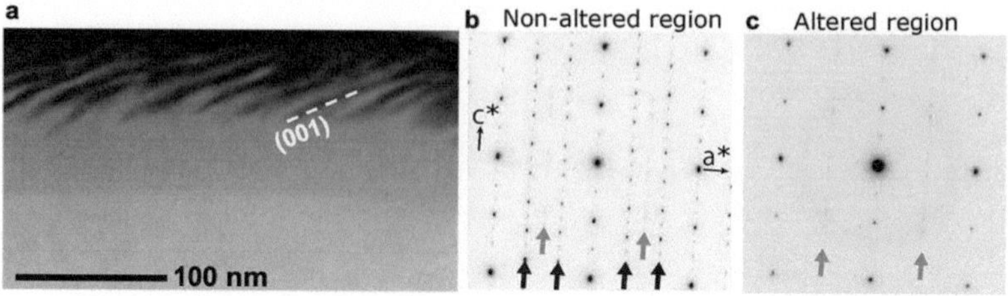


**Fig. 3.4** Space weathered iron sulfides ($Fe_{1-x}S$). **a** scanning transmission electron microscope (STEM) image of the sulfide surface of an Itokawa particle. Open vesicles parallel to the c-plane (001) are indicated by a dotted line. **b, c** Selected area electron diffraction (SAED) patterns of the iron sulfide of lunar grain from the Apollo 17 landing site. The SAED pattern in **b** is obtained from non-altered region of the iron sulfide and that in **c** is obtained from altered region, which is affected by space weathering. Both patterns are indexed by the 1C cell. Black arrows in **b** show diffraction spots of 2C superstructure, which is two time larger than the 1C cell. Gray arrows in **c** show diffraction spots of non-integral superstructure. Note that 2C superstructure reflections disappear in (**c**)

## 3.4 Vesiculation and Blistering by Accumulation of Solar Wind Ions

The partially or completely amorphous layers of silicate minerals, Fe-Ti oxides, and iron sulfides in lunar regolith and Itokawa particles often include vesicles with a typical size of several tens of nm (Fig. 3.2b). [23]. They have a lenticular shape and are elongated parallel to the surface. Some vesicles lift up the surface and form the blistering structure (Fig. 3.5). This structure is similar to the ion-irradiated materials used in the walls of nuclear reactors [24]. The formation of vesicles in the irradiated materials may be due to the accumulation of implanted atoms with the saturation limit in crystal structures. The concentration of implanted ions may have been promoted by the development of dislocation loops and the migration of vacancies and implanted atoms into the loops [25]. The electron energy-loss spectroscopy (EELS) technique has been used to visualize the distribution of helium accumulated within lunar regolith grains [26]. The EELS analysis showed that helium is concentrated in vesicles and in the crystallographically oriented defects along the (001) planes in ilmenite, suggesting that the movement of atoms into the defects drives the concentration of helium and the formation of vesicles. Open vesicles aligned along the (001) planes have been observed on the space-exposed surface of iron sulfides in Itokawa particles and lunar regolith [27] (Fig. 3.4a). Dislocation loops may preferentially form along the (001) plane of hexagonal iron sulfides due to its minimum surface energy, attracting vacancies and implanted atoms that may cause vesiculation. It has been proposed that heating due to nearby micrometeoroid bombardment is necessary to form vesicles in silicate minerals by releasing solar wind gases trapped in the crystals [28]. Blisters can be non-destructively identified by surface observation using scanning electron microscopy, allowing the distribution of space weathered surfaces on regolith particles to be examined. The distribution of blisters on grain

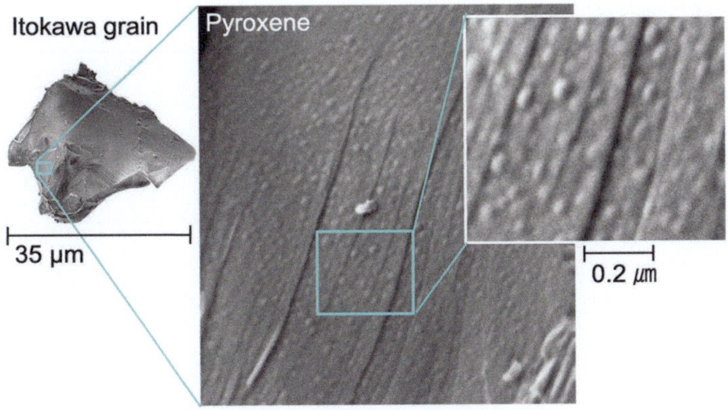


**Fig. 3.5** SEM image of blistering structures on the surface of Itokawa particle. Numerous bright spots on the right image correspond to the blisters formed by the accumulation of solar wind gases

surfaces suggests the rotation and fragmentation of particles by regolith activity in microgravity [29].

## 3.5 Micrometeoroid Bombardment

Materials on airless surfaces are also modified by the impacts of small dust from the interplanetary space [1]. Micrometeoroids are dust particles less than a few tens of millimeters in size, and originate from asteroids, comets, planets and their moons and rings, and the interstellar environment. They are transported from their sources by gravitational forces of the Sun and planets, radiation pressures, and magnetic fields. Micrometeoroid bombardment causes the formation of small impact craters (Fig. 3.6) [30], vaporization and subsequent condensation of the vaporized materials, and transient heating of nearby mineral surfaces by the hot vapors. Minerals exposed to space are often modified by the combined effects of micrometeoroid impacts and solar wind irradiation.

## 3.6 Condensation Products

In regolith grains, vapor deposition layers, or also called redeposition rims, of various thickness are observed above the amorphous rims formed by solar wind implantation (Fig. 3.7). The redeposition rims consist of amorphous materials containing elements not present in the substrate. They are products of sputtering by solar wind irradiation and/or condensation of impact-generated vapors from neighboring mineral grains.

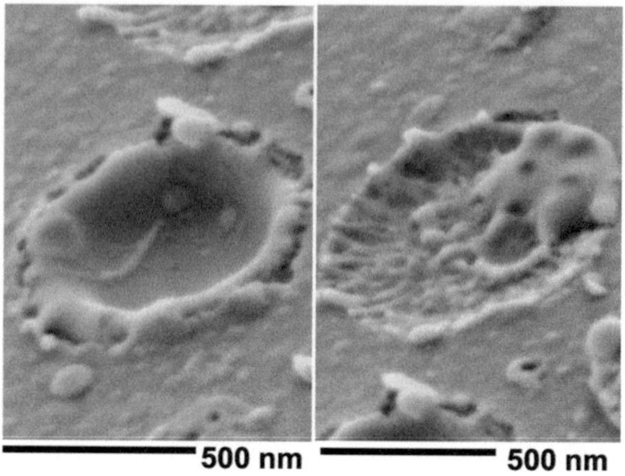


**Fig. 3.6** SEM images of small impact craters on the surface of silicate minerals of an Itokawa particle. The crater floors are covered by melted materials

The thickness of the redeposition rims is several nm in Itokawa grains [18] and up to 200 nm in lunar regolith grains [23].

Vapor condensates with a distinct morphology have been identified in lunar regolith samples. A number of stalks with bulbs at their tips have been found on lunar regolith grains by SEM study [31]. The bulbs have diameter up to 0.2 μm and the length of stalks varies from 0.015 to 0.15 μm. The bulbs are probably a mixture of iron and sulfur, while the stalks are composed of metallic iron. The chemical composition suggests that the stalks and bulbs are the result of the vapor–liquid-solid (VLS) growth mechanism operating in vapors produced during an impact event. The iron stalks may have been formed from the melted bulbs where iron atoms are incorporated as impurities and were expelled as metallic iron stalks. Lunar samples also contain graphite whiskers as condensates [32]. The graphite whiskers embedded in lunar rocks were identified by Raman imaging spectroscopy. They could have been

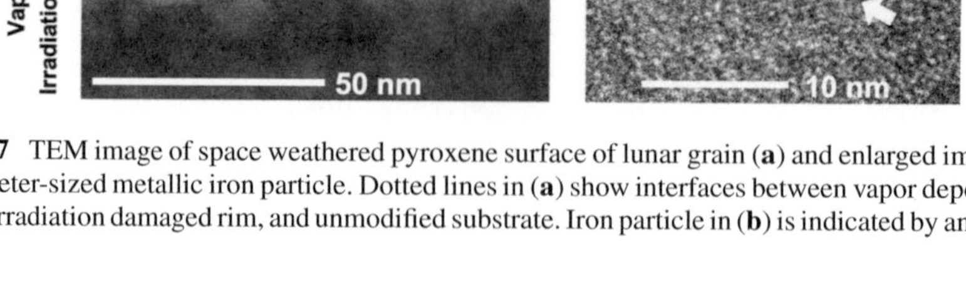


**Fig. 3.7** TEM image of space weathered pyroxene surface of lunar grain (**a**) and enlarged image of nanometer-sized metallic iron particle. Dotted lines in (**a**) show interfaces between vapor deposition layer, irradiation damaged rim, and unmodified substrate. Iron particle in (**b**) is indicated by an arrow

formed by condensation of carbon-rich gas released during impacts of carbonaceous materials.

## 3.7 Formation of Nanoparticles of Iron Compounds

Nanometer-sized metallic iron particles (nanophase $Fe^0$ or $npFe^0$) [1] are dispersed in impact-generated glasses called agglutinates, redeposition rims produced by sputtering and/or micrometeoroid impacts, and amorphous rims formed by solar wind implantation (Fig. 3.7). These nanoparticles are the product of space weathering and are mainly derived from iron-containing silicates, sulfides and oxides, which contain ferrous ($Fe^{2+}$) iron or ferric ($Fe^{3+}$) iron in their crystal structures. In addition to metallic iron, nanoparticles of (Mg,Fe)S are present in the redeposition rims of Itokawa particles [18]. The $npFe^0$ found in the deposition rims in lunar regolith is a range of ~1–10 nm in diameter and an average of ~3 nm, while the iron particles in lunar agglutinates have a wider range of sizes, up to hundreds of nanometers in diameter [1]. Ryugu samples contain nanoparticles of iron sulfide, $(Fe, Ni)_{1-x}S$, in irradiated amorphous rims of phyllosilicates and melted layers produced by micrometeoroid impacts [19]. Nanoparticles of metallic iron are rare in Ryugu samples compared to Itokawa grains and lunar grains, probably because Ryugu samples are rich in water and therefore iron is surrounded by highly oxidized environment.

The formation mechanisms of $npFe^0$ include elemental fractionation during vapor-deposition process, reduction of iron-bearing minerals by solar wind implantation [1], and iron charge disproportionation reaction [33]. The vapor-deposition process refers to vaporization of FeO by impacts or sputtering and subsequent dissociation of FeO to elemental $Fe^0$ prior to its condensation on the surface of regolith particles. The reduction by solar wind implantation refers to the reduction of ferric or ferrous iron in mineral by ion implantation. Reduction by solar hydrogen is represented as the reaction: $FeO + H_2 = Fe^0 + H_2O$. The iron charge disproportionation reaction occurs in silicate phase and is described as $3Fe^{2+} = Fe^0 + 2Fe^{3+}$. The reaction is based on the presence of ferric iron in irradiation-induced amorphous rims in silicate minerals [33], suggesting that iron cations in the ferrous state in silicate minerals are segregated to condense as $npFe^0$, and the ferric iron from the host silicate grain remains in the silicate phase.

Iron nanoparticles have been of particular interest among the products of space weathering because they alter the reflectance spectra in the visible to infrared wavelengths. Lunar regolith grains show a darkening and reddening of the spectra compared to lunar rocks [34], while the surface of asteroid Itokawa is darker and redder compared to meteorites [35]. The spectral change can be explained by the presence of iron nanoparticles (<40 nm) based on Rayleigh scattering [34]. In addition, abundant large particles (>40–50 nm) in lunar samples likely cause Rayleigh scattering and Mie scattering [36]. Pulse laser irradiation of silicate minerals simulating micrometeoroid impacts reproduced impact-induced vaporization, the formation of

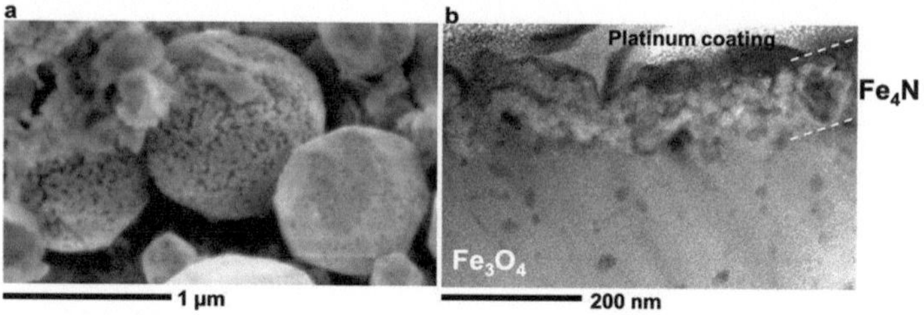


**Fig. 3.8** Space weathered magnetite ($Fe_3O_4$) grains of Ryugu sample. **a** SEM image of magnetite grains showing porous surface features that correspond to modifications by space weathering. **b** STEM image of the surface of the space weathered magnetite grain. The surface layer of iron nitride is indicated by dotted lines. The iron nitride layer consists of nanoparticles. The surface layer is covered with platinum during sample preparation

iron nanoparticles, and optical properties of the surface of the Moon and Itokawa [37].

Transformation of metallic iron nanoparticles has also been identified on the surfaces of magnetite ($Fe_3O_4$) grains of samples from the asteroid Ryugu [38]. The magnetite surfaces are coated with nanoparticles of iron nitrides ($Fe_4N$) (Fig. 3.8). Solar wind irradiation and micrometeorite impacts may have caused selective loss of oxygen atoms from the magnetite, leading to the formation of metallic iron on the magnetite surface. The iron nitride is likely a product of nitridation of the iron metals by the chemical reaction of reactive ammonia vapors that could have been produced by micrometeorite impacts.

## 3.8 Formation of Iron Whiskers

Whisker formation is the most prominent crystal growth caused by space weathering. Metallic iron whiskers up to three micrometers in length have been found on the surface of iron sulfide in Itokawa grains and lunar regolith grains (Fig. 3.9) [27]. Whiskers have a striated texture and are thinner at their tips. Curved or bending whiskers are often observed. TEM observations showed that the whiskers are composed of body-centered cubic iron. Some are composed of a set of several sub-whiskers elongated from the sulfide surface. The surfaces of the iron sulfide show open vesicles with crystallographic orientation, suggesting the accumulation of solar wind gases (Fig. 3.9). These surface structures of iron sulfide are similar between lunar and Itokawa particles, indicating that whisker growth on sulfide grains is common on the surfaces of airless bodies. The ratio of iron atom to sulfur atom in the iron sulfides increases toward the surface. This suggests that sulfur atoms were selectively released from the sulfide surface. The selective release of sulfur may have been caused by the reduction of iron sulfide by chemical reaction of solar wind hydrogen and by thermal effects associated with micrometeorite impacts. The excess iron atoms at the surface likely led to the nucleation of metallic iron by rapid diffusion of iron atoms through defects and by their surface diffusion. Whisker growth may have occurred within the solar wind exposure age of several hundred years for the Itokawa grains. A well-known whisker growth in engineering materials is tin whiskers, which grow from tin surfaces. The tin-whisker growth leads to electronic system failures by bridging closely-spaced semiconductor circuits and often cause problems in electronic components and satellites [39]. The mechanism of whisker growth is not fully understood, despite a quarter of a century of research. It has been proposed that whisker formation is stimulated by stress accumulation and thermal cycling. The formation of iron whiskers from iron oxides has been reported experimentally, but their growth mechanism is poorly discussed [40]. In the natural environment on Earth, silver whiskers, also called wire silver, have been identified in silver deposits where silver sulfides are oxidized by water [41]. Recent experimental studies of the synthesis of wire silver on the surface of silver sulfides suggest that the wire growth is induced by the solid-state ion conduction, by which metal cations can be rapidly transported through a crystal lattice in response to a thermoelectric potential, as opposed to the slower diffusion of cations [41]. This cation migration process can occur in superionic-conducting material including silver sulfides, but would not occur in iron sulfides, which do not have this property. Considering the proposed mechanisms of whisker growth at tin surface, growth of iron whiskers in airless surfaces could have been related to the relaxation of the strain field associated with excess interstitial $Fe^{2+}$ as a compensation for the increase in the surface energy of metals due to whisker growth. Another possible factor for whisker growth is the relaxation of the compressive stress in the metal caused by the difference in thermal expansion between troilite and metal induced by thermal cycling between day and night on airless surfaces. Numerous dislocations within the whiskers are observed in lunar regolith and could be related to the whisker growth process (Fig. 3.9). Further detailed observations are needed to constrain the poorly understood mechanism of whisker growth in the space environment.

## 3.9 Conclusion

The returned samples from the Moon and asteroids record the modifications in the microstructures of crystals that have experienced solar wind implantation and micrometeoroid impacts. The harsh space environment not only causes crystallographic damage to crystalline structures, but also promotes crystal growth, such as whiskers and nanoparticles. Further research into space weathering processes, focusing on the optical, magnetic and mechanical properties of their products, have potential applications in the development of nanoengineering materials.

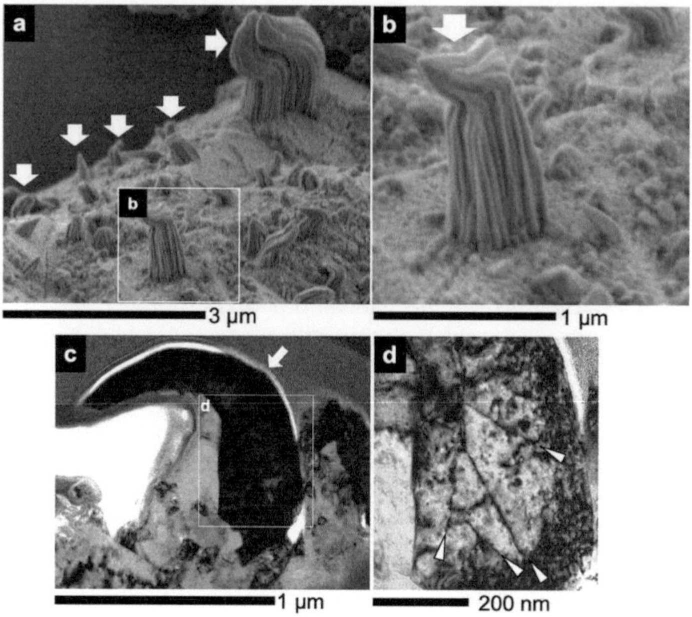


**Fig. 3.9** Iron whiskers found at the surface of iron sulfides of Itokawa particles. **a**, **b** Scanning electron image of iron whiskers (arrowed). The sulfide surface shows porous feature. **c**, **d** TEM image of an iron whisker. Uniform contrast of the whisker in **c** indicates that the whisker consists mainly of a single crystal. Triangles in **b** show possible dislocations dispersed in the iron whisker

## References

1. C.M. Pieters, S.K. Noble, Space weathering on airless bodies. J. Geophys. Res. Planets **121**(10), 1865–1884 (2016)
2. H.J. Melosh, *Planetary Surface Processes,* vol. 13 (Cambridge University Press, 2011)
3. J. Papike, L. Taylor, S. Simon, Lunar minerals, in *Lunar Sourcebook*, pp. 121–181 (1991)
4. T. Nakamura, T. Noguchi, M. Tanaka, M.E. Zolensky, M. Kimura, A. Tsuchiyama, A. Nakato, T. Ogami, H. Ishida, M. Uesugi, T. Yada, K. Shirai, A. Fujimura, R. Okazaki, S.A. Sandford, Y. Ishibashi, M. Abe, T. Okada, M. Ueno, J. Kawaguchi, et al.: Itokawa dust particles: a direct link between S-type asteroids and ordinary chondrites. Science **333**(6046), 1113–1116 (2011). https://doi.org/10.1126/science.1207758
5. T. Yokoyama, K. Nagashima, I. Nakai, E.D. Young, Y. Abe, J. Aléon, C.M.O. Alexander, S. Amari, Y. Amelin, K. Bajo, Samples returned from the asteroid Ryugu are similar to Ivuna-type carbonaceous meteorites. Science, eabn7850 (2022)
6. J.P. Bradley, Early solar nebula grains-interplanetary dust particles. Meteor. Cosmochem. Process. **1**, 287–308 (2014)
7. A. Rubin, C. Ma, *Meteorite Mineralogy,* vol. 26 (Cambridge University Press, 2021)
8. J.N. Goswami, I.D. Hutcheon, J.D. Macdougall, Microcraters and solar flare tracks in crystals from carbonaceous chondrites and lunar breccias, in *Lunar and Planetary Science Conference Proceedings*, vol. 7 (1976), pp. 543–562
9. S.K. Noble, L. Keller, Evidence of space weathering in regolith breccias I : lunar regolith breccia. Meteorit. Planet. Sci. **40**(3), 397–408 (2005). https://doi.org/10.1111/j.1945-5100.2005.tb00390.x

10. M. Zolensky, T. Mikouchi, K. Hagiya, K. Ohsumi, M. Komatsu, A. Cheng, L. Le, Evidence for impact shock and regolith transportation on CM, CI, and CV chondrite parent asteroids. Meteorit. Planet. Sci. **57**(10), 1902–1919 (2022)
11. J.P. Bradley, D.E. Brownlee, P. Fraundorf, Discovery of nuclear tracks in interplanetary dust. Science **226**(4681), 1432–1434 (1984)
12. D.B. Reisenfeld, D.S. Burnett, R.H. Becker, A.G. Grimberg, V.S. Heber, C.M. Hohenberg, A.J.G. Jurewicz, A. Meshik, R.O. Pepin, J.M. Raines, Elemental abundances of the bulk solar wind: analyses from genesis and ACE. Space Sci. Rev. **130**(1–4), 79–86 (2007)
13. D.B. Reisenfeld, R.C. Wiens, B.L. Barraclough, J.T. Steinberg, M. Neugebauer, J. Raines, T.H. Zurbuchen, Solar wind conditions and composition during the genesis mission as measured by in situ spacecraft. Space Sci. Rev. **175**(1), 125–164 (2013)
14. M. Neugebauer, The quasi-stationary and transient states of the solar wind. Science **252**(5004), 404–409 (1991)
15. C. Jäger, D. Fabian, F. Schrempel, J. Dorschner, T. Henning, W. Wesch, Structural processing of enstatite by ion bombardment. Astron. Astrophys. **401**(1), 57–65 (2003). https://doi.org/10.1051/0004-6361:20030002
16. D. Harries, F. Langenhorst, The mineralogy and space weathering of a regolith grain from 25143 Itokawa and the possibility of annealed solar wind damage. Earth Planets Space **66**(1), 1–11 (2014). https://doi.org/10.1186/s40623-014-0163-1
17. J.P. Bibring, J.P. Duraud, L. Durrieu, C. Jouret, M. Maurette, R. Meunier, Ultrathin amorphous coatings on lunar dust grains. Science **175**(4023), 753–755 (1972)
18. T. Noguchi, T. Nakamura, M. Kimura, M.E. Zolensky, M. Tanaka, T. Hashimoto, M. Konno, A. Nakato, T. Ogami, A. Fujimura, M. Abe, T. Yada, T. Mukai, M. Ueno, T. Okada, K. Shirai, Y. Ishibashi, R. Okazaki, Incipient space weathering observed on the surface of Itokawa dust particles. Science **333**(6046), 1121–1125 (2011). https://doi.org/10.1126/science.1207794
19. T. Noguchi, T. Matsumoto, A. Miyake, Y. Igami, M. Haruta, Y. Seto, M. Miyahara, N. Tomioka, A dehydrated space weathered skin cloaks the hydrated interior of Ryugu. Nat. Astron., 1–12 (2022)
20. P. Carrez, K. Demyk, P. Cordier, L. Gengembre, J. Grimblot, L. D'hendecourt, A.P. Jones, H. Leroux, Low-energy helium ion irradiation-induced amorphization and chemical changes in olivine: Insights for silicate dust evolution in the interstellar medium. Meteorit. Planet. Sci. **37**(11), 1599–1614 (2002)
21. L.P. Keller, E.L. Berger, A transmission electron microscope study of Itokawa regolith grains. Earth Planets Space **66**(1), 1–7 (2014). https://doi.org/10.1186/1880-5981-66-71
22. R. Christoffersen, L.P. Keller, Space radiation processing of sulfides and silicates in primitive solar systems materials: comparative insights from in situ TEM ion irradiation experiments. Meteorit. Planet. Sci. **46**(7), 950–969 (2011). https://doi.org/10.1111/j.1945-5100.2011.01203.x
23. L. P. Keller, D.S. McKay, The nature and origin of rims on lunar soil grains. Geochimica Cosmochima Acta **61**(11), 2331–2341 (1997). https://doi.org/10.1016/S0016-7037(97)00085-9
24. S.J. Zinkle, Effect of H and He irradiation on cavity formation and blistering in ceramics. Nucl. Instrum. Methods Phys. Res. Sect. B **286**, 4–19 (2012)
25. N. Sasajima, T. Matsui, S. Furuno, K. Hojou, H. Otsu, Damage accumulation in $Al_2O_3$ during H2+ or He+ ion irradiation. Nucl. Instrum. Methods Phys. Res. Sect. B **148**(1–4), 745–751 (1999)
26. K.D. Burgess, R.M. Stroud, PHase-dependent space weathering effects and spectroscopic identification of retained helium in a lunar soil grain. Geochim. Cosmochim. Acta **224**, 64–79 (2018). https://doi.org/10.1016/j.gca.2017.12.023
27. T. Matsumoto, D. Harries, F. Langenhorst, A. Miyake, T. Noguchi, Iron whiskers on asteroid Itokawa indicate sulfide destruction by space weathering. Nat. Commun. **11**(1), 1–8 (2020). https://doi.org/10.1038/s41467-020-14758-3
28. L.P. Keller, E.L. Berger, S. Zhang, R. Christoffersen, Solar energetic particle tracks in lunar samples: A transmission electron microscope calibration and implications for lunar space

weathering. Meteorit. Planet. Sci. **56**(9), 1685–1707 (2021). https://doi.org/10.1111/maps.13732
29. T. Matsumoto, A. Tsuchiyama, K. Uesugi, T. Nakano, M. Uesugi, J. Matsuno, T. Nagano, A. Shimada, A. Takeuchi, Y. Suzuki, T. Nakamura, M. Nakamura, A. Gucsik, K. Nagaki, T. Sakaiya, T. Kondo, Nanomorphology of Itokawa regolith particles: application to space-weathering processes affecting the Itokawa asteroid. Geochim. Cosmochim. Acta **187**, 195–217 (2016). https://doi.org/10.1016/j.gca.2016.05.011
30. T. Matsumoto, S. Hasegawa, S. Nakao, M. Sakai, H. Yurimoto, Population characteristics of submicrometer-sized craters on regolith particles from asteroid Itokawa. Icarus **303**, 22–33 (2018). https://doi.org/10.1016/j.icarus.2017.12.017
31. J.L. Carter, VLS (vapor-liquid-solid): newly discovered growth mechanism on the lunar surface? Science **181**(4102), 841–842 (1973). https://doi.org/10.1126/science.181.4102.841
32. A. Steele, F.M. McCubbin, M. Fries, M. Glamoclija, L. Kater, H. Nekvasil, Graphite in an apollo 17 impact melt breccia. Science **329**(5987), 51 (2010). https://doi.org/10.1126/science.1190541
33. L.J. Hicks, J.C. Bridges, T. Noguchi, A. Miyake, J.D. Piercy, S.H. Baker, Fe-redox changes in Itokawa space-weathered rims. Meteorit. Planet. Sci. **55**(12), 2599–2618 (2020)
34. B. Hapke, Space weathering from mercury to the asteroid belt. J. Geophys. Res. **106**(NO. E5), 10039–10073 (2001). https://doi.org/10.1029/2000JE001338
35. M. Abe, Y. Takagi, K. Kitazato, S. Abe, T. Hiroi, F. Vilas, B.E. Clark, P.A. Abell, S.M. Lederer, K.S. Jarvis, Near-infrared spectral results of asteroid Itokawa from the Hayabusa spacecraft. Science **312**(5778), 1334–1338 (2006)
36. P.G. Lucey, M.A. Riner, The optical effects of small iron particles that darken but do not redden: evidence of intense space weathering on Mercury. Icarus **212**(2), 451–462 (2011)
37. S. Sasaki, K. Nakamura, Y. Hamabe, E. Kurahashi, T. Hiroi, Production of iron nanoparticles by laser irradiation in a simulation of lunar-like space weathering. Nature **410**(6828), 555–557 (2001)
38. T. Matsumoto, T. Noguchi, A. Miyake, Y. Igami, M. Haruta, Y. Seto, M. Miyahara, N. Tomioka, H. Saito, S. Hata, Influx of nitrogen-rich material from the outer Solar System indicated by iron nitride in Ryugu samples. Nat. Astron., **8**(2), 207–215 (2024)
39. J. Brusse, G. Ewell, J. Siplon, Tin whiskers: attributes and mitigation. Carts Europe, 16
40. R. Nicolle, A. Rist, The mechanism of whisker growth in the reduction of wüstite. Metall. Trans. Bans. B **10**(3), 429–438 (1979). https://doi.org/10.1007/BF02652516
41. C.J. Anderson, R. Mathur, J. Rakovan, A.S. Tremsin, Natural solid-state ion conduction induces metal isotope fractionation. Geology **47**(7), 617–621 (2019)

# Part II
# Nucleation and Growth of Minerals

# Chapter 4
# The Chemistry of Phase Separation in Aqueous Hydrolyzing Metal Systems: Aluminium(III) and Iron(III)

**Miodrag J. Lukić and Denis Gebauer**

**Abstract** Hydrolysis of aqueous metal ion systems is fundamental to the phase separation of (oxy)(hydr)oxides, the understanding of which can help controlling, e.g., the formation of specific crystal structures, and advance various applications. Unlike the predominant, simplified physical notions implemented in the framework of classical nucleation theory, recent advances in the understanding of nucleation processes have highlighted the role of subtle chemical changes in precursors species and of the solvent, water. All of this has demonstrated the importance of "non-classical" concepts when discussing phase separation in hydrolyzing metal systems, beyond the classical ideas of the size of critical nuclei and the interpretation of supersaturation. Keeping in mind the vital importance of the chemistries of aluminium(III) and iron(III) ions in geochemical, physiological, and engineering processes, but also in the medical and renewable energy fields, herein, we summarize recent findings on the mechanisms of (oxy)(hydr)oxide formation from homogeneous aqueous solutions. Specific attention is paid to the role of water in understanding the nucleation behavior, an area where significant further advances are expected in the future. Also, we review the role of organic additives in controlling the nucleation of aluminium(III) and iron(III) (oxy)(hydr)oxides, including biomineralization processes.

Climate change, water shortage, environmental pollution, pandemic outbreaks, and fossil fuel dependency are just some of the most striking problems that we are facing

M. J. Lukić · D. Gebauer (✉)
Institute of Inorganic Chemistry, Gottfried Wilhelm Leibniz University, Hannover, Callinstr. 9, 30167 Hannover, Germany
e-mail: gebauer@acc.uni-hannover.de

M. J. Lukić
e-mail: miodrag.lukic@vin.bg.ac.rs; lukic@acc.uni-hannover.de

M. J. Lukić
Laboratory of Physics, "Vinča" Institute of Nuclear Sciences, National Institute of the Republic of Serbia, The University of Belgrade, Mike Petrovića Alasa 12-14, 11351 Belgrade, Republic of Serbia



F. Ishikawa et al. (eds.), *Engineering Crystal Habit*, Topics in Applied Physics 152,
https://doi.org/10.1007/978-981-96-0266-7_4

nowadays. Modern societies rely on constantly developing industrial and technological improvements toward sustainable living environments and balanced harvesting of natural resources. Aluminum and iron are among the most abundant elements in Earth's crust, playing a central role in shaping the humankind as we know it today. The aqueous chemistry of hydrolyzing Al(III) and Fe(III) systems is literally omnipresent, having tremendous importance for industrial processes but also everyday human life. The importance and relevance of Al(III) chemistry spans from climate change [1], soil chemistry [2, 3], agriculture [4], catalysis [5, 6], treatment of industrial wastewaters [7], nuclear technology [8], renewable energy sources [9], medicine [10, 11], pharmacy [12], and multifunctional hybrid materials to the military industry [13]. Fe(III) chemistry is crucial in (bio)geochemical cycles [14], ecological applications [15], magnetic materials [16], medical imaging [17], theranostic devices [18], steel industry [19], pigments [20], catalysis [21], but is also extensively present and has the utmost importance in many biomineralization processes, from bacteria to humans [22, 23]. Besides their historical application areas, both systems attract increasing attention due to novel findings made in the course of technological development. For example, we are experiencing an expansion in the vaccine development as a global response to the COVID pandemic, where Al(III) compounds, mostly Al(III) (oxy)hydroxides and Al phosphates, are used as particulate adjuvants of the first generation, or as the next generation of adjuvants involving particulate matter with immune potentiators for precise delivery to immune cells [24]. Some of the recently reported data indicate that Al(III) species are conducive, in contrast to many other metals, to the chemical protection of black phosphorus as an attractive post-graphene 2D nanomaterial [25]. A hot-debated topic is the Al(III)-ion rechargeable battery, which currently suffers, among other issues, from the strong interaction of the high-charge-density Al(III) species with the host electrode materials, leading to unwanted crystallization and impairing capacity and reversibility during cycling [26]. Iron-oxide nanoparticles can be used to improve the intercellular mitochondrial transfer for disease treatments [27], suggesting their partial degradation in cells and formation of Fe(II/III) ions. Understanding nucleation and crystallization in these chemical systems has been attracting great research attention and has been continuously improved due to the development of advanced instrumental techniques, also facilitated by computational studies. Altogether, this led to the development of novel theoretical frameworks for the description of nucleation mechanisms. Indeed, vast data on the nucleation of Al(III) and Fe(III) chemical systems has been interpreted from the point of view of classical nucleation theory (CNT), but there is an increasing number of recent studies that advocate for mechanistically different pathways involving, e.g., pre-nucleation cluster (PNC) species [28, 29]. The latter approach offers an improved explanatory power to consider nucleation as a chemical process consisting of several subsequent steps, where highly dynamic PNCs as precursor species in solution are the starting point. The "non-classical" nucleation pathway via PNCs is described for hydrolyzing systems in more detail elsewhere [30]. In brief, PNCs behave as solutes without a phase boundary, initially yielding amorphous intermediates on the pathway to distinct solid materials. It avoids the "capillary assumption", which is inevitable in the CNT approach, and broadens the understanding of nucleation beyond the notions of supersaturation ratio and interfacial free energies. The PNC pathway considers a change in the dynamics of highly dynamic PNC species, rather than their size, as a determining factor toward phase separation [28, 29], providing a framework for understanding chemical changes that govern the nucleation process besides the purely "physical" properties considered within CNT.

Besides nucleation from aqueous Al(III) and Fe(III) systems, here, we pay special attention to the role of water in the nucleation of the corresponding solid materials. Furthermore, we address recent knowledge obtained on the interactions of Al(III) and Fe(III) chemical species with spectator species, i.e., the influence of organic additives, which may affect the nucleation process on different levels and at different stages. Finally, this allows us to discuss biomineralization processes mainly from the point of view of Fe chemistry. Keeping in mind our recent review paper on non-classical approaches to the nucleation in Al(III) and Fe(III) solutions [30], we here focus on more recent data, however, referring back to milestone achievements published earlier. This chapter exclusively focuses on the nucleation and crystallization of Al(III) and Fe(III) (oxy)(hydr)oxides in aqueous systems, without consideration of the nucleation of (oxy)(hydr)oxides containing Fe(II) (except a few examples of magnetite formation), nucleation from the melt, related redox processes, or other non-aqueous solutions.

## 4.1 Nucleation of Al(III) and Fe(III) (Oxy)(Hydr)oxides from Homogeneous Aqueous Solutions

Chemical pathways from aqueous Al(III) and Fe(III) systems toward their solid (oxy)(hydr)oxide compounds involve numerous hydrolysis and condensation steps that proceed on different timescales, yielding monomeric species, small oligomeric species, molecular clusters, nanoscopic solids, and amorphous phases, which further aggregate and/or crystallize, leading to distinct solid materials. These two systems are extremely versatile, reflecting the richness of chemical environments, and especially Fe(III) represents a prime example of diversity in inorganic chemistry [31]. The Al(III) and Fe(III) systems are similar at first sight, but actually, the underlying chemistry is rather distinct, which originates from their different polarizing power and different electronegativity [32]. When dissolved in very acidic solutions, both metals form hexa-aqua coordinated monomeric species, $[Al(H_2O)_6]^{3+}$ and $[Fe(H_2O)_6]^{3+}$, which undergo hydrolysis upon increasing pH depending on the total concentration of the corresponding metal. $[Fe(H_2O)_6]^{3+}$ is more reactive than $[Al(H_2O)_6]^{3+}$, and its hydrolysis starts already at pH ~ 1–2 (Baes and Mesmer 1976), while $[Al(H_2O)_6]^{3+}$ hydrolyzes at pH ~ 3–4 (Jolivet 2019). Besides their different formal acidities, the hydrolysis behavior of Fe(III) and Al(III) is extraordinarily distinct at later hydrolysis stages. The hydrolytic species, $[Fe(H_2O)_{3-x}(OH)_x]^{3-x}$, occur essentially successively in a wide pH range from 2 to 9 (Baumgartner and Faivre 2015), while all

$[Al(H_2O)_{3-x}(OH)_x]^{3-x}$ hydrolytic species can simultaneously exist between pH 4 and 6. Two types of chemical reactions that lead to condensation dominate the hydrolyzing Fe(III) and Al(III) systems: olation, i.e., removal of a water molecule between hydroxo- and/or mixed aqua-hydroxo complexes, yielding M-OH-M bonds, and oxolation, i.e., removal of a proton between hydroxo-bridged species, typically released as a water molecule together with a hydroxo ligand, yielding stronger M–O-M bonds. The high reactivity of Fe(III) yields structurally and chemically distinct solids compared to Al(III), e.g., $Fe(OH)_3$ does not readily form compared to $Al(OH)_3$ because oxo ($O^{2-}$) species are more dominant than hydroxo ($OH^-$) species in the Fe(III) system. Consequently, the formation of Fe(III) oxides can be realized at lower temperatures, as opposed to that of Al(III) oxides. In the alkaline pH range, Al(III) species exist as tetrahedrally coordinated $Al(OH)_4^-$ complexes, while the formation of similar $Fe(OH)_4^-$ species does not occur; instead, the tetrahedrally coordinated anionic complex $[FeO_4]^{5-}$ that does not condense to higher-order compounds is found (Jolivet 2019). In the entire pH range of interest for both systems, the formation of solid compounds can in principle be considered from chemically distinct starting conditions, i.e., acidic or alkaline environments, proceeding via different formation mechanisms toward solid materials. Here, we focus on nucleation in the Al(III) and Fe(III) systems starting from (mildly) acidic solutions, but will also provide an overview of the current knowledge on the plausible mechanism of nucleation from alkaline aqueous solutions.

Strictly speaking, in the existing nucleation and crystallization literature on the aqueous Fe(III) system, assuming certain given physico-chemical starting conditions (pH, concentration, temperature, pressure), we can define two main approaches: the first is phase separation from homogeneous solution followed by the initial formation of amorphous/low-crystalline ferrihydrite/akageneite phases, and the second is the nucleation of other stable crystalline phases from already formed amorphous precursors, i.e., heterogeneous systems. Results from both approaches have been discussed from the perspectives of classical and non-classical nucleation pathways. Considering final applications and technological aspects, the latter approach could seem more important; however, striving to understand biomineralization, biochemical, geochemical, and environmental processes, the former approach is undoubtedly also critical. Indeed, both approaches are interdependent and can be neither physically nor chemically separated. Certainly, structural and spatial (pre)-arrangements in pre-nucleation species are determined by chemistry, and may play decisive roles in the final crystallization and polymorph selection process. At mildly acidic to alkaline pH, ambient pressure and temperature, amorphous, highly-disordered, and hydrated nanosized ferrihydrite, β-FeOOH, was reported to be the first forming solid in the Fe(III) system [33]. In solutions with high chloride concentrations, the akageneite (oxy)(hydr)oxide phase precipitates first, which can then transform into iron oxide [32, 34]. Figure 4.1 shows the crystal structures of ferrihydrite (A), with a domain size of 2 nm, and akageneite (B). While ferrihydrite consists of a combination of tetrahedral and octahedral environments, Fe(III) in akageneite is octahedrally coordinated, possessing structural channels in which structure-stabilizing chloride ions are situated. Structurally similar to ferrihydrite is the Al(III) mineral

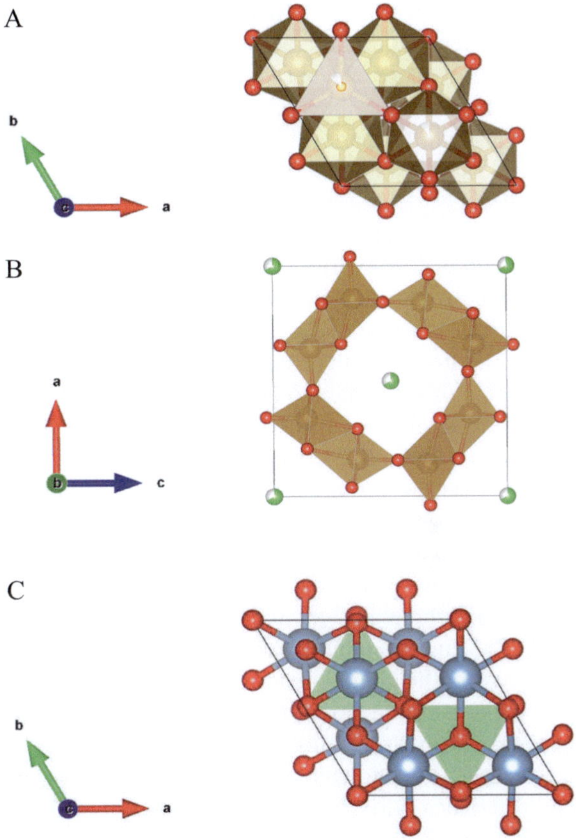


**Fig. 4.1** Crystal structures of **A** ferrihydrite with a 2-nm domain [35] and **B** akaganeite [36]; **C** akdalaite, an Al mineral structurally similar to ferrihydrite [37]. Brown atoms and polyhedra correspond to Fe(III), red atoms are oxygen, and green atoms are chloride. Crystal structures are plotted using the Vesta software [38]

akdalaite, $Al_{10}O_{14}(OH)_2$, which can be produced in different morphologies using Al (oxy)(hydr)oxide flocks or nanocrystalline boehmite, Fig. 4.1(C), and it contains $Al_{13}O_{40}$ Keggin features. The structural model of akdalaite was precisely determined, and it can be used as a reference for studying ordered/disordered ferrihydrite nanoparticles and the related structural transformations [39].

The condensation of Fe(III) species in solution can be an extremely fast process, depending on concentration, and proceeds mainly via olation reactions, while phase-separated (solid) compounds do contain oxolated bonds [32]. The existence of small polymeric species preceding phase separation is highly debated due to the high reactivity of Fe(III) and experimental difficulties to measure them, and indeed, they can play crucial roles in the formation of solids, as recently discussed elsewhere [30]. Bottero et al. reported that at an OH/Fe ratio of 1.5, the hydrolysis of $FeCl_3$ solutions initially yields dimers with edge-sharing octahedra, then trimers with edge- and corner-sharing octahedra after 50 min, finally leading to kinetically stable β-FeOOH-like polycations after 1 h [40]. Considering acidic solutions, at pH 3, the hydrolysis of Fe(III) can be very rapid given sufficient iron concentrations (1 mM), immediately (within 1 s) yielding larger polycations of several nanometers in size. This implies that the interactions between smaller (not strictly monomeric) species are plausibly more relevant for nucleation than $[Fe(OH)_3]^0$ [41]. Indeed, it was previously proposed that the aggregation of $[Fe(OH)_3]^0$ species was critically important in ferrihydrite formation [32]. The crystal structure of ferrihydrite remains highly debated due to its nanocrystalline nature, and it was suggested that it consists of approximately 20% Fe in tetrahedral and 80% Fe in octahedral orientation [35], although the exact Fe(III) coordination still remains elusive. Zhu et al. studied the nucleation of ferrihydrite from acidic concentrated Fe(III) nitrate solutions [42]. In their model of ferrihydrite nucleation at $1.8 < pH < 3$, they assumed the formation of μ-oxo dimers, which undergo structural recombination as a rate-limiting step, yielding aqua or hydrolyzed dihydroxo dimers, and finally, the system evolves quickly via the generation of oligomers to ferrihydrite. This mechanism also seemed more evident than the formation of ferrihydrite from neutral $[Fe(OH)_3]^0$ species, due to the low Fe(III) and hydroxide concentrations at the investigated experimental conditions. However, as the formation of Keggin "$Fe_{13}$" ($Fe_{13}O_{40}$) species that contain tetrahedral sites were already reported [43], the model cannot explain their particular role. The explanation and mechanistic role of the occurrence of Keggin species is an important feature of nucleation pathways, especially in the context of non-classical nucleation where clusters and aggregates are involved in the condensation process [44]. The kinetic stabilization of Keggin $Fe_{13}$ species in the pH range of 1–3 was achieved by the simultaneous employment of trifluoro-acetate ligands and Bi ions, and without stabilization, the process readily yields ferrihydrite. The relationship between ferrihydrite in its hydrated and dehydrated forms is hard to depict, but the ferrihydrite formation pathway may critically influence its final structural and surface characteristics [45]. Based on the number of XRD reflections, 2-line (the least crystallized) and 6-line (the most crystallized) ferrihydrite structures were reported [46], but synchrotron-based X-ray total scattering studies indicated that this interpretation rather originates from the limited size of scattering domains due to small crystallite sizes than from different intermediate-range atomic arrangements, which is nearly the same for ferrihydrite with average crystalline domains of 2, 3, and 6 nm [47]. However, surface properties of ferrihydrite samples may differ depending on the preparation method, e.g., between those prepared by alkalization and acidification approaches to neutral pH 7, influencing its distinct application behavior [48].

We recently commented on the occurrence of larger (oligomeric) olated species in the Fe(III) system, implying that changes in the structural conformation within highly-dynamic olated prenucleation clusters (PNCs), the fast exchange of water molecules between the solvent and PNCs, the specific acid–base behavior, or interactions with counter ions, may enable further hydrolysis and growth of oligomeric species/clusters [30, 49], despite the previously suggested charge accumulation hindrance that could suppress the growth of oligomers. Indeed, at low driving force for phase separation in the pH range from 2.1 to 2.85, using $FeCl_3{\cdot}6H_2O$ as a precursor, the chemical transition from olation to oxolation within PNCs, accompanied by reduced dynamics due to the formation of stronger bonds, governs the phase separation event from PNCs. The existence of olation Fe(III) PNCs was experimentally proved by means of a combination of potentiometric titrations, analytical ultracentrifugation, high-resolution transmission electron microscopy, and SAXS measurements [49]. These experimental conditions lead to the formation of akaganeite [50, 51]. Keeping in mind the extremely high reactivity of Fe(III) species, experimental approaches using high driving force for phase separation, i.e., high chemical potential/supersaturation ratio (with respect to the solid phase with the lowest solubility)/high metal concentrations and strong bases, might lead to very fast condensation and almost immediate formation of oxolated phase-separated species, thus circumventing the above-described pathway via solute olation PNCs. This is important to understand when considering the experimental conditions employed in numerous studies of nucleation and crystallization in the Fe(III) system, since some of them might already be far beyond the phase separation limit via local oversaturation (e.g., due to pH gradients, etc.).

Ferrihydrite is a metastable phase at ambient conditions, and may evolve into different, more stable phases, such as hematite, goethite, and magnetite. The evolution from ferrihydrite may thereby proceed via dehydration and local rearrangement (at low solubility conditions), dissolution-recrystallization (at high solubility conditions), but also oriented attachment of primary particles [32, 52]. Baumgartner et al. used cryo-transmission electron microscopy to show that magnetite nucleation from solution proceeds through rapid agglomeration of primary nanoparticles, which may or may not correspond to PNCs [53], without intermediate amorphous bulk precursor phases, which can be interpreted in the framework of CNT [54]. They reported a molecule-like behavior of disordered primary particles possibly due to their small size of 1–2 nm (in the range of large molecules). Furthermore, Baumgartner et al. investigated the self-confined nucleation of magnetite from amorphous ferrihydrite-like precursors prepared by instantaneous mixing of appropriate amounts of Fe(III) and Fe(II). The concentration of the acidic iron dispersion was in the range from 10 to 100 mMat pH 1, and it was mixed with 1 M NaOH in a microfluidic reactor at a total flow rate of 60 mL min$^{-1}$. As determined by WAXS, amorphous particles were formed initially, and then crystallized to magnetite. The study implied that the self-confinement by amorphous nanostructures, besides their thermodynamics, determined the crystal growth rate of magnetite, decreasing it by 2 orders of magnitude when the crystal size reaches the size of the amorphous precursor particles [55]. Montes-Hernandez used time-resolved Raman spectroscopy to study magnetite nucleation and also reported the absence of amorphous intermediate phases during the magnetite synthesis, by adding a mixed-valent iron solution into an alkaline one, or vice versa, with a total starting concentration of ferro- and ferric-species of 0.5 M [56]. In this set of experiments, the mixed-valent iron species, at rather high concentrations, were added to NaOH at pH 14, or a NaOH solution of high pH was added into iron solutions at acidic pH values. These conditions generate a high driving force for phase separation and can give rise to the occurrence of pH inhomogeneities and strong local oversaturation. Indeed, as mentioned above, this can alter the pathway and/or lead to highly transient early stages of nucleation in the system [49], such as amorphous intermediates, which may thus be overseen experimentally.

In a recent report, the initially forming clusters were assumed to consist of crystalline ferrihydrite based on cryo-TEM measurements [57], describing the process of nucleation and crystallization of magnetite ($Fe_3O_4$) particles via formation and aggregation of primary crystalline ferrihydrite particles and the thermodynamics and the kinetics of the process in terms of colloidal assembly. A later study indicated that magnetite can form directly from amorphous mixed valent iron phosphate precursors, which critically influence magnetite formation, shifting it to a rather high pH of 11 [58]. Indeed, the evolution of ferrihydrite via a dissolution-recrystallization mechanism and the subsequent formation of magnetite have been mostly interpreted from the point of view of CNT. However, nucleated (hetero-phase) amorphous/low-crystalline ferrihydrite particles subjected to dissolution at conditions differing from their formation, e.g., higher pH values or temperatures, are probably not the same species as Fe(III) prenucleation species that form in homogeneous solutions before the formation of the pristine solid. Solutes re-formed from an already precipitated amorphous solid, i.e., ferrihydrite, upon dissolution may adopt the chemical form, i.e., the ratio of aqua/hydroxo/oxo ligands in the first hydration shell, and possibly the structure of the second hydration shell, corresponding to the new surrounding conditions—provided sufficient time for equilibration. However, the structural characteristics, i.e., bond lengths, bond angles, and coordination environments of the re-forming solute may contain features reminiscent of pre-dissolution structural arrangements in the solids underneath since this process inevitably comprises of a dissolution front, where many oxo or hydroxo chemical bonds of the metal centers are cleaved. Dissolution itself necessarily causes structural changes at the solid/liquid interface, which are accompanied by a continuous internal structural reorganization of the remaining solid directly beneath the first dissolving layer. Thus, the rate of dissolution, determined by the characteristics of the employed dissolution medium, and the rate of the internal structural reorganization, determined by the chemistry and (meta)stability of the amorphous precursor, compete, and the specific balance of these processes may ultimately determine the chemical/structural characteristics of newly formed solute species. This is also important when considering the growing number of evidence toward the central role of water in nucleation and crystallization mechanisms, since the water structure around newly formed solutes can play a critical role in the dynamics and also thermodynamics of these solute species and, from the point of view of the PNC pathway, thus, also in the formation of the final oxide(s). This non-classical nucleation pathway can thus rationalize the formation of different transient structures, e.g., isolated oligomers, prenucleation clusters, amorphous structures, and proto-minerals. Furthermore, the notions of the PNC pathway inherently account for the role of chemical processes in the early stages of precipitation, e.g., condensation, ligand exchange reactions, structural rearrangements, dissolution, and growth of amorphous precursors, leading to final crystalline solids. In CNT, on the other hand, the main determining parameters toward nucleation are the reduction of interfacial/surface free energy and supersaturation, while the role of chemistry is largely neglected, as recently discussed in detail for the hydrolyzing Fe(III) and Al(III) systems elsewhere [30].

Indeed, crystallization by particle attachment or oriented attachment was reported in many studies of iron oxides [57, 59]. Remarkably, though, crystallization by particle attachment or oriented attachment has been rather rarely reported for the aqueous Al(III) system, to the authors' best knowledge, attracting attention due to practical applications [60]. It was reported that boehmite nanoplatelets undergo oriented aggregation with crystallographic alignment [61], where the process is influenced by the morphology of the aggregating particles, and pH and the ionic strength of the solution. Considering the forces involved in the (oriented) aggregation processes, at a short-range, atomic structures, i.e., the chemical nature, determine the structure of hydration layers and their overlapping, while the long-range effects are controlled by the solution parameters (pH, concentration, ionic strength, the type of electrolyte) [62]. The role of the interfacial water structure in oriented attachment still needs to be explored [63], but it certainly becomes critical considering hydrolysing systems and the role of water for their nucleation and crystallization. In the case of Al(III) (oxy)(hydr)oxides, this crystallization pathway is likely chemically driven, and keeping in mind that the Al(III) system requires higher temperatures for the formation of Al-O bonds and oxide phases than Fe(III) due to its lower electronegativity, it seems fair to assume that oxolated metal centers could play a significant role in particle-mediated, non-classical crystallization mechanisms in hydrolyzing systems. Indeed, Sun et al. have recently reported a non-classical mechanism of magnetite formation based on the alignment of small (<2 nm) amorphous iron oxide clusters and polymer-protected amorphous iron oxide clusters via larger amorphous particles, where the former were synthesized in ethanol to prevent them from further oxolation and crystallization [64]. The magnetite phase was obtained by thermal treatment of the amorphous iron oxide clusters, and pre-alignment of the clusters played a critical role. Herein, the roles of dehydration, dehydroxylation, water binding, electric dipole interactions, and entropy effects probably need to be better understood for elucidating this non-classical crystallization pathway further. On the other hand, Scheck et al. have explained a non-classical mechanism of hematite nucleation at low temperatures and low driving force for phase separation by a change in the reaction mechanism, rationalizing the formation of akaganeite, ferrihydrite, and hematite from solution under different conditions based on the chemistry of olation PNCs [34]. The process of PNC formation is entropy-driven due to the significant release of water molecules from condensation reactions, the magnitude of which could be rationalized only by the formation of olation clusters. The first traces of hematite appeared already at 45 °C, and it was shown that the formation mechanism at higher supersaturation ratios (i.e., corresponding to hydrothermal conditions) differs from that in the titration approach at low driving force and constant temperature. In fact, olation PNCs play an important role in hematite formation, and their shortage was reflected in the titration curves, which might be alleviated under hydrothermal conditions.

Similar to the Fe(III) system, the aqueous Al(III) system may yield various solid Al(III) compounds. The initial precipitation typically yields amorphous $Al(OH)_3$, transforming later into crystalline gibbsite $\alpha$-$Al(OH)_3$ (stable at room temperature) upon appropriate treatment [65], or bayerite $\gamma$-Al(OH)$_3$ (stable at higher temperatures). Two dehydrated forms also appear at higher temperatures, $\alpha$-AlO(OH), diaspore, and $\gamma$-AlO(OH), boehmite [66]. The nucleation of Al(III) (oxy)(hydr)oxide from homogeneous solutions is commonly explained using "core-links", "cage-like" Keggin-Al$_{13}$, and "continuous" models, and each of them is applicable in certain particular situations. The "core-links" model is reportedly useful for explaining the Al(III) polymerization and gibbsite formation by the coalescence of octahedrally coordinated single hexamer rings $(Al_6(OH)_{12}(H_2O)_{12})^{6+}$, where the structure of the Al polymeric species in solution is the same as that of the solid Al(OH)$_3$ [67]. It was assumed that dioctahedral sheets, characteristic for gibbsite, form from monomeric species aggregating into $Al_6(OH)_{12}(H_2O)_{12}{}^{6+}$ [68], but such species were not experimentally observed [69]. Still, the "core-links" model can rationalize the formation of different small polymeric species [70] but cannot explain the formation of Keggin-Al$_{13}$ species that are experimentally detected. The "cage-like" Al$_{13}$ model recognizes the formation of only monomers, dimers, Keggin Al$_{13}$, and larger species (determined by $^{27}$Al-NMR) in solution, which can directly transform from one into the other; it has gained wide acceptance since the proposed species can be observed experimentally. However, the "cage-like" model cannot explain the polymerization process itself and the formation of metastable medium-sized transient species/oligomers in solution (experimentally observed by other methods, e.g., ESI–MS [70]), but also the formation of Keggin-Al$_{13}$ is not clarified and it assumes undefined transformation and reshuffling mechanisms (generically speaking, mostly by aging and treatments at high temperatures). The "continuous" model reportedly unites the previous two models accounting for the base addition rate, aluminum and base concentrations, and aging time. The model intends to unite the results from spectrophotometric studies, potentiometric titrations, and existing models of hydrolysis-polymerization, and assumes continuous change from monomeric to large polymeric species, which can, via self-assembly, transform into more stable small Al and Keggin-Al$_{13}$ species upon, e.g., aging or addition of specific counter-anions. The "continuous" model can be considered rather as a kinetic than a thermodynamic one, since it advocates for the flux of base addition (alkali neutralization) determining the evolution of different hydrolytic polymeric species [67]. It was suggested that under moderate base addition rates, the Al(III) species change in a transient manner and that the existence of any of the transient (polynuclear) species is granted under specific experimental conditions [67]. Figure 4.2 schematically depicts the main steps in the hydrolysis-polymerization-nucleation cascade according to "core-links" and "cage-like" models.

Thus, one of the most critical questions in the nucleation of Al(III) (oxy)(hydr)oxides is the formation of the Al$_{13}$ Keggin structure, which contains one tetrahedrally coordinated Al(III) center bonded to four trimers of octahedrally coordinated Al species. Another critical question that logically follows at this point is how Keggin structures transform into gibbsite, which does not have any tetrahedrally coordinated Al species in its crystal structure. Indeed, currently existing models still cannot explain these critical questions. Experimental results showed that in the pH range from 4.2 to 4.9, X-ray amorphous or microcrystalline gibbsite can form [71]. Furthermore, natural flocs are assumed to consist of aggregated Keggin-$Al_{13}$ clusters with a tetrahedrally coordinated $AlO_4$ unit [71]. Furrer et al. reported the existence of Keggin-$Al_{13}$ species when the total Al(III) concentration is 20 μMat pH values between 5 and 5.5 [71]. It is assumed that the formation of this structure originates from smaller oligomeric Al(III) species upon aging [67]. Jolivet has explained the formation of Keggin $Al_{13}$ cations by the condensation of four planar trimers and one hexacoordinated monomer, facilitated by heating up to 80 °C [32]. At low and circumneutral pH values, Al(III) centers are reported

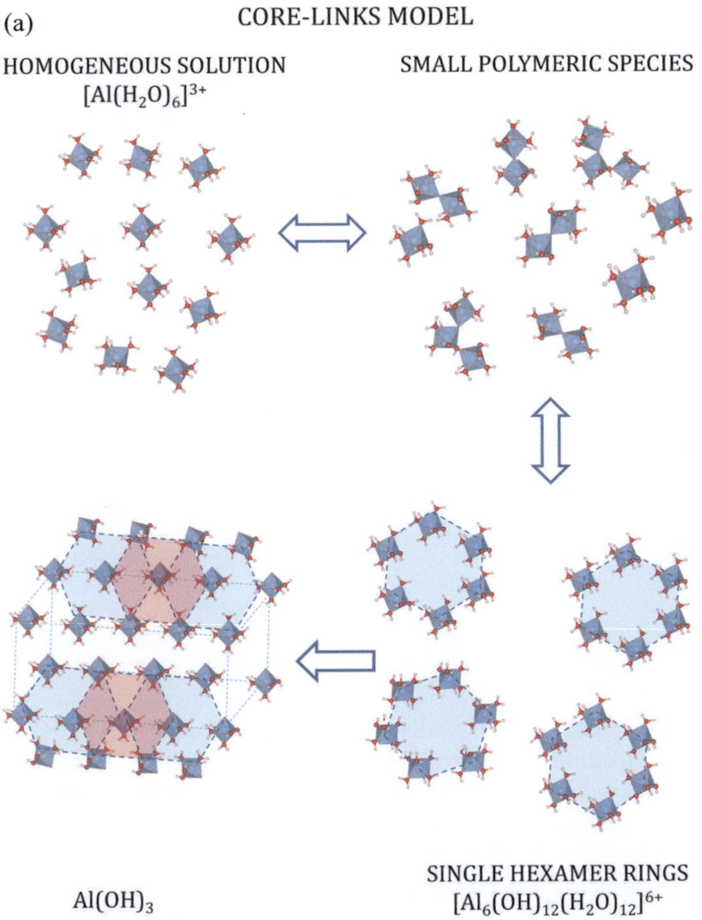


**Fig. 4.2** Schematic representation of the main steps in the hydrolysis-polymerisation-nucleation cascade of Al(III) from aqueous solutions according to the **a** "core-links" and **b** "cage-like" Keggin-$Al_{13}$ models. (Al octahedra are represented in blue, oxygen is given as red balls, and H as small white balls; Al tetrahedra are given in yellow, and yellow circles in Keggin-$Al_{13}$ represent Al in tetrahedral coordination). The crystal structures were plotted using the Vesta software [38]

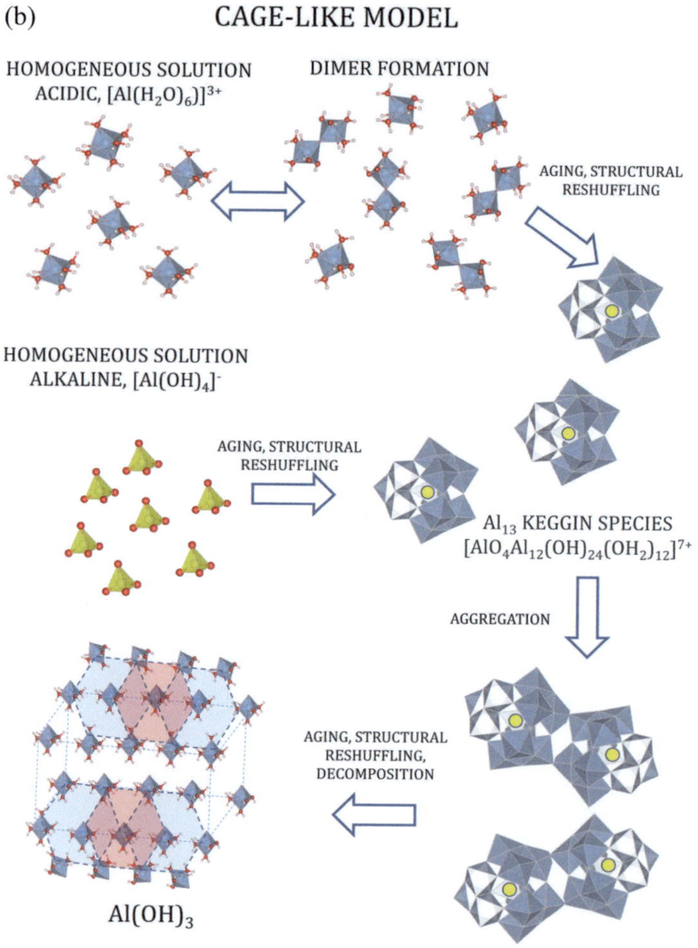


**Fig. 4.2** (continued)

to be octahedrally coordinated, while tetrahedrally coordinated $Al(OH)_4^-$ is present under (highly) alkaline conditions. The formation of gibbsite with octahedrally coordinated Al(III) species is thus thought to occur via amorphous $Al(OH)_3$ precursors at near neutral pH conditions [65]. Consequently, the formation mechanism of gibbsite from tetrahedrally coordinated environments at high pH values remains unclear, although is extremely important from an industrial point of view, e.g., for the formation of alumina ceramics via the Bayer process. Similar to the Fe(III) system, two basic chemical processes that compete in these systems are olation and oxolation reactions, but the water exchange reaction from Al(III) species is slower compared to Fe(III) [72], which can be expected to be the case also for the corresponding olation reactions. The removal of water molecules upon interaction between aqua-coordinated species yields entities with highly dynamic olated $-(Al–OH-Al)_n-$ and less dynamic oxolated -Al–O–Al- bonds, respectively. Indeed, CNT neglects critical chemical aspects of nucleation mentioned above for the formation of solid Al(III) compounds from solution, leaving questions toward the importance of underlying chemical parameters such as pH and speciation, and their role in the nucleation of distinct forms without answer. Figure 4.3 shows the sheet structures of (A) gibbsite [73] and (B) bayerite [74] $Al(OH)_3$, while the (C) panel illustrates boehmite (oxy)(hydr)oxide phase, AlO(OH) [75].

New insight into the nucleation mechanism in the aqueous Al(III) system at mild pH levels and low driving force for phase separation was obtained recently [76]. A pH-constant titration assay was used, and the hydrolysis was induced by increasing the total concentration of Al(III) while compensating for even the slightest oscillations in pH by automatic counter-titration. This approach is different from most used alkalization methods, where a base is added to a solution at a fixed starting total Al(III) concentration. The early stages of nucleation in the pH range from 3.9 to 4.6 were studied by the constant pH approach, which is the regime characteristic for Al(III)

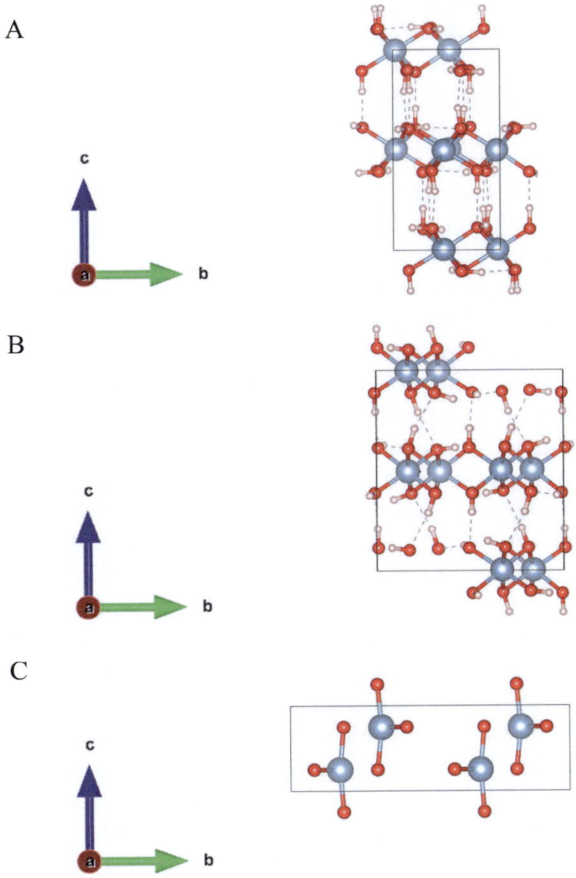


**Fig. 4.3** Crystal structures of **A** gibbsite [73], **B** bayerite [74], and **C** boehmite [75]. Blue atoms correspond to Al(III), red atoms are oxygen, and white atoms are hydrogen. Crystal structures are plotted using the Vesta software [38]

hydrolysis. The obtained titration curves revealed at least two distinct hydrolysis regimes, the first one with a low and linear base consumption, passing through a transient regime with a strongly increased base consumption, and reaching a second linear regime characterized by a stronger base consumption. Further analyses by means of "stop-experiments", liquid-state $^{27}$Al-NMR spectroscopy, and HR-TEM showed that in the first linear regime, highly-dynamic olation Al(III) PNCs dominate the speciation, while the formation of oxolated $AlO_4$ environments initially occurs at the transition point between the hydrolysis regimes. That point represents the early onset of the phase separation event in the Al(III) system. The role of chemical changes that trigger phase separation is particularly important. The chemical changes are related to structural rearrangements, i.e., a decrease in the coordination number within Al(III) PNC species from octahedrally to tetrahedrally coordinated Al(III) centers. The former continue to exist in the solution together with tetrahedral species, but the total amount of the latter linearly increases with the total amount of added Al(III) upon entering the second hydrolysis stage. With increasing total Al(III) concentration, the amount of released protons per added Al(III) asymptotically approaches 2, implying a change from octahedral to tetrahedral coordination and a steady progress in the hydrolysis reactions, which was corroborated by NMR spectroscopy. In the proposed model of Al(III) hydrolysis at low driving force for phase separation and under mildly acidic conditions, the nucleation mechanism in the aqueous system can be understood as follows: after dosing Al(III) in the solution at pH values above 4.3 (for the investigated range of Al(III) concentrations, i.e., up to 10 mM), the initial $[Al(H_2O)_6]^{3+}$ complexes deprotonate, yielding $[Al(H_2O)_5(OH)]^{2+}$ hydrolytic species existing in the solution in equilibrium with $[Al(H_2O)_6]^{3+}$. As reported previously [77], these species may lose water ligands, forming $[Al(H_2O)_4(OH)]^{2+}$ species with pentahedral coordination. These Al(III) species form highly-dynamic chain-like olation Al(III) PNCs, exhibiting similar dynamics as monomeric complexes, and there is formally no phase boundary between the clusters of various sizes and the solution. As the olation PNCs behave like solutes, the formation of transient sites with tetrahedral geometry is possible and seems to be more likely to occur in larger PNCs. Such sites may then undergo deprotonation of a tetrahedral OH bridge and protonation of an adjacent hydroxo ligand toward releasing a water molecule and yielding the first oxolation species (O-bridging). Consequently, a decrease in the Al-O bond length of Al–OH and Al-$OH_2$ and an increase of the Al-O–H bond angles [78] may occur, decreasing the dynamics at the tetrahedral site and leading to the first, irreversible formation of tetrahedrally coordinated $AlO_4$ centers. In this view, the formation of tetrahedral olation sites within PNCs is a prerequisite for phase separation, yielding less dynamic phase-separated entities, that is, olation PNCs transform into partially oxolated post-nucleation clusters. The tetrahedral $AlO_4$ environments are kinetically much more stable, and the phase-separated species may further aggregate to decrease their surface area and total interfacial free energy, leading eventually to the formation of distinct solid species [76]. Figure 4.4 shows nucleation in the aqueous Al(III) solution according to the PNC pathway [76].

Altogether, the above-described findings and notions strongly suggest that $Al_{13}$ Keggin species are phase-separated entities, and that they cannot be considered as

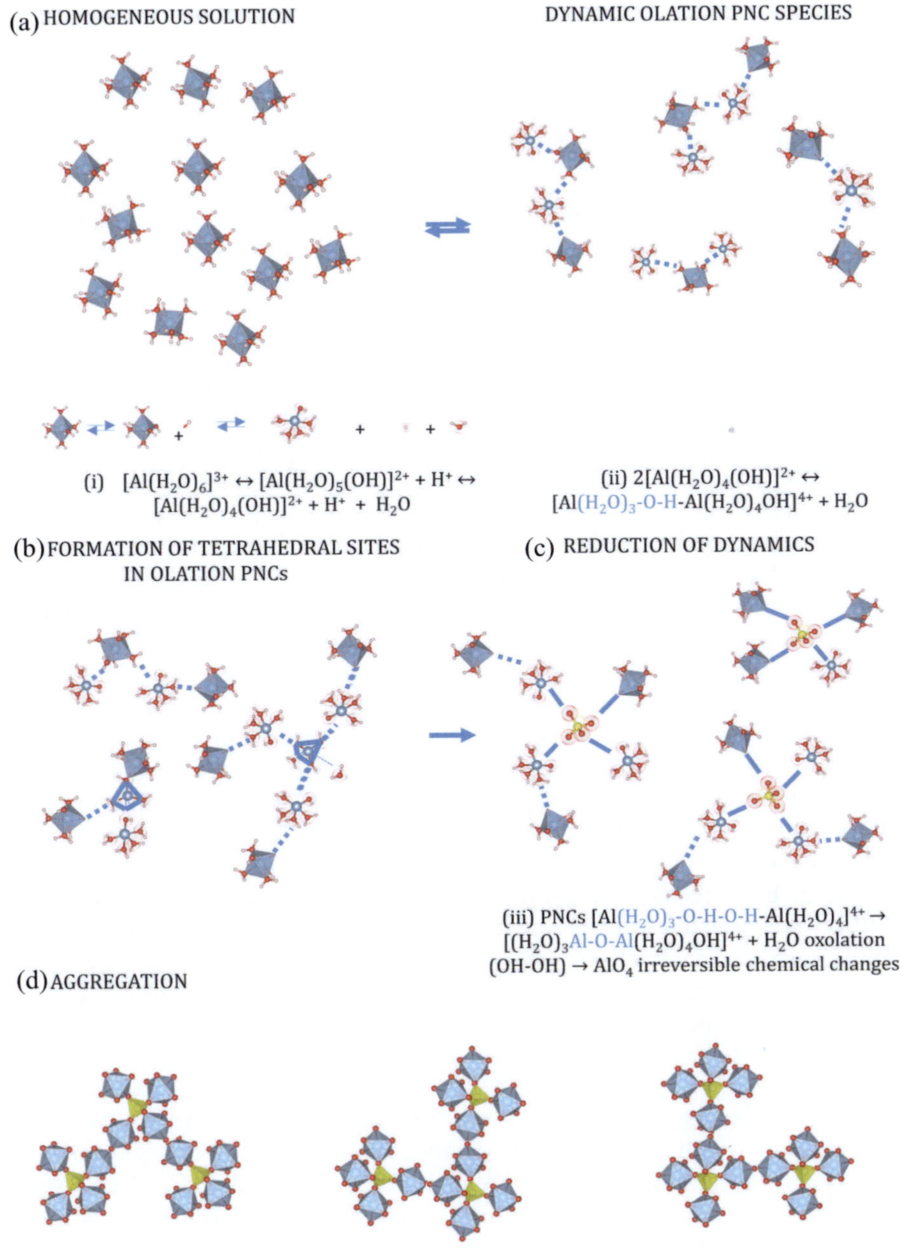
(a) HOMOGENEOUS SOLUTION
DYNAMIC OLATION PNC SPECIES
(i) $[Al(H_2O)_6]^{3+} \leftrightarrow [Al(H_2O)_5(OH)]^{2+} + H^+ \leftrightarrow$
$[Al(H_2O)_4(OH)]^{2+} + H^+ + H_2O$
(ii) $2[Al(H_2O)_4(OH)]^{2+} \leftrightarrow$
$[Al(H_2O)_3\text{-O-H-}Al(H_2O)_4OH]^{4+} + H_2O$
(b) FORMATION OF TETRAHEDRAL SITES IN OLATION PNCs
(c) REDUCTION OF DYNAMICS
(iii) PNCs $[Al(H_2O)_3\text{-O-H-O-H-}Al(H_2O)_4]^{4+} \rightarrow$
$[(H_2O)_3Al\text{-O-}Al(H_2O)_4OH]^{4+} + H_2O$ oxolation
(OH-OH) $\rightarrow AlO_4$ irreversible chemical changes
(d) AGGREGATION

◂**Fig. 4.4** Nucleation in the aqueous Al(III) solution according to the PNC pathway [76]. **a** Prenucleation phase—deprotonation reactions of Al(III) (represented by Al octahedrons (where pale blue, red, and white colors represent Al, O, and H atoms, respectively) in homogeneous solution, yielding dynamic olation Al(III) PNC species, including hexa- and penta-coordinated Al(III) environments [penta-coordinated species are represented as a ball-and-stick model with dot surface, while blue dotted lines represent olation (-OH-) bridging]. **b** Tetrahedral (pale blue triangles) sites formed in the olation PNCs lead to an irreversible oxolation reaction. **c** Phase separation in the Al(III) system due to the reduction of dynamics and the formation of stronger bonds in $AlO_4$ tetrahedral environments [solid blue lines—oxolation (–O) bridging; oxolated while Al atoms in tetrahedral coordination—yellow]. **d** Post-nucleation phase—as-formed $AlO_4$ tetrahedrally coordinated species (yellow tetrahedra) take part in further aggregation (H atoms are not shown for clarity)

PNCs due to their slow dynamics [29] but rather as nanoscopic solid precursors to Al (oxy)(hydroxide) [79]. The formation of Al PNCs allows rationalizing the formation of Al species with different coordination environments in solution and of small oligomeric forms as outlined above. The event of phase separation in the Al(III) solution can thereby be considered as liquid–liquid phase separation, and the subsequent post-nucleation species with $AlO_4$ centers may represent a dense liquid stage in the PNC crystallization pathway [80]. The findings align well with a previous study of natural Al flocs, which showed that these are formed by aggregation of $Al_{13}$-Keggin polyoxocationic species, possessing a central $AlO_4$ unit [71], implying that the proposed Al PNC pathway may be relevant for the mechanism of their formation. Moreover, the formation of tetrahedrally coordinated species was reported for pH values as low as pH 4.3 [66], implying that the proposed mechanism is general. The formation of the layered gibbsite $Al(OH)_3$ structure without tetrahedrally coordinated species may occur upon structural rearrangement of the detected tetrahedral positions, or via an alternative mechanism. Recently, the formation of gibbsite nanoplates in acidic and neutral environments was reported, exhibiting local differences in interlayer distances [81]. 7–8% vol. % of the initial gibbsite particles exhibited this defect. It was plausibly ascribed to $Al_{13}$ aluminum hydroxide clusters, almost isostructural to gibbsite, which could be trapped during the growth process. The source of disorder of gibbsite synthesized at these conditions is well explained by a cluster-mediated growth mechanism, based on the interlayer incorporation of remaining flat $Al_{13}$ oligomers.

In highly-alkaline environments, Al(III) (oxy)(hydr)oxides can undergo phase-transformation into gibbsite, bayerite, and boehemite depending on the concentrations of Al and NaOH, and temperature, proceeding via dissolution-reprecipitation mechanisms [8]. Dembowski et al. recently studied nucleation from alkaline solutions, starting from $Al(OH)_4^-$ monomers or possibly $Al_2O(OH)_6^-$ dimers. They suggested that the transformation from tetrahedral toward octahedral coordination governed the precipitation process of octahedral gibbsite. The results implied that the transformation occurs in the thin water layer on the surface of the solid phase, determining the selection toward gibbsite or dawsonite by altering the humidity level via changing the carbonate/bicarbonate chemical equilibrium; higher humidity levels favored the formation of gibbsite [82]. Similar as in the case of the formation of different Fe(III) oxides, such as hematite and magnetite, from amorphous phases like ferrihydrite, an often employed method for synthesizing gibbsite nanoparticles is using amorphous $Al(OH)_3$ and subjecting it to increased temperatures for prolonged times. A so-called two-step synthesis route was applied to obtain gibbsite with controlled sizes and morphology [83]. An amorphous $Al(OH)_3$ precursor was aged at room temperature for long times (7 days) and exhibited mainly tetra- and penta-coordinated Al(III), and it was found that more tetrahedrally coordinated Al(III) centers appeared in gibbsite prepared via the dissolution-precipitation mechanism of amorphous $Al(OH)_3$ aged in water at room temperature (and crystallized at 50 °C for 7 days) and at 50 °C (and crystallized at 25 °C for 3 months) than in those aged at higher temperatures [83]. The morphology of the synthesized gibbsite nanoplates changed from truncated triangles to hexagons, pseudo-hexagons, and rhombic morphologies, which was explained based on CNT. Baccarella et al. reported a cluster-mediated conversion of amorphous $Al(OH)_3$ to $\gamma$-AlOOH via hydrothermal treatment at 473 K in the presence of water, where six-, five-, and four-coordinated Al-based Keggin species existed in gels during the initial period of the conversion reaction [84], similar to Al-flocs discovered in nature [71]. The speciation of Al(III) in alkaline solutions is critical for the precipitation behavior. The formation of both oxo-bridged $Al_2O(OH)_6^{2-}$ or dihydroxo-bridged $Al_2(OH)_8^{2-}$ dimers under alkaline environments was reported [85]. Simulation results showed that ion-pairing of aluminate species with either $Li^+$, $Na^+$, or $K^+$ cations favors the formation of oxo-bridged dimers, while the first interaction between, reportedly, monomeric units occurs via one hydroxo ligand, yielding a labile $Al_2(OH)_8^{2-}$ dimer. Remarkably, cation–anion interactions are regarded as being critical for both dimerization pathways. Medium-range ion-pair interactions, i.e., solvent-separated $Al(OH)_4^-$-$Na^+$ ion pairs, were proposed to critically influence the kinetics of gibbsite crystallization in a dissolution-precipitation mechanism in alkaline solutions [86]. Recently, a study on the structure of alumina glass also revealed unusual behavior, where many tri- and tetra-clusters occurred alongside edge-sharing penta- and hexa-coordinated $AlO_n$ polyhedra, as opposed to conventional glass-forming rules [87].

Considering the above discussions, it is clear that significant progress in understanding nucleation processes in the Al(III) and Fe(III) systems has been achieved. Indeed, an increasing number of experimental and theoretical studies on the nucleation of Fe(III) and Al(III) reveals the complexity of both systems, and notions of classical and non-classical nucleation theories are used to explain the different observations. However, profound insights into chemical/structural changes, i.e., proton transfer reactions, dehydration, and coordinative adaptation, are needed to further our understanding of non-classical mechanisms. This is also true for many other chemical systems nucleating from solutions or even the gas phase. For example, clustering of $I_xO_y$ was recently proposed as a plausible pathway in the formation of iodine oxide particles from the gas phase [88]. Eventually, new rationales of nucleation from solution are required to consider chemical landscapes and explain non-classical, cluster-mediated pathways. Figure 4.5 schematically represents the formation of Fe(III) and Al(III) (oxy)(hydr)oxides from acidic and alkaline solutions, not intending to provide a holistic picture of all possible solid formation mechanisms.

## 4.2 The Role of Water in the Nucleation of Al(III) and Fe(III) (Oxy)(Hydr)oxides from Solution

The discussions in the previous section show that nucleation events in Fe(III) and Al(III) aqueous systems cannot be fully understood without considering the role of water as a ligand in initial hexacoordinated complexes, as the hydration medium or the bulk solvent, or even as a humidity determining factor on the surface of mineral phases. For studying the role of water, the pH-constant approach is particularly useful [34, 49, 51, 76, 92]. It enables controlling the early stages of nucleation in the sense that all non-hydrolyzed metal species introduced as solutes from acidic solutions

(a)

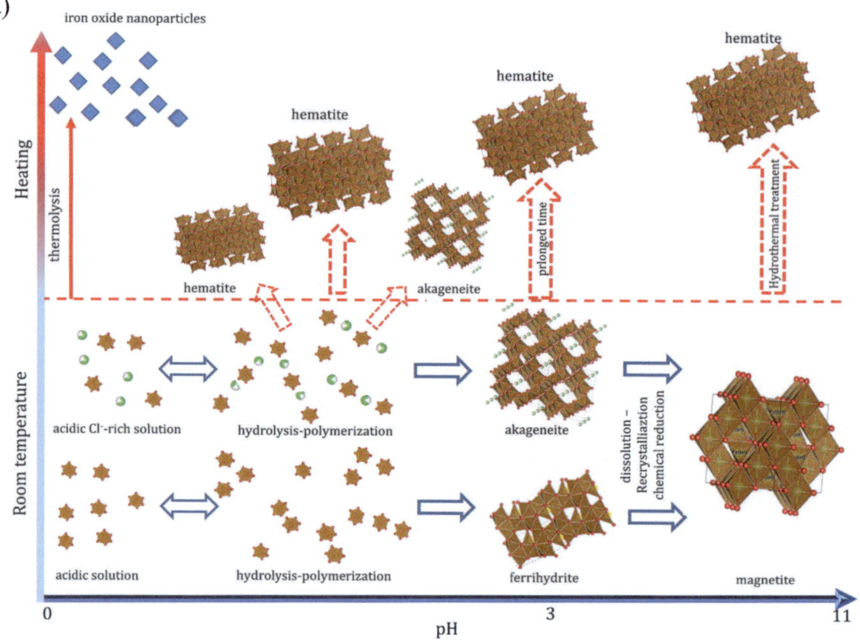


**Fig. 4.5** Schematic representation of the formation of **a** Fe(III) and **b** Al(III) (oxy)(hydr)oxides from solution. It should be noted that these schemes do not represent the holistic approach, i.e., do not account for all possible phases and mechanisms. The crystal structures were plotted using the Vesta software [38], and the following cif files were used: gibbsite [73], boehmite [74], gamma alumina [89], ferrihydrite [37], akaganeite [36], magnetite [90], and hematite [91]

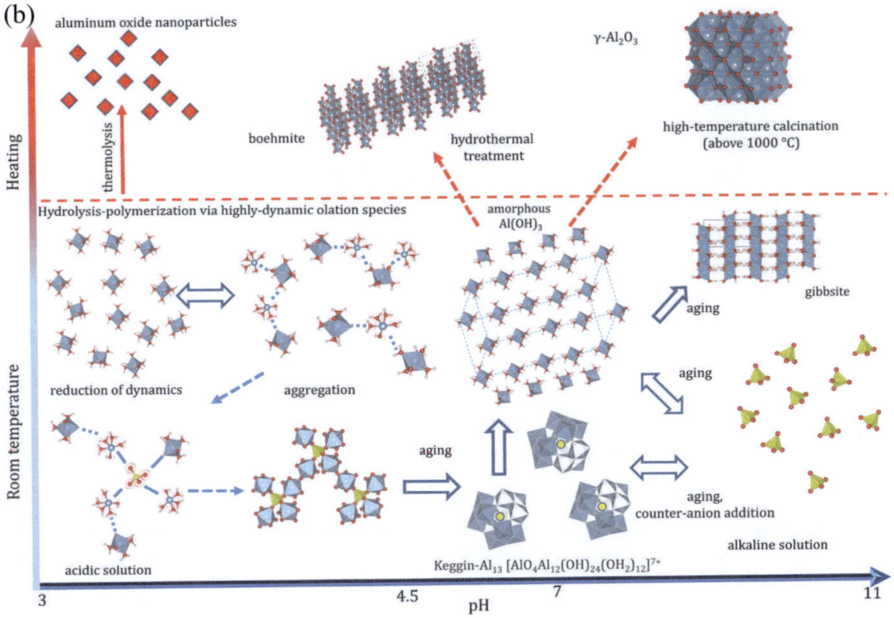


**Fig. 4.5** (continued)

equilibrate with corresponding hydrolytic forms according to the externally-adjusted pH value (given that other parameters like temperature are constant). In that way, all subsequent changes in the system can be considered from a precisely defined chemical reference point. Furthermore, a low driving force for hydrolysis as realized by the very slow mixing of dilute solutions prevents the occurrence of abrupt local chemical changes, which, in turn, inevitably occur when using externally-controlled hydrolysis driving agents, such as highly alkaline solutions, as in the case of the often-employed alkalization approaches. In the pH-constant approach, the diffusion of Al(III) and Fe(III) ions can be considered to be a limiting process, which might be closer to natural processes where precipitation of Fe(III) and Al(III) (oxy)(hydr)oxides occurs. Another important consequence of this approach is that a dissociative hydrolysis mechanism is favored over an associative one, i.e., externally added $OH^-$ ions, usually at high concentrations (as in the case of alkalinization methods) associate with metal ions, behaving as nucleophilic complexing ligands, which is apparent in the potentiometric pH vs. OH/Al curves (i.e., the initial base addition does not significantly increase the pH value when the OH/Al ratio changes from 0.2 to 2.5, pH increases from 4.1 to 4.6, while at an OH/Al ratio of 2.8, the pH sharply raises to more than 9 since $OH^-$ ions are not used anymore for hydrolysis-oligomerization/polymerization reactions, and are free in solution) [67]. In contrast, in the case of pH-constant titrations, hexa-aqua Al complexes dissociate (hydrolyze) due to the addition of Al(III), and $H^+$ ions released due to Al(III) hydrolysis are compensated by the base addition. In this case, initially formed Al(III) aqua-hydroxo complexes possess a nucleophilic OH group and a leaving $H_2O$ group, so the reaction has a dissociative mechanism and $H_2O$ can leave (forming an intermediate with reduced coordination) before the formation of Al–OH-Al bonds occurs in the olation reaction [32]. After all, due to the realization of a low driving force (in sense of supersaturation) for phase separation in the pH-constant approach, the role of water can arguably be better studied than in other experimental settings.

The role of water in nucleation was reported for different systems, e.g., calcium carbonate [93], while the release of water molecules from monomeric ions upon the formation of PNC species, i.e., increasing the solvent entropy, was reported to be a thermodynamic driving force underlying PNC formation in non-classical nucleation [94]. As already discussed above, water contributes to both building and solvating Al(III) and Fe(III) (oxy)(hydr)oxide PNCs [30]. Olation and oxolation reactions comprise the core chemical events, determining the onset and early stages of phase separation in both systems. The starting point in the consideration of the role of water for hydrolyzing systems is the polarizing power of the metal ions. It depends on the ionic charge and radius, but also electronegativity, explaining a stronger polarization of water molecules in the first coordination shell of Fe(III) than Al(III), although the former has a larger ionic radius than the latter [32]. This leads to a lower stability of Fe(III) hexaaqua complexes, their higher reactivity at low pH values, and the more frequent occurrence of oxo- than hydroxo-bridges in distinct ferric compounds (they rather form oxides than oxyhydroxides) compared to Al(III) hexaaqua species [32]. The rapid water exchange from Fe(III) was considered to be crucial for the condensation and nucleation from iron nitrate solutions at pH 3 [41].

One of the main characteristics of Fe(III) and Al(III) (oxy(hydr)oxide nucleation from homogeneous solution is that these ions strongly bind water molecules in the first hydration shell, so they should be actually considered as ligands rather than hydrating species. Thus, anions that constitute Al(III) and Fe(III) hydrolytic species, $OH^-/O^{2-}$, actually originate from water ligands within existing complexes rather than from free ions in solution [30]. Due to their strong polarization, water molecules complexed by these metal ions are very sensitive to any pH variation, so these systems in a specific pH range act as cationic acids with limited capacity. These render the chemistry of nucleation of hydrolyzing systems even more complex than that of ionic crystals, where both reactants, cations and anions, can be externally controlled, in principle. As pointed out repeatedly, this chemical aspect is completely overseen in CNT that mainly considers the physical parameters of nucleation, while chemical aspects of hydrolyzing systems cannot be adequately interpreted. The first solvation shell of both Al(III) and Fe(III) was investigated in detail, and these ions are coordinated by six water molecules in the first layer, while the second hydration shell of both ions consists of 12 water molecules [95]. Based on theoretical calculations, the water molecules in the first coordination shell of Al(III) form rather strong hydrogen bonds with the water molecules of the second hydration shell [96]. The presence of water can stabilize Fe-OH bonds characteristic for oxy-hydroxides, e.g., ferrihydrite, so its further transformation into iron oxides inevitably correlates with the disruption of the structural water. Indeed, a recent report on the high-temperature phase transformations and crystallization of ferrihydrite [97] indicated the critical role of water vapor as an external factor that facilitates crystallization of ferrihydrite to hematite at lower temperatures (350 °C) than usually required (above 500 °C) via the generation of additional nucleation sites. Dehydration processes that are necessary for the phase transformation proceed via proton transfer reactions between OH groups in the presence of water. Furthermore, the interaction of additives requires a ligand exchange reaction with water molecules [98], which determines the rate and the nature of the interaction. The role of water, particularly the release of water, was recently demonstrated in the non-classical nucleation of hematite at rather low temperatures. The formation of PNCs induces the release of many water molecules from the hydration shell, and based on the magnitude of the entropic contribution, it is estimated that ~12 water molecules are released per Fe(III) ion. The cluster formation thus disturbs both the first and second hydration shells, causing a greater entropic contribution than one that can be rationalized based upon monomeric hydrolyzed entities [34].

The properties of hydration and bulk water may play critical roles in the change of the coordination number of metal ions during the early stages of non-classical nucleation as discussed above. Indeed, in the case of the aqueous Al(III) system, it was reported that the metal-water coordination number could change due to deprotonation reactions [99] or by the growth of polymeric species [100]. Dynamic exchanges between Al(III) environments with different coordination environments are critically important for phase separation, and there is increasing evidence for the role of penta-coordinated species, such as $[Al(H_2O)_4(OH)]^{2+}$ [77]) and $[Al(H_2O)_2(OH)_3]^o$ [101], as well as tetra-coordinated metal centers, such as $[Al(H_2O)(OH)_3]^o$ [101] and Al(III) (oxy)(hydr)oxide PNCs [76] in solution. Dembowski et al. explored the role of atmospheric water and carbon dioxide in dawsonite vs. gibbsite precipitation from highly alkaline solutions [82] and showed that the presence of carbonates favors the formation of gibbsite, where the presence of the water layer plays a critical role. Water ordering on magnetite surfaces and its influence on the adsorption of the second water layer was also experimentally observed using microcalorimetry and temperature-programmed desorption [102]. Dong et al. [101] considered tetrahedral $[Al(H_2O)(OH)_3]^o$ species in amorphous sols as precursors of Keggin-$Al_{13}$, while at low driving force for phase separation, tetrahedral sites form within solute Al(III) (oxy)(hydr)oxide PNCs species without aging treatment or specific interactions with anions [76]. Cooperative structural and chemical changes within highly dynamic olation PNCs, where the rate of water exchange from $[Al(OH)(H_2O)_{n-1}]^{2+}$ is on the order of $10^4$ $s^{-1}$ [77], certainly imply that water does play a central role in the early stage of nucleation of hydrolyzing systems. A better understanding of water dynamics and the hydration characteristics of cationic and anionic species accompanying the hydrolyzing metal thus seems extremely important to further advance the notions of non-classical nucleation pathways.

## 4.3 Nucleation in the Presence of Organic Additives and Biomineralization

It is well established that different anions can influence the formation of Al(III) and Fe(III) (oxy)(hydr)oxide solids as spectator species, i.e., by changing the chemical environment, the characteristics of the hydration shell, or the structure of bulk water. Indeed, all of this is likely also the case for other species, including organic molecules, present during nucleation. Nucleation in the aqueous Al(III) and Fe(III) systems in the presence of organic additives is a topic with tremendous scientific and practical importance. Organic additives, and generally, organic matter, closely couple with the chemistry of Fe(III) and Al(III), and they can influence the precipitation of metal (oxy)(hydr)oxides by interacting with the forming metal species, depending on the physical–chemical conditions. These chemical systems in the environment and the living nature are constantly exposed to many organic molecules, ranging from small (bio)molecules to peptides and proteins. Fe(III) is involved in numerous biomineralization processes in living species, from magnetotactic bacteria to humans, and the process of biomineralization starts from the characteristic molecular interactions within physiological environments. Besides living systems, Al(III) and Fe(III) are involved in many processes of environmental remediation, including treatments of nuclear wastes and industrial wastewaters, as well as soil acidification, and geochemical and also extraterrestrial cycles [103]. Al(III) species are occasionally reported as being related to different neurodegenerative disorders, reporting them to be critical in protein denaturation and the formation of insoluble amyloid-like deposits in the human brain [104]. It is generally assumed that organically-complexed Al(III) is less toxic than its aqua or hydroxo forms [105]. One of the contemporary examples is the importance of understanding the chemistry of the interactions of Al adjuvants with proteinaceous antigens used in modern vaccines to improve their compatibility and render the vaccination procedure more beneficial for patients [106]. Indeed, a rather distinct behavior of Al(III) and Fe(III) in contact with organic matter was reported [107], where the reactivity of metal species is sensitive to both pH and oxygen level.

Keeping in mind the above elaborations on the aqueous chemistry of Fe(III) and Al(III), and also the very different pH ranges where Al(III) and Fe(III) compounds usually form in nature or are used in industry, e.g., acidic soils, aquatic acidification, natural wastewater, physiological environments, or the Bayer process, the interaction of organic ligands with metal species can occur in rather distinct chemical environments. This concerns the formation of the corresponding solids, influencing nucleation and crystallization events at different stages and yielding solid phases with specific chemical/morphological features. Organic additives, especially macromolecules and polymers, may bind to the metal ions, prenucleation clusters, or aggregates [80], governing, on the one hand, the nucleation and crystallization behavior, while, in turn, the conformation and stability of macromolecules may be influenced by the ion binding as well. Hydrolyzing species can interact with organic molecules in different ways, ranging from electrostatic interactions, depending on the charge of the ligand, i.e., the pH value of the solution in case of acid/base equilibria, to hydrogen bonding and van der Waals interactions [105]. All these possibilities for interactions of metal species with organic additives can serve as a basis for the organization into specific hybrid structures depending on the ligand conformation, potentially facilitating their crystallization toward specific polymorphs or forms. Elucidating the primary interaction mechanisms in Al(III) and especially Fe(III) systems with model organic macromolecules thus seems to be a necessary step forward, toward understanding biomineralization processes and the formation of natural biostructures with precisely tuned size, uniformity, organization, and morphological and functional characteristics.

As discussed in detail elsewhere [80], CNT fails to provide adequate explanations for many observations made for nucleation processes in the presence of organic additives, especially having in mind opposite effects that the same additive may exert on different minerals. For example, polyaspartic acid (pAsp) strongly suppresses the nucleation of calcium carbonate, while it promotes the nucleation of Fe(III) (oxy)(hydr)oxide [92]. In the latter case, the effect was explained by the promotion of oxolation within Fe(III) (oxy)(hydr)oxide PNCs upon interaction with pAsp, and it was shown that the effect was present only in the case of pAsp, while monomeric Asp did not show any effects. Furthermore, the specific fashion how organic additives may influence the early stages of mineralization in the Fe(III) system also depends on the structure, molecular weight, and length of the additive sidechain. It was recently shown in a theoretical study that Asp and Glu peptide sequences, differing only in one $-CH_2-$ group more in the sidechain of Glu, display a distinct influence on mineralization in the presence of calcium and chloride ions, showing that the Asp sequence attracts chemical species more strongly, forming different molecular clusters on the surface, unlike the Glu sequence [108]. Indeed, nucleation in the aqueous Fe(III) system was investigated experimentally in the presence of poly-L-glutamic (pGlu) acid with 20 monomeric units using potentiometric titrations under precisely controlled and constant pH conditions between 2 and 3 and low driving force for phase separation. The obtained results point toward several steps in the interactions of pGlu with nascent Fe(III) (oxy)(hydr)oxide species, namely, initial binding, stabilization of Fe(III) PNCs by pGlu 20, and phase separation, yielding the akageneite phase at pH 3.0. [51]. Unlike pAsp [92], there was no sudden precipitation, and the nucleation was even slightly postponed in the presence of pGlu20, although the effect was not dominant at the investigated pGlu concentration, corroborating the presence of different effects of structurally similar organic additives. Akagenite was the only XRD-detectable precipitated phase independent of the presence of the additive at pH 3.0. However, although pGlu20 did not influence the phase composition, it yielded an akageneite phase that exhibited a somewhat higher degree of crystallinity. The presence of pGlu residues in the final material was confirmed, implying that pGlu incorporation positively influenced the crystalline arrangement of the forming Fe(III) (oxy)(hydr)oxide, affecting the final characteristics. A similar effect was found by Chang et al. who reported a continuous growth of iron oxide nanoparticles from iron-oleate precursors, which consisted of tri-iron-oxo cores instead of monomeric complexes [109]. These tri-iron-oxo clusters served as starting points for the growth of nanoparticles by condensation between hydroxyl groups of the iron-oxo clusters formed after the esterification of the oleate ligands. The authors proposed a continuous growth model induced by the controlled reaction of clusters and strong binding of organic ligands and added alcohol in the solution, implying certain commonalities in the nature of interaction of Fe(III) with organic molecules. Indeed, the role of the polymeric iron-oleate network being formed prior to nucleation seems to be important [110]. Remarkably, the formation of tri-iron carboxylation clusters as an iron mineral nucleation site was also reported in the case of biomineralization in the presence of ferritin [111], where anchoring of the metal centers was realized through glutamate amino-acid residues. The formation of tetra- and octa-iron clusters bound to L-subunits of ferritin was experimentally confirmed using electrospray ionization mass spectrometry, indicating the importance of Glu residues [112]. Surprisingly, no binding of ferric species to Asp residues was reported in single amino acid substituted MMS6, although there were no critical conformational preferences of the ferric binding domain, indicating the critical role of cooperative iron binding by the sequence-specific motif at the C-terminus [113]. These observations point toward a cooperative nature of the interaction between Fe(III) and (bio)macromolecules during biomineralization, where biomacromolecules, depending on their chemistry, i.e., amino-acid sequence, configuration, and conformation, play a vital role in controlling the early stages of mineralization. Accordingly, organic ligands can also be used to isolate or stabilize metal-oxo clusters [44].

Unlike Al(III), Fe(III) is abundant in many living organisms and is involved in biomineralization processes. Studying iron biomineralization is important regarding the synthesis of advanced materials with unique structures using similar biological systems (e.g., proteins, protein subunits, peptides), which can find potential applications in energy storage, biomedicine, catalysis, and other relevant fields. Iron biomineralization can also be used for an integrated detection and treatment of lung tumors in their early stages, avoiding the use of toxic drugs or any invasive treatments [114]. Remarkably, strategies identified in biomineralization research are not limited to Fe but can be transferred to many other metals and chemical systems. Thus, findings on the mineralization mechanisms in living systems serve as inspiration for creating syntheses toward tailor-made particles on the laboratory scale, and beyond. Iron is stored in many biological systems by ferritin and hemosiderin proteins. Ferritin is the major iron storage molecule, and it comprises a spherical protein shell around an inner cavity, accumulating up to 4,500 iron atoms in the form of ferrihydrite [115, 116]. This form of iron is favorable under biological conditions, i.e., as a backup form containing both hydroxo and oxo groups and an undefined structure, facilitating the conversion into different iron oxide forms when necessary, also buffering any changes in the Fe concentration. Looking back at the above-mentioned debate on amorphous, 2-line, and 6-line ferrihydrite, ferrihydrite in natural systems exhibits mostly an amorphous/non-crystalline structure, and it was found by means of transmission electron microscopy and electron energy loss spectroscopy that ferrihydrite exists in hydrated human spleen ferritin and human heart ferritin, while stable hematite occurred in dehydrated samples [117]. The ferrihydrite transformation can be mediated by extracellular polymeric substances [118], and the effect of extracellular polymer substances on the ferrihydrite transformation largely depends on the chemical properties of these compounds, namely, the presence of polar carboxylic and phosphate functional groups. The mechanisms of iron biomineralization, and particularly those occurring during nucleation and crystallization events, have been extensively studied. One of the most known biomineralization examples is the formation of magnetite nanoparticles by magnetotactic bacteria. Such-formed magnetic nanoparticles are highly crystalline, consisting of a single magnetic domain with narrow size distribution [119]. In the magnetotactic biomineralization (MTB) of iron, magnetite nucleation and crystal growth occur after several preceding processes, including compartmentalization, cellular arrangement, formation of the required internal physico-chemical conditions, and protein coordination. The MTB formation of magnetite nanoparticles is assumed to proceed via co-precipitation of Fe(II) and Fe(III) species or via a solid precursor, i.e., ferrihydrite. The MMS6 protein, the most abundant one in magnetosome membranes, is probably also the most important protein for iron nucleation [113], and it was proposed that magnetite formation proceeds via the fusion of many prenucleation clusters [119]. Recently, a plausible interaction between highly-charged unstructured regions of MMS6 proteins involved in the biomineralization process and artificially prepared iron oxide nanoparticles was investigated using cryo-EM and X-ray diffraction studies. It was found that proteins are in direct contact with nascent mineral particles and can respond by changing their conformation, where the intrinsically unstructured regions become more ordered, which is probably applicable for many biomineralization systems [120, 121]. Such an orchestrated scenario in biomineralization processes would not only involve the part of the protein that directly interacts with the mineral phase but also the rest of the biomacromolecule. A recent study found that the localization of iron imported into magnetotactic bacteria, which points towards such orchestration, precedes the nucleation and growth of magnetite [122].

Lid et al. recently simulated the ferritin-promoted nucleation of Fe(III) (oxy)(hydr)oxides and assessed potential anchoring sites for Fe(III) clusters on the protein subunits [98]. The results showed that Fe(III) mainly interacts with amino acid sidechains and not extensively with the protein backbone. The interaction with Fe(III) nanoclusters can differ even among the amino acids of the ferritin subunits, so the oxygen atoms of carboxylic acid groups form bidentate bonds with the Fe(III) oxyhydroxide surface, while other amino-acid residues can interact via hydrogen or even van der Waals bonds. The formation of naturally occurring unusual morphologies of iron oxides seems to be related to the interfacially-driven nucleation process due to changes in chemical potential gradients [59].

Turning to the interactions of Al(III) with polypeptides, the initial position of organic–inorganic interactions and the existence of cooperative interactions with carboxylic acid groups and the polypeptide backbone lead to irreversible denaturation [123], although this requires further investigation as there are opposite results evidencing dominant side-chain interactions [124]. It was reported that Al(III) greatly influences biological molecules, often inducing their compaction and precipitation from solution, e.g., of DNA, and exhibiting gene silencing ability [125]. Al(III) ions exhibit a strong chaotropic effect and destabilize the collagen protein structure, behaving completely different than, e.g., Cr(III) [126]. One of the prime examples of the interactions of Al(III) species and organic molecules utilized in practical applications is the synthesis and formation of Al-MOF compounds that are applied in various engineering contexts, e.g., gas storage [127], drug delivery [128], controlled nucleation [129], adsorption cooling [130], etc. In general, the synthesis of these compounds is performed by mixing certain amounts of Al(III) salts and organic molecules in specific ratios, followed by treatments at increased temperatures for prolonged times, empirically searching for the optimal conditions for the preparation of crystalline materials. Although these empirical approaches are generally successful, we still lack detailed knowledge about the nature of the interactions, which often prevents us from rationally controlling the synthesis outcome beyond trial-and-error approaches. Thus, there is a potentially enormous, yet untapped potential to further optimize the final MOF materials' properties. Recent insights into the synthesis process suggest that the formation of prenucleation units are important during MOF synthesis [131], and it was shown that MOF synthesis exhibits complex dynamics, a proper understanding of which can likely enable the improvement of their functional properties [132]. Interestingly, anisotropic nanostructures with their dynamic interphases dispersed in battery electrolytes improve the electrochemical performance of multi-valent-ion batteries [133], and indeed, Al-organic compounds were used as electrolytes in Al-ion batteries [134]. A combination of Al species, possibly Al(III) (oxy)(hydr)oxide PNC species, stabilized by specific additives that can provide charge transfer but suppress long-term precipitation, might resolve some of the current issues of Al-based batteries while offering considerable electrochemical performance.

Altogether, the recent years have brought novel insights into the chemistry and nucleation mechanisms of the hydrolyzing aqueous Al(III) and Fe(III) systems. Furthermore, advances were made in the understanding of their chemical behavior also in the presence of organic species and in biomineralizing environments. As these systems are extremely important for industry and everyday life, they have been extensively studied for more than a century, yet, many open questions remain. Switching the paradigm of classical nucleation theory toward non-classical approaches allows thinking and developing new concepts toward more direct, chemical controls over nucleation processes, enabling us to potentially target particular reaction steps. The introduction of chemical, especially organic, additives at specific and controlled hydrolysis-polymerization-nucleation steps could allow the formation of hybrid structures with unique morphological and chemical properties. The prenucleation cluster pathway was shown to be useful in interpreting the nucleation mechanisms in these hydrolyzing systems, with and without the presence of (organic) additives, and may also be transferred to other relevant metal systems. Remarkably, the role of water seems to become increasingly more recognized as being important not only for the fundamental understanding of nucleation (and biomineralization) mechanisms but also for the vast practical applications of the metal ions discussed herein.

**Acknowledgements** MJL and DG thank the Deutsche Forschungsgemeinschaft (DFG, German Research Foundation)—Project number: 455689363. The authors declare no conflict of interest.

## References


1. L. Zhou, Y. Tan, L. Huang, C. Fortin, P.G.C. Campbell, Aluminum effects on marine phytoplankton: implications for a revised iron hypothesis (iron–aluminum hypothesis). Biogeochemistry **139**, 123–137 (2018). https://doi.org/10.1007/s10533-018-0458-6
2. K. Hagvall, P. Persson, T. Karlsson, Speciation of aluminum in soils and stream waters: the importance of organic matter. Chem. Geol. **417**, 32–43 (2015). https://doi.org/10.1016/j.chemgeo.2015.09.012
3. M.A. Rahman, S.H. Lee, H.C. Ji, A.H. Kabir, C.S. Jones, K.W. Lee, Importance of mineral nutrition for mitigating aluminum toxicity in plants on acidic soils: current status and opportunities. Int. J. Mol. Sci. **19**, 3073 (2018). https://doi.org/10.3390/ijms19103073
4. K.K.I.U. Arunakumara, B.C. Walpola, M.-H. Yoon, Aluminum toxicity and tolerance mechanism in cereals and legumes—a review. J. Korean Soc. Appl. Biol. Chem. **56**, 1–9 (2013). https://doi.org/10.1007/s13765-012-2314-z
5. M. Mohammadi, M. Khodamorady, B. Tahmasbi, K. Bahrami, A. Ghorbani-Choghamarani, Boehmite nanoparticles as versatile support for organic–inorganic hybrid materials: synthesis, functionalization, and applications in eco-friendly catalysis. J. Ind. Eng. Chem. **97**, 1–78 (2021). https://doi.org/10.1016/j.jiec.2021.02.001
6. A.N. Mukhamed'yarova, B.I. Gareev, D.K. Nurgaliev, F.A. Aliev, A.V. Vakhin, A review on the role of amorphous aluminum compounds in catalysis: avenues of investigation and potential application in petrochemistry and oil refining. Processes **9**, 1811 (2021). https://doi.org/10.3390/pr9101811
7. E. Kumar, A. Bhatnagar, W. Hogland, M. Marques, M. Sillanpää, Interaction of anionic pollutants with Al-based adsorbents in aqueous media—a review. Chem. Eng. J. **241**, 443–456 (2014). https://doi.org/10.1016/j.cej.2013.10.065
8. H. Zhang, X. Zhang, T.R. Graham, C.I. Pearce, H. Hlushko, J.A. LaVerne et al., Crystallization and phase transformations of aluminum (oxy)hydroxide polymorphs in caustic aqueous solution. Inorg. Chem. **60**, 9820–9832 (2021). https://doi.org/10.1021/acs.inorgchem.1c01111
9. H. Yang, H. Li, J. Li, Z. Sun, K. He, H.M. Cheng et al., The rechargeable aluminum battery: opportunities and challenges. Angew. Chem. Int. Ed. **58**, 11978–11996 (2019). https://doi.org/10.1002/anie.201814031
10. B. Sun, T. Xia, Nanomaterial-based vaccine adjuvants. J. Mater. Chem. B. **4**, 5496–5509 (2016). https://doi.org/10.1039/C6TB01131D
11. Z. Liang, X. Wang, G. Yu, M. Li, S. Shi, H. Bao et al., Mechanistic understanding of the aspect ratio-dependent adjuvanticity of engineered aluminum oxyhydroxide nanorods in prophylactic vaccines. Nano Today **43**, 101445 (2022). https://doi.org/10.1016/j.nantod.2022.101445
12. R.M. Peres, J.M.L. Sousa, M.O. de Oliveira, M.V. Rossi, R.R. de Oliveira, N.B. de Lima et al., Pseudoboehmite as a drug delivery system for acyclovir. Sci. Rep. **11**, 15448 (2021). https://doi.org/10.1038/s41598-021-94325-y
13. T. Lertvanithphol, C. Limnonthakul Phom-on, P. Jaroenapibal, C. Chananonnawathorn, S. Limwichean, M. Horprathum, Facile fabrication and optical characterization of nanoflake aluminum oxide film with high broadband and omnidirectional transmittance enhancement. Opt. Mater. **111**, 110567 (2021). https://doi.org/10.1016/j.optmat.2020.110567
14. A. Tagliabue, A.R. Bowie, P.W. Boyd, K.N. Buck, K.S. Johnson, M.A. Saito, The integral role of iron in ocean biogeochemistry. Nature **543**, 51–59 (2017). https://doi.org/10.1038/nature21058
15. A.G. Leonel, A.A.P. Mansur, H.S. Mansur, Advanced functional nanostructures based on magnetic iron oxide nanomaterials for water remediation: a review. Water Res. **190**, 116693 (2021). https://doi.org/10.1016/j.watres.2020.116693
16. H. Li, Y.J. Zhu, Liquid-phase synthesis of iron oxide nanostructured materials and their applications. Chemistry **26**, 9180–9205 (2020). https://doi.org/10.1002/chem.202000679

17. E. Alphandéry, Iron oxide nanoparticles as multimodal imaging tools. RSC. Adv. **9,** 40577–40587 (2019). https://doi.org/10.1039/C9RA08612A
18. M.G. Montiel Schneider, M.J. Martín, J. Otarola, E. Vakarelska, V. Simeonov, V. Lassalle et al., Biomedical applications of iron oxide nanoparticles: current insights progress and perspectives. Pharmaceutics **14**, 204 (2022). https://doi.org/10.3390/pharmaceutics14010204
19. D. Spreitzer, J. Schenk, Reduction of iron oxides with hydrogen—a review. Steel Res. Int. **90**, 1900108 (2019). https://doi.org/10.1002/srin.201900108
20. G.P. Mastrotheodoros, K.G. Beltsios, Pigments—iron-based red, yellow, and brown ochres. Archaeol. Anthropol. Sci. **14**, 35 (2022). https://doi.org/10.1007/s12520-021-01482-2
21. B. Singh, M.B. Gawande, A.D. Kute, R.S. Varma, P. Fornasiero, P. McNeice et al., Single-atom (iron-based) catalysts: synthesis and applications. Chem. Rev. **121**, 13620–13697 (2021). https://doi.org/10.1021/acs.chemrev.1c00158
22. J.L. Jambor, J.E. Dutrizac, Occurrence and constitution of natural and synthetic ferrihydrite, a widespread iron oxyhydroxide. Chem. Rev. **98**, 2549–2586 (1998). https://doi.org/10.1021/cr970105t
23. D. Faivre, *Iron Oxides: From Nature to Applications* (John Wiley & Sons, Chichester, UK, 2016)
24. R.N. Lodaya, S. Gregory, M.M. Amiji, D.T. O'Hagan, Overview of vaccine adjuvants, in *Practical Aspects of Vaccine Development,* 2 edn, ed. by P. Kolhe, S. Ohtake (Academic Press, Cambridge, MA), pp. 9–25
25. X. Liu, L. Xiao, J. Weng, Q. Xu, W. Li, C. Zhao et al., Regulating the reactivity of black phosphorus via protective chemistry. Sci. Adv. **6** (2020). https://doi.org/10.1126/sciadv.abb4359
26. Z. Lin, M. Mao, C. Yang, Y. Tong, Q. Li, J. Yue et al., Amorphous anion-rich titanium polysulfides for aluminum-ion batteries. Sci. Adv. **7** (2021). https://doi.org/10.1126/sciadv.abg6314
27. T. Huang, T. Zhang, X. Jiang, A. Li, Y. Su, Q. Bian et al., Iron oxide nanoparticles augment the intercellular mitochondrial transfer–mediated therapy. Sci. Adv. **7**, eabj0534 (2021). https://doi.org/10.1126/sciadv.abj0534
28. D. Gebauer, A. Völkel, H. Cölfen, Stable prenucleation calcium carbonate clusters. Science **322**, 1819–1822 (2008). https://doi.org/10.1126/science.1164271
29. D. Gebauer, M. Kellermeier, J.D. Gale, L. Bergström, H. Cölfen, Pre-nucleation clusters as solute precursors in crystallisation. Chem. Soc. Rev. **43**, 2348–2371 (2014)
30. M.J. Lukić, D. Gebauer, A. Rose, Nonclassical nucleation towards separation and recycling science: iron and aluminium (Oxy)(hydr)oxides. Curr. Opin. Colloid Interface Sci. **46**, 114–127 (2020). https://doi.org/10.1016/j.cocis.2020.03.010
31. J.P. Jolivet, C. Chanéac, E. Tronc, Iron oxide chemistry. from molecular clusters to extended solid networks. Chem. Commun. (Camb). 481–483, 481–487 (2004). https://doi.org/10.1039/B304532N
32. J.P. Jolivet, *Metal Oxide Nanostructures Chemistry: Synthesis from Aqueous Solutions* (Oxford University Press, Oxford, 2019)
33. U. Schwertmann, J. Friedl, H. Stanjek, From Fe(III) ions to ferrihydrite and then to hematite. J. Colloid Interface Sci. **209**, 215–223 (1999). https://doi.org/10.1006/jcis.1998.5899. (PubMed: 9878155)
34. J. Scheck, L.M. Fuhrer, B. Wu, M. Drechsler, D. Gebauer, Nucleation of hematite: a nonclassical mechanism. Chemistry **25**, 13002–13007 (2019). https://doi.org/10.1002/chem.201902528
35. F.M. Michel, L. Ehm, S.M. Antao, P.L. Lee, P.J. Chupas, G. Liu et al., The structure of ferrihydrite, a nanocrystalline material. Science **316**, 1726–1729 (2007). https://doi.org/10.1126/science.1142525
36. J.E. Post, V.F. Buchwald, Crystal structure refinement of akaganéite. Am. Miner. **76**, 272–277 (1991)
37. G. Yamaguchi, M. Okumiya, S. Ono, Refinement of the structure of tohdite $5Al_2O_3$ $H_2O$. Bull. Chem. Soc. Jpn **42**, 2247–2249 (1969). https://doi.org/10.1246/bcsj.42.2247

38. K. Momma, F. Izumi, VESTA 3 for three-dimensional visualization of crystal, volumetric and morphology data. J. Appl. Crystallogr. **44**, 1272–1276 (2011). https://doi.org/10.1107/S0021889811038970
39. J.B. Parise, B. Xia, J.W. Simonson, W.R. Woerner, A.M. Plonka, B.L. Phillips et al., Structural chemistry of akdalaite, $Al_{10}O_{14}(OH)_2$, the isostructural aluminum analogue of ferrihydrite. Crystals **9**, 246 (2019). https://doi.org/10.3390/cryst9050246
40. J.Y. Bottero, A. Manceau, F. Villieras, D. Tchoubar, Structure and mechanisms of formation of iron oxide hydroxide (chloride) polymers. Langmuir **10**, 316–319 (1994). https://doi.org/10.1021/la00013a046
41. A.L. Rose, M.W. Bligh, R.N. Collins, T.D. Waite, Resolving early stages of homogeneous iron(III) oxyhydroxide formation from iron(III) nitrate solutions at ph 3 using time-resolved SAXS. Langmuir **30**, 3548–3556 (2014). https://doi.org/10.1021/la404712r
42. M. Zhu, C. Frandsen, A.F. Wallace, B. Legg, S. Khalid, H. Zhang et al., Precipitation pathways for ferrihydrite formation in acidic solutions. Geochim. Cosmochim. Acta **172**, 247–264 (2016). https://doi.org/10.1016/j.gca.2015.09.015
43. J.S. Weatherill, K. Morris, P. Bots, T.M. Stawski, A. Janssen, L. Abrahamsen et al., Ferrihydrite formation: the role of Fe13 Keggin clusters. Environ. Sci. Technol. **50**, 9333–9342 (2016). https://doi.org/10.1021/acs.est.6b02481
44. O. Sadeghi, C. Falaise, P.I. Molina, R. Hufschmid, C.F. Campana, B.C. Noll et al., Chemical stabilization and electrochemical destabilization of the iron Keggin ion in water. Inorg. Chem. **55**, 11078–11088 (2016). https://doi.org/10.1021/acs.inorgchem.6b01694
45. J. Baumgartner, D. Faivre, Iron solubility, colloids and their impact on iron (oxyhydr)oxide formation from solution. Earth Sci. Rev. **150**, 520–530 (2015). https://doi.org/10.1016/j.earscirev.2015.09.003
46. E. Jansen, A. Kyek, W. Schäfer, U. Schwertmann, The structure of six-line ferrihydrite. Appl. Phys. A **74**, s1004–s1006 (2002). https://doi.org/10.1007/s003390101175
47. F.M. Michel, L. Ehm, G. Liu, W.Q. Han, S.M. Antao, P.J. Chupas et al., Similarities in 2- and 6-line ferrihydrite based on pair distribution function analysis of X-ray total scattering. Chem. Mater. **19**, 1489–1496 (2007). https://doi.org/10.1021/cm062585n
48. H. Liu, X. Li, Y. Wang, X. Yang, Z. Zhen, R. Chen et al., New insight into the effect of the formation environment of ferrihydrite on its structure and properties. RSC Adv. **4**, 11451–11458 (2014). https://doi.org/10.1039/C4RA00696H
49. J. Scheck, B. Wu, M. Drechsler, R. Rosenberg, A.E.S. Van Driessche, T.M. Stawski et al., The molecular mechanism of iron(III) oxide nucleation. J. Phys. Chem. Lett. **7**, 3123–3130 (2016). https://doi.org/10.1021/acs.jpclett.6b01237
50. J. Scheck, T. Lemke, D. Gebauer, The role of chloride ions during the formation of akaganéite revisited. Minerals **5**, 778–787 (2015). https://doi.org/10.3390/min5040524
51. M.J. Lukić, F. Lücke, T. Ilić, K. Petrović, D. Gebauer, On the role of poly-glutamic acid in the early stages of iron(III) (oxy)(hydr)oxide formation. Minerals **11**, 715 (2021). https://doi.org/10.3390/min11070715
52. D. Li, M.H. Nielsen, J.R.I. Lee, C. Frandsen, J.F. Banfield, J.J. De Yoreo, Direction-specific interactions control crystal growth by oriented attachment. Science **336**, 1014–1018 (2012). https://doi.org/10.1126/science.1219643
53. J. De Yoreo, Crystal nucleation: more than one pathway. Nat. Mater. **12**, 284–285 (2013). https://doi.org/10.1038/nmat3604
54. J. Baumgartner, A. Dey, P.H.H. Bomans, C. Le Coadou, P. Fratzl, N.A.J.M. Sommerdijk et al., Nucleation and growth of magnetite from solution. Nat. Mater. **12**, 310–314 (2013). https://doi.org/10.1038/nmat3558
55. J. Baumgartner, R.K. Ramamoorthy, A.P. Freitas, M.A. Neouze, M. Bennet, D. Faivre et al., Self-confined nucleation of iron oxide nanoparticles in a nanostructured amorphous precursor. Nano Lett. **20**, 5001–5007 (2020). https://doi.org/10.1021/acs.nanolett.0c01125
56. G. Montes-Hernandez, N. Findling, F. Renard, Direct and indirect nucleation of magnetite nanoparticles from solution revealed by time-resolved Raman spectroscopy. Cryst. Growth Des. **21**, 3500–3510 (2021). https://doi.org/10.1021/acs.cgd.1c00282

57. G. Mirabello, A. Ianiro, P.H.H. Bomans, T. Yoda, A. Arakaki, H. Friedrich et al., Crystallization by particle attachment is a colloidal assembly process. Nat. Mater. **19**, 391–396 (2020). https://doi.org/10.1038/s41563-019-0511-4
58. G. Mirabello, M. GoodSmith, P.H.H. Bomans, L. Stegbauer, D. Joester, G. de With, Iron phosphate mediated magnetite synthesis: a bioinspired approach. Chem. Sci. **12**, 9458–9465 (2021). https://doi.org/10.1039/D0SC07079C
59. G. Zhu, M.L. Sushko, J.S. Loring, B.A. Legg, M. Song, J.A. Soltis et al., Self-similar mesocrystals form via interface-driven nucleation and assembly. Nature **590**, 416–422 (2021). https://doi.org/10.1038/s41586-021-03300-0
60. T.A. Ho, L.J. Criscenti, Molecular-level understanding of gibbsite particle aggregation in water. J. Colloid Interface Sci. **600**, 310–317 (2021). https://doi.org/10.1016/j.jcis.2021.05.016
61. E. Nakouzi, J.A. Soltis, B.A. Legg, G.K. Schenter, X. Zhang, T.R. Graham et al., Impact of solution chemistry and particle anisotropy on the collective dynamics of oriented aggregation. ACS Nano **12**, 10114–10122 (2018). https://doi.org/10.1021/acsnano.8b04909
62. L.M. Anovitz, X. Zhang, J. Soltis, E. Nakouzi, A.J. Krzysko, J. Chun et al., Effects of ionic strength, salt, and ph on aggregation of boehmite nanocrystals: tumbler small-angle neutron and X-ray scattering and imaging analysis. Langmuir **34**, 15839–15853 (2018). https://doi.org/10.1021/acs.langmuir.8b00865
63. S.N. Kerisit, J.J. De Yoreo, Effect of hydrophilicity and interfacial water structure on particle attachment. J. Phys. Chem. C **124**, 5480–5488 (2020). https://doi.org/10.1021/acs.jpcc.9b12053
64. S. Sun, D. Gebauer, H. Cölfen, Alignment of amorphous iron oxide clusters: a non-classical mechanism for magnetite formation. Angew. Chem. Int. Ed. **56**, 4042–4046 (2017). https://doi.org/10.1002/anie.201610275
65. J.Z. Hu, X. Zhang, N.R. Jaegers, C. Wan, T.R. Graham, M. Hu et al., Transitions in al coordination during gibbsite crystallization using high-field $^{27}$Al and $^{23}$Na MAS NMR spectroscopy. J. Phys. Chem. C **121**, 27555–27562 (2017). https://doi.org/10.1021/acs.jpcc.7b10424
66. C.F. Baes, R.E. Mesmer, *The Hydrolysis of Cations* (Wiley, Chichester, UK, 1976)
67. S. Bi, Studies on the mechanism of hydrolysis and polymerization of aluminum salts in aqueous solution: correlations between the "core-links" model and "cage-like" Keggin-Al13 model. Coord. Chem. Rev. **248**, 441–455 (2004). https://doi.org/10.1016/j.ccr.2003.11.001
68. R.W. Smith, J.D. Hem, *Effect of Aging on Aluminum Hydroxide Complexes in Dilute Aqueous Solutions* (1972)
69. P.M. Bertsch, D.R. Parker, Aqueous polynuclear aluminum species, in *The Environmental Chemistry of Aluminum,* 2nd edn. (CRC Press, Boca Raton, FL)
70. H. Zhao, H. Liu, J. Qu, Effect of ph on the aluminum salts hydrolysis during coagulation process: formation and decomposition of polymeric aluminum species. J. Colloid Interface Sci. **330**, 105–112 (2009). https://doi.org/10.1016/j.jcis.2008.10.020
71. G. Furrer, B.L PHillips, K.U. Ulrich, R. Pöthig, W.H. Casey, The origin of aluminum flocs in polluted streams. Science **297**, 2245–2247 (2002)
72. L. Helm, A.E. Merbach, Water exchange on metal ions: experiments and simulations. Coord. Chem. Rev. **187**, 151–181 (1999). https://doi.org/10.1016/S0010-8545(99)90232-1
73. E. Balan, M. Lazzeri, G. Morin, F. Mauri, First-principles study of the OH-stretching modes of gibbsite. Am. Miner. **91**, 115–119 (2006). https://doi.org/10.2138/am.2006.1922
74. E. Balan, M. Blanchard, J.-F. Hochepied, M. Lazzeri, Surface modes in the infrared spectrum of hydrous minerals: the OH stretching modes of bayerite. Phys. Chem. Miner. **35**, 279–285 (2008). https://doi.org/10.1007/S00269-008-0221-Y
75. X. Bokhimi, J.A. Toledo-Antonio, M.L. Guzmán-Castillo, F. Hernández-Beltrán, Relationship between crystallite size and bond lengths in boehmite. J. Solid State Chem. **159**, 32–40 (2001). https://doi.org/10.1006/jssc.2001.9124
76. M.J. Lukić, E. Wiedenbeck, H. Reiner, D. Gebauer, Chemical trigger toward phase separation in the aqueous Al(III) system revealed. Sci. Adv. **6**, eaba6878 (2020b). https://doi.org/10.1126/sciadv.aba6878

77. T.W. Swaddle, J. Rosenqvist, P. Yu, E. Bylaska, B.L. Phillips, W.H. Casey, Kinetic evidence for five-coordination in $AlOH(aq)^{2+}$ ion. Science **308**, 1450–1453 (2005)
78. J.M. Ruiz, M.H. McAdon, J.M. Garcés, Aluminum complexes as models for Broensted acid sites in zeolites: structure and energetics of $[Al(OH)_4]^-$, $[Al(H_2O)_6]^{3+}$, and intermediate monomeric species $[Al(OH)_x(H_2O)_{n-x}\ mH_2O]^{3-x}$ obtained by hydrolysis. J. Phys. Chem. B **101**, 1733–1744 (1997). https://doi.org/10.1021/jp961987n
79. D. Reusser, W.H. Casey, A. Navrotsky, Energetic insight into the formation of solids from aluminum Polyoxocations. Angew. Chem. Int. Ed. **54**, 9253–9256 (2015). https://doi.org/10.1002/anie.201503544
80. D. Gebauer, How can additives control the early stages of mineralisation? Minerals **8**, 179 (2018). https://doi.org/10.3390/min8050179
81. S.T. Mergelsberg, M. Dembowski, M.E. Bowden, T.R. Graham, M. Prange, H.W. Wang et al., Cluster defects in gibbsite nanoplates grown at acidic to neutral ph. Nanoscale **13**, 17373–17385 (2021). https://doi.org/10.1039/D1NR01615F
82. M. Dembowski, J.S. Loring, M.E. Bowden, J.G. Reynolds, T.R. Graham, K.M. Rosso et al., The controlling role of atmosphere in dawsonite versus gibbsite precipitation from tetrahedral aluminate species. Dalton Trans. **50**, 13438–13446 (2021). https://doi.org/10.1039/D1DT02081A
83. S. Wang, X. Zhang, T.R. Graham, H. Zhang, C.I. Pearce, Z. Wang et al., Two-step route to size and shape controlled gibbsite nanoplates and the **crystal growth** mechanism. CrystEngComm **22**, 2555–2565 (2020). https://doi.org/10.1039/D0CE00114G
84. A.M. Baccarella, R. Garrard, M.L. Beauvais, U. Bednarksi, S. Fischer, A.M.M. Abeykoon et al., Cluster mediated conversion of amorphous $Al(OH)_3$ to $\gamma$-AlOOH. J. Solid State Chem. **301**, 122340 (2021). https://doi.org/10.1016/j.jssc.2021.122340
85. M. Pouvreau, E. Martinez-Baez, M. Dembowski, C.I. Pearce, G.K. Schenter, K.M. Rosso et al., Mechanisms of Al3+ dimerization in alkaline solutions. Inorg. Chem. **59**, 18181–18189 (2020). https://doi.org/10.1021/acs.inorgchem.0c02660
86. T.R. Graham, M. Dembowski, E. Martinez-Baez, X. Zhang, N.R. Jaegers, J. Hu et al., In situ $^{27}$Al NMR spectroscopy of aluminate in sodium hydroxide solutions above and below saturation with respect to gibbsite. Inorg. Chem. **57**, 11864–11873 (2018). https://doi.org/10.1021/acs.inorgchem.8b00617
87. H. Hashimoto, Y. Onodera, S. Tahara, S. Kohara, K. Yazawa, H. Segawa et al., Structure of alumina glass. Sci. Rep. **12**, 516 (2022). https://doi.org/10.1038/s41598-021-04455-6
88. J.C. Gómez Martín, T.R. Lewis, M.A. Blitz, J.M.C. Plane, M. Kumar, J.S. Francisco et al., A gas-to-particle conversion mechanism helps to explain atmospheric particle formation through clustering of iodine oxides. Nat. Commun. **11**, 4521 (2020). https://doi.org/10.1038/s41467-020-18252-8
89. L. Smrcok, V. Langer, J. Krestan, Gamma-alumina: a single-crystal X-ray diffraction study. Acta Crystallogr. C **62**, i83–i84 (2006). https://doi.org/10.1107/S0108270106026850
90. C. Haavik, S. Stølen, H. Fjellvåg, M. Hanfland, D. Häusermann, Equation of state of magnetite and its high-pressure modification: thermodynamics of the Fe-O system at high pressure. Am. Mineral. **85**, 514–523 (2000). https://doi.org/10.2138/am-2000-0413
91. R.L. Blake, R.E. Hessevick, T. Zoltai, L.W. Finger, Refinement of the hematite structure. Am. Mineral. **51**, 123–129 (1966)
92. J. Scheck, M. Drechsler, X. Ma, M.T. Stöckl, J. Konsek, J.B. Schwaderer et al., Polyaspartic acid facilitates oxolation within iron(III) oxide pre-nucleation clusters and drives the formation of organic–inorganic composites. J. Chem. Phys. **145**, 211917 (2016). https://doi.org/10.1063/1.4963738
93. H. Lu, Y.C. Huang, J. Hunger, D. Gebauer, H. Cölfen, M. Bonn, Role of water in $CaCO_3$ biomineralization. J. Am. Chem. Soc. **143**, 1758–1762 (2021). https://doi.org/10.1021/jacs.0c11976
94. M. Kellermeier, P. Raiteri, J.K. Berg, A. Kempter, J.D. Gale, D. Gebauer, Entropy drives calcium carbonate ion association. ChemPhysChem **17**, 3535–3541 (2016). https://doi.org/10.1002/cphc.201600653

95. H. Ohtaki, T. Radnai, Structure and dynamics of hydrated ions. Chem. Rev. **93**, 1157–1204 (1993). https://doi.org/10.1021/cr00019a014
96. W.W. Rudolph, R. Mason, C.C. Pye, Aluminium(III) hydration in aqueous solution: a Raman spectroscopic investigation and an ab initio molecular orbital study of aluminium(III) water clusters. Phys. Chem. Chem. Phys. **2**, 5030–5040 (2000). https://doi.org/10.1039/B003764H
97. A. Arinchtein, R. Schmack, K. Kraffert, J. Radnik, P. Dietrich, R. Sachse et al., Role of water in phase transformations and crystallization of ferrihydrite and hematite. ACS Appl. Mater Interfaces **12**, 38714–38722 (2020). https://doi.org/10.1021/acsami.0c05253
98. S. Lid, D. Carmona, M. Maas, L. Treccani, L. Colombi Ciacchi, Anchoring of iron oxyhydroxide clusters at H and L ferritin subunits. ACS Biomater. Sci. Eng. **4**, 483–490 (2018). https://doi.org/10.1021/acsbiomaterials.7b00814
99. R.B. Martin, $Fe^{3+}$ and $Al^{3+}$ hydrolysis equilibria: cooperativity in $Al^{3+}$ hydrolysis reactions. J. Inorg. Biochem. **44**, 141–147 (1991). https://doi.org/10.1016/0162-0134(91)84026-6
100. P.L. Brown, R.N. Sylva, G.E. Batley, J. Ellis, The hydrolysis of metal ions part 8. Aluminium(III). J. Chem. Soc. Dalton Trans. (9), 1967 (1985). https://doi.org/10.1039/dt9850001967
101. S. Dong, W. Shi, J. Zhang, S. Bi, $^{27}$Al NMR chemical shifts and relative stabilities of aqueous monomeric $Al^{3+}$ hydrolytic species with different coordination structures. ACS Earth Space Chem. **3**(7), 1353–1361 (2019). https://doi.org/10.1021/acsearthspacechem.9b00102
102. E. Zaki, Z. Jakub, F. Mirabella, G.S. Parkinson, S. Shaikhutdinov, H.J. Freund, Water ordering on the magnetite $Fe_3O_4$ surfaces. J. Phys. Chem. Lett. **10**, 2487–2492 (2019). https://doi.org/10.1021/acs.jpclett.9b00773
103. T.S. Peretyazhko, M.J. Pan, D.W. Ming, E.B. Rampe, R.V. Morris, D.G. Agresti, Reaction of akaganeite with Mars-relevant anions. ACS Earth Space Chem. **3**, 314–323 (2019). https://doi.org/10.1021/acsearthspacechem.8b00173
104. C. Exley, The coordination chemistry of aluminium in neurodegenerative disease. Coord. Chem. Rev. **256**, 2142–2146 (2012)
105. G. Sposito, *The Environmental Chemistry of Aluminum*, 2nd edn. (CRC Press, Boca Raton, FL, 1995)
106. D. Bhattacharya, C. Wiley, A. Latal, V. Krishna, Antigen–adjuvant formulations—key considerations, in *Practical Aspects of Vaccine Development*, 7. ed. by P. Kolhe, S. Ohtake (Academic Press, Cambridge, MA, 2022), pp. 207–224
107. K.N. Nitzsche, Z.E. Kayler, K. Premke, A. Gessler, R. Wagai, Divergent roles of iron and aluminum in sediment organic matter association at the terrestrial–aquatic interface. Biogeochemistry **157**, 355–378 (2022). https://doi.org/10.1007/s10533-021-00878-5
108. T. Lemke, M. Edte, D. Gebauer, C. Peter, Three reasons why aspartic acid and glutamic acid sequences have a surprisingly different influence on mineralization. J. Phys. Chem. B **125**, 10335–10343 (2021). https://doi.org/10.1021/acs.jpcb.1c04467
109. H. Chang, B.H. Kim, H.Y. Jeong, J.H. Moon, M. Park, K. Shin et al., Molecular-level understanding of continuous growth from iron-oxo clusters to iron oxide nanoparticles. J. Am. Chem. Soc. **141**, 7037–7045 (2019). https://doi.org/10.1021/jacs.9b01670
110. A. Feld, A. Weimer, A. Kornowski, N. Winckelmans, J.P. Merkl, H. Kloust et al.**,** Chemistry of shape-controlled iron oxide nanocrystal formation. ACS Nano **13**, 152–162 (2019). https://doi.org/10.1021/acsnano.8b05032
111. C. Pozzi, S. Ciambellotti, C. Bernacchioni, F.D. Di Pisa, S. Mangani, P. Turano, Chemistry at the protein–mineral interface in L-ferritin assists the assembly of a functional (μ3-oxo)Tris[(μ2-peroxo)] triiron(III) cluster. Proc. Natl. Acad. Sci. USA **114**, 2580–2585 (2017). https://doi.org/10.1073/pnas.1614302114
112. L. Massai, S. Ciambellotti, L. Cosottini, L. Messori, P. Turano, A. Pratesi, Direct detection of iron clusters in L ferritins through ESI-MS experiments. Dalton Trans. **50**, 16464–16467 (2021). https://doi.org/10.1039/D1DT03106F
113. A.E. Rawlings, P. Liravi, S. Corbett, A.S. Holehouse, S.S. Staniland, Investigating the ferric ion binding site of magnetite biomineralisation protein Mms6. Plos One **15**, e0228708 (2020). https://doi.org/10.1371/journal.pone.0228708

114. K. Zhang, J. Wu, X. Zhao, J. Qin, Y. Xue, W. Zheng et al., Prussian Blue/calcium peroxide nanocomposites-mediated tumor cell iron mineralization for treatment of experimental lung adenocarcinoma. ACS Nano **15**, 19838–19852 (2021b). https://doi.org/10.1021/acsnano.1c07308
115. P. Dusek, M. Dezortova, J. Wuerfel, *Chapter nine, imaging of iron*, ed. by K.P. Bhatia, S.A. Schneider. International Review of Neurobiology (Academic Press, Cambridge, MA, 2013), pp. 195–239. https://doi.org/10.1016/B978-0-12-410502-7.00010-7
116. C. Dehner, N. Morales-Soto, R.K. Behera, J. Shrout, E.C. Theil, P.A. Maurice et al., Ferritin and ferrihydrite nanoparticles as iron sources for Pseudomonas aeruginosa. J. Biol. Inorg. Chem. **18**, 371–381 (2013). https://doi.org/10.1007/s00775-013-0981-9
117. S. Narayanan, E. Firlar, M.G. Rasul, T. Foroozan, N. Farajpour, L. Covnot et al., On the structure and chemistry of iron oxide cores in human heart and human spleen ferritins using graphene liquid cell electron microscopy. Nanoscale **11**, 16868–16878 (2019). https://doi.org/10.1039/C9NR01541H
118. W. Yan, W. Guo, L. Wang, C. Jing, Extracellular polymeric substances from Shewanella oneidensis MR-1 biofilms mediate the transformation of ferrihydrite. Sci. Total. Environ. **784**, 147245 (2021). https://doi.org/10.1016/j.scitotenv.2021.147245
119. S. Ben-Shimon, D. Stein, R. Zarivach, Current view of iron biomineralization in magnetotactic bacteria. J. Struct. Biol. X **5**, 100052 (2021). https://doi.org/10.1016/j.yjsbx.2021.100052
120. J.S. Evans, Polymorphs, proteins, and nucleation theory: a critical analysis. Minerals **7**, 62 (2017). https://doi.org/10.3390/min7040062
121. G. Davidov, G. Abelya, R. Zalk, B. Izbicki, S. Shaibi, L. Spektor et al., Folding of an intrinsically disordered iron-binding peptide in response to sedimentation revealed by cryo-EM. J. Am. Chem. Soc. **142**, 19551–19557 (2020). https://doi.org/10.1021/jacs.0c07565
122. J. Werckmann, J. Cypriano, C.T. Lefèvre, K. Dembelé, O. Ersen, D.A. Bazylinski et al., Localized iron accumulation precedes nucleation and growth of magnetite crystals in magnetotactic bacteria. Sci. Rep. **7**, 8291 (2017). https://doi.org/10.1038/s41598-017-08994-9
123. B. Song, Q. Sun, H. Li, B. Ge, J.S. Pan, A.T.S. Wee et al., Irreversible denaturation of proteins through aluminum-induced formation of backbone ring structures. Angew. Chem. Int. Ed. **53**, 6358–6363 (2014). https://doi.org/10.1002/anie.201307955
124. J.I. Mujika, G. Dalla Torre, E. Formoso, R. Grande-Aztatzi, S.J. Grabowski, C. Exley et al., Aluminum's preferential binding site in proteins: sidechain of amino acids versus backbone interactions. J. Inorg. Biochem. **181**, 111–116 (2018). https://doi.org/10.1016/j.jinorgbio.2017.10.014
125. W.J. Lukiw, Evidence supporting a biological role for aluminum in chromatin compaction and epigenetics. J. Inorg. Biochem. **104**, 1010–1012 (2010). https://doi.org/10.1016/j.jinorgbio.2010.05.007
126. L. He, S. Cai, B. Wu, C. Mu, G. Zhang, W. Lin, Trivalent chromium and aluminum affect the thermostability and conformation of collagen very differently. J. Inorg. Biochem. **117**, 124–130 (2012). https://doi.org/10.1016/j.jinorgbio.2012.08.017
127. D. Alezi, Y. Belmabkhout, M. Suyetin, P.M. Bhatt, ŁJ. Weseliński, V. Solovyeva et al., MOF crystal chemistry paving the way to gas storage needs: aluminum-based soc-MOF for $CH_4$, $O_2$, and $CO_2$ storage. J. Am. Chem. Soc. **137**, 13308–13318 (2015). https://doi.org/10.1021/jacs.5b07053
128. X. Zhong, Y. Zhang, L. Tan, T. Zheng, Y. Hou, X. Hong et al., An aluminum adjuvant-integrated Nano-MOF as antigen delivery system to induce strong humoral and cellular immune responses. J. Control. Release **300**, 81–92 (2019). https://doi.org/10.1016/j.jconrel.2019.02.035
129. M.R. Mani, R. Chellaswamy, Y.N. Marathe, V.K. Pillai, Enhanced nucleation of polypropylene by metal–organic frameworks (MOFs) based on aluminium dicarboxylates: influence of structural features. R.S.C. Adv. **6**, 1907–1912 (2016). https://doi.org/10.1039/C5RA22764J
130. N. Tannert, S.-J. Ernst, C. Jansen, H.-J. Bart, S.K. Henninger, C. Janiak, Evaluation of the highly stable metal–organic framework MIL-53(Al)-TDC (TDC = 2,5-thiophenedicarboxylate) as a new and promising adsorbent for heat transformation applications. J. Mater. Chem. A **6**, 17706–17712 (2018). https://doi.org/10.1039/C8TA04407D

131. G. Zahn, H.A. Schulze, J. Lippke, S. König, U. Sazama, M. Fröba et al., A water-born Zr-based porous coordination polymer: modulated synthesis of Zr-fumarate MOF. Micropor. Mesopor. Mater. **203**, 186–194 (2015). https://doi.org/10.1016/j.micromeso.2014.10.034
132. X. Lv, P. Sullivan, D. Feng, MOFs the movie: molecule to nuclei evolution during metal–organic framework formation. Matter **5**, 11–13 (2022). https://doi.org/10.1016/j.matt.2021.11.009
133. W. Zhang, Q. Zhao, Y. Hou, Z. Shen, L. Fan, S. Zhou et al., Dynamic interphase–mediated assembly for deep cycling metal batteries. Sci Adv. 7, eabl3752 (2021c). https://doi.org/10.1126/sciadv.abl3752
134. G.A. Elia, K.V. Kravchyk, M.V. Kovalenko, J. Chacón, A. Holland, R.G.A. Wills, An overview and prospective on Al and Al-ion battery technologies. J. Power. Sources **481**, 228870 (2021). https://doi.org/10.1016/j.jpowsour.2020.228870

# Chapter 5
# Visualization of pH and Ionic Concentration for Dissolving/Forming Crystals

## Visualization of pH and Ionic Concentration

**Jun Kawano and Takashi Toyofuku**

**Abstract** To elucidate the formation and dissolution mechanisms of crystals in aqueous solution, monitoring local pH or ionic concentrations at or near the solid–fluid interface of crystals is critical. Using fluorescent probes, which produce different fluorescent intensities depending on pH or the ionic concentration of the solution containing the probes, a two- or three-dimensional visualization of the local distribution of pH or ionic concentration around the forming/dissolving crystal is realized. In this chapter, the technique of visualizing ionic concentrations under a microscope using a fluorescent probe is described, along with examples of visualized pH and calcium ion concentrations during the growth and dissolution of calcium carbonate. Using this technique, ion migration on the surface of growing/dissolving crystals can be directly measured, and it is expected to generate new insights into the mechanisms of their growth and dissolution.

## 5.1 Visualization Using Fluorescent Probes

During crystallization, it is important to control polymorphism and morphology, and understanding elementary processes at the crystal–solution interface is indispensable. The recent development of modern analytical instruments, such as atomic force microscopy (AFM), has advanced the understanding of nanoscale phenomena on

---

J. Kawano (✉)
Department of Earth and Planetary Sciences, Faculty of Science, Hokkaido University, N10 W8, Kita-Ku, Sapporo 060-0810, Japan
e-mail: j-kawano@sci.hokudai.ac.jp

T. Toyofuku
Institute for Extra-Cutting-Edge Science and Technology Avant-Garde Research (X-Star), Japan Agency for Marine-Earth Science and Technology (JAMSTEC), Natsushima-Cho 2-15, Yokosuka 237-0061, Japan
e-mail: toyofuku@jamstec.go.jp

 
F. Ishikawa et al. (eds.), *Engineering Crystal Habit*, Topics in Applied Physics 152,
https://doi.org/10.1007/978-981-96-0266-7_5

crystal surfaces [1, 2], but a complete understanding of the atomic-scale dynamic processes of these surface reactions is far from complete.

To clarify specific ionic migration during reactions, monitoring the pH and ionic concentration of bulk solutions is still a commonly used method. Even with modern AFM, which can observe phenomena at the atomic level, it is difficult to directly determine the type of atoms being observed. Moreover, specific ionic behaviors at the material–solution interface related to crystallization and dissolution are not well understood.

Recently, to determine ion dynamics near the crystal–solution interface during crystal precipitation and dissolution, visualization of the distribution of pH and ionic concentration using a fluorescent probe has been attempted, whereby the fluorescent intensity of reagents changes depending on the ion concentration or pH when the solution in which they are dissolved is irradiated with light of a certain wavelength [3]. This method is less accurate than electrode measurements, but it is suitable for tracking fast chemical reactions because of its quick response to changes in the environment and allows the two-dimensional observation of reactions in solution along with local distributions and changes in pH and ionic concentration. This technique was originally developed in the field of biology to determine the behavior of pH and ions in cells, where it has been observed, for example, that pH increases where calcification is occurring in foraminifera (Fig. 5.1) [4, 5]. By applying this technique to reactions in aqueous solution, it is possible to quantify the movement of specific ions in and out of the solid–liquid interface. This method has great potential for clarifying the behavior of ions at the crystal–solution interface and for understanding polymorphism and morphogenesis on time and spatial scales directly related to actual material development applications.

In addition to describing a practical method for visualizing pH and ionic concentration under a microscope using such a fluorescent probe, we present an example of the application of this technique to the dissolution and growth reactions of calcium

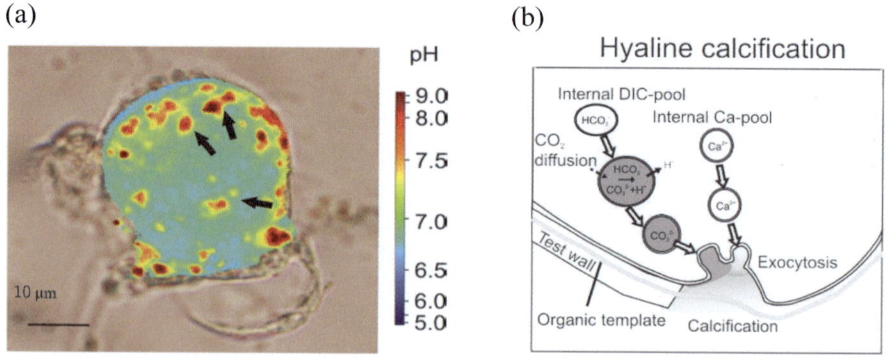


**Fig. 5.1** **a** Intracellular pH distribution in foraminifera containing high-pH vesicles correlated with the location of crystallites (black arrows). **b** Schematic illustration of the calcification pathway. Reproduced from [4]. Copyright 2009

carbonate ($CaCO_3$). The formation and dissolution mechanisms of $CaCO_3$ have attracted attention because of its importance as an industrial material and biomaterial [6–12]. In particular, phenomena occurring on its surfaces have been actively investigated for a long time using AFM [13–15], surface X-ray scattering [16, 17], computational simulation [18, 19], etc. Furthermore, $CaCO_3$ has several polymorphs, and the selection of these polymorphs remains controversial [10, 20]. The methods introduced in this chapter could provide new insights into these issues.

## 5.2 Selection of Fluorescent Probes

Fluorescent probes can measure the concentration of a dissolved ion by changing its fluorescence intensity by binding to it. As previously mentioned, because this method has been developed in the field of biology, various probes have been developed to visualize the behavior of pH and $Ca^{2+}$ ions, which are important when considering the function of cells. To apply this method to the crystallization of inorganic materials, the selected probe depends on the target to be visualized. Table 5.1 presents examples of typical pH and $Ca^{2+}$ ion probes. The following points should be considered when selecting and using these probes:

1. Acetoxymethyl group derivatives, which are permeable to cell membranes, are usually used for in vivo visualization, whereas potassium and other salts are used for observations in solutions. However, for some probes, only one is available.
2. In fluorescent probes, the fluorescence intensity at a specific wavelength varies with the ionic concentration when the probe is irradiated with excitation light with a wavelength specific to the probe [21–23]. Table 5.1 presents these excitation/

**Table 5.1** Examples of typical pH and $Ca^{2+}$ probes. The excitation and fluorescence wavelengths and the applicable ranges and dissociation constants ($K_d$) are shown. (Note that these values may vary depending on the experimental conditions.)

| pH probe | Excitation (nm) | Emission (nm) | Application range for pH | Refs. |
|---|---|---|---|---|
| HPTS | 405/460 | 520 | 6.0–10.0 | [3, 4] |
| BCECF | 440/490 | 530 | 6.5–7.5 | [21] |
| SNARF | 488 | 580/640 | 6–8 | [21] |
| $Ca^{2+}$ probe | Excitation (nm) | Emission (nm) | $K_d$ for $Ca^{2+}$ (nM) | Refs. |
| Fluo-4 | 494 | 516 | 345 | [23] |
| Fura-2 | 335/363 | 512 | 145 | [23] |
| Indo-1 | 488 | 405/485 | 230 | [23] |
| Fura-5F | 336/363 | 512 | 0.4 | [23] |
| Fura-FF | 335/364 | 510 | 5,500 | [23] |
| Rhod-FF | 552 | 580 | 19,000 | [23] |
| Rhod-5N | 552 | 580 | 320,000 | [23] |

fluorescence wavelengths. If the target sample emits fluorescence or phosphorescence, wavelengths that do not overlap should be selected. Some probes use only one excitation/fluorescence wavelength (i.e., single-wavelength excitation or single-wavelength fluorescence), whereas others use two wavelengths (i.e., dual-wavelength excitation or dual-wavelength fluorescence).

3. When accurate quantification is required, it is desirable to use probes that use two different wavelengths and normalize them by ratiometry, using the ratio of fluorescence intensities. For example, pyranine, (8-hydroxypyrene-1,3,6-trisulfonic acid; HPTS), a typical pH indicator, is a probe with two-wavelength excitation and one-wavelength fluorescence. The higher the pH of the solution, the weaker the fluorescence intensity near 520 nm when irradiated with ultraviolet light near 405 nm as the excitation light, and the stronger the fluorescence intensity at the same wavelength when irradiated with excitation light near 460 nm. Therefore, the ratio of the fluorescence intensities at a wavelength of approximately 520 nm when irradiated with these two types of excitation light can be used for quantification.
4. Quantification is accomplished by preparing solutions with different ionic concentrations and pH values in the experimental system, measuring the fluorescence intensity at a specific wavelength when irradiated with the appropriate excitation light, and creating a calibration curve. However, the fluorescence intensity may be affected by dissolved ions that are not the measurement target. Therefore, particular attention should be paid to the presence of ions considered to behave similarly to the measurement target with respect to the fluorescent probe, such as ions of the same valency. An example of the pH calibration curve for HPTS is presented in Fig. 5.2. The concentration of the probe should be optimized according to this system.
5. Because the applicable concentration (or pH) range differs depending on the probe, the range should be selected according to the system and purpose of the experiment. For example, for pH, the HPTS calibration curve shown in Fig. 5.2 indicates that the quantifiable range is approximately 6–10, depending on the conditions. When quantifying ionic concentrations, it should be noted that the dissociation constant ($K_d$) differs between probes; therefore, the range of concentrations that can be measured also differs. In general, the higher the $K_d$, the higher the concentration that can be measured, but the lower the sensitivity to minute changes. Therefore, a suitable probe that can detect the concentration range expected in the target system should be selected. For example, Rhod-5N has a large $K_d$, making it suitable for measuring relatively high concentrations (several hundred μM).
6. If two or more probes are used simultaneously, it is possible, in principle, to visualize and measure multiple elements simultaneously, and efforts are currently being made to realize this [24]. In this case, the fluorescence/excitation wavelengths of the probes used should not overlap or affect each other. In addition, it should be noted that the fluorescence intensity emitted by the ion concentration indicator often differs depending on the pH, even for the same ion concentration. When simultaneously quantifying and visualizing pH and ion concentration, it

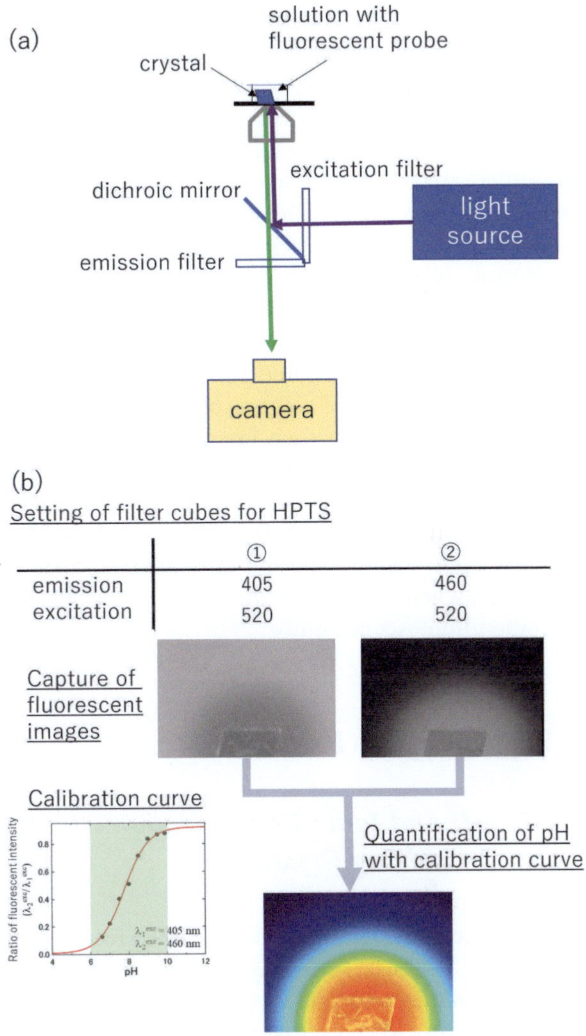


**Fig. 5.2** **a** Experimental setting and **b** procedure for pH imaging using the fluorescent HPTS probe

is necessary to accurately estimate the effects of pH and ion concentration on the fluorescence intensity emitted by each probe.

## 5.3 Imaging

For imaging, conventional fluorescence microscopes or laser confocal microscopes can be used. Figure 5.2 shows an overview of the observation setup when using an inverted fluorescence microscope and the imaging process. For fluorescence microscopy, a mercury, halogen, or LED lamp is applied as the light source, and a mirror cube with an excitation filter, a fluorescence filter, and a dichroic mirror are used to extract excitation light and fluorescence at the desired wavelengths. Fluorescence images are captured by a monochrome or color camera. Some recently developed laser confocal microscopes are equipped with a high-sensitivity spectroscopic detector with variable fluorescence wavelengths. When using a dual-wavelength excitation/fluorescence probe or when simultaneously imaging two or more elements, two-dimensional images showing the fluorescence intensity at different excitation light wavelengths can be acquired by automatically and instantaneously switching the mirror cubes. Using the spectroscopic detector, spectroscopic observation with no time lag is possible when using a probe with dual-wavelength fluorescence.

The fluorescence intensity for each pixel of an image acquired in this way is converted to pH and ionic concentration according to the calibration curve to obtain two-dimensional images of the pH and ionic concentration distributions. Furthermore, by taking such images at certain time intervals, a time-lapse image sequence can be obtained that shows the time variation of the respective distributions. However, because the solution containing the fluorescent probe fades when irradiated with strong light, it is necessary to thoroughly check that the exposure time for each imaging and the time interval during time-lapse imaging are under conditions where fading does not occur, especially when using a single-wavelength excitation/fluorescence probe.

## 5.4 Application Examples

### *5.4.1 pH and Ionic Concentration Changes Around Dissolving Crystals*

$CaCO_3$ is not only the largest stable carbon reservoir but also one of the most reactive minerals in the crust. The dissolution of $CaCO_3$ plays a critical role in regulating the chemistry of the ocean and atmosphere and has attracted the attention of researchers for a long time. Recent progress in research using state-of-the-art instruments has shown that the behavior and reactivity of ions at or near the solid–fluid interface of $CaCO_3$ is often more complex than previously thought [1, 2] and it differs from the case of the bulk phase. Moreover, the details of this behavior and reactivity are not fully understood. The visualization technique described above is a powerful tool for determining local atomic behavior at the crystal–solution interface during $CaCO_3$ dissolution.

When $CaCO_3$ dissolves in an aqueous solution, the pH is expected to increase because of the following reaction:

$$CaCO_3 + H_2O \rightarrow Ca^{2+} + HCO_3^- + OH^- \quad (5.1)$$

When calcite, the stable polymorph of $CaCO_3$, is immersed in an aqueous solution containing 40 μM HPTS as the pH probe, a rapid concentric rise in pH around the crystal was observed. In Fig. 5.3, both the time variation of pH distribution and the line profile of pH along the white dotted transect line shown in Fig. 5.3a are shown. In the case of aragonite, which is a metastable polymorph of $CaCO_3$ at room temperature and pressure, the pH around the crystal rose even more rapidly (Fig. 5.4a). The pH distribution aragonite dissolution is more in line with the crystal outline, which may reflect the polymorphs and crystal surface conditions. Furthermore, some biological hard tissues made of calcium carbonate, such as coral and foraminifera, are also observed to dissolve easily in water, as shown in Fig. 5.4b. In the case of foraminifera, immediately after immersion in water, the pH changes followed the shape of the hard tissue protrusions; however, with time, the distribution became more concentric. The imaging of the three-dimensional distribution of pH can also be realized using a laser confocal microscope. Figure 5.5 shows that a cross-section view of the three-dimensional pH distribution during calcite powder dissolution was successfully observed.

Furthermore, if a $Ca^{2+}$ indicator is used with HPTS, it is possible to visualize pH and $Ca^{2+}$ ion concentrations simultaneously. When observing phenomena such as the dissolution of $CaCO_3$ crystals, in which the $Ca^{2+}$ ion concentration is expected to become somewhat high, Rhod-FF or Rhod-5N, which have relatively large dissociation constants, should be selected [22, 23]. In this case, the fluorescence intensity emitted by the HPTS indicates the exact pH regardless of the $Ca^{2+}$ concentration, whereas the fluorescence intensity emitted by Rhod-FF or Rhod-5N varies with pH, even at the same $Ca^{2+}$ concentration. Therefore, by first determining the pH and then correcting for $Ca^{2+}$ concentration based on a prepared calibration curve considering the pH effect, both can be simultaneously quantified and visualized. This makes it possible to observe differences in the trends of changes in the distribution of pH and $Ca^{2+}$ concentration around dissolving crystals and to analyze the dissolution reaction based on the direct measurement of the change in the concentration of ions and pH just above the surface.

### 5.4.2 *Visualization of pH Changes During Crystal Precipitation in Gels*

One method to precipitate crystals by controlling the diffusion of ions is the gel double-diffusion method. During the synthesis of $CaCO_3$, $Ca^{2+}$ ions and $CO_3^{2-}$ or $HCO_3^-$ ions are diffused from both sides of a gel-like material, such as agar gel, and crystals are deposited in the aggregation zone. This method has long been used to synthesize large crystals [25, 26] and has recently received renewed attention because it mimics biomineralization [27–32]. In this method, the polymorphism and morphology of precipitated crystals synthesized often vary depending on their location in the gel. However, it is difficult to obtain direct chemical information about

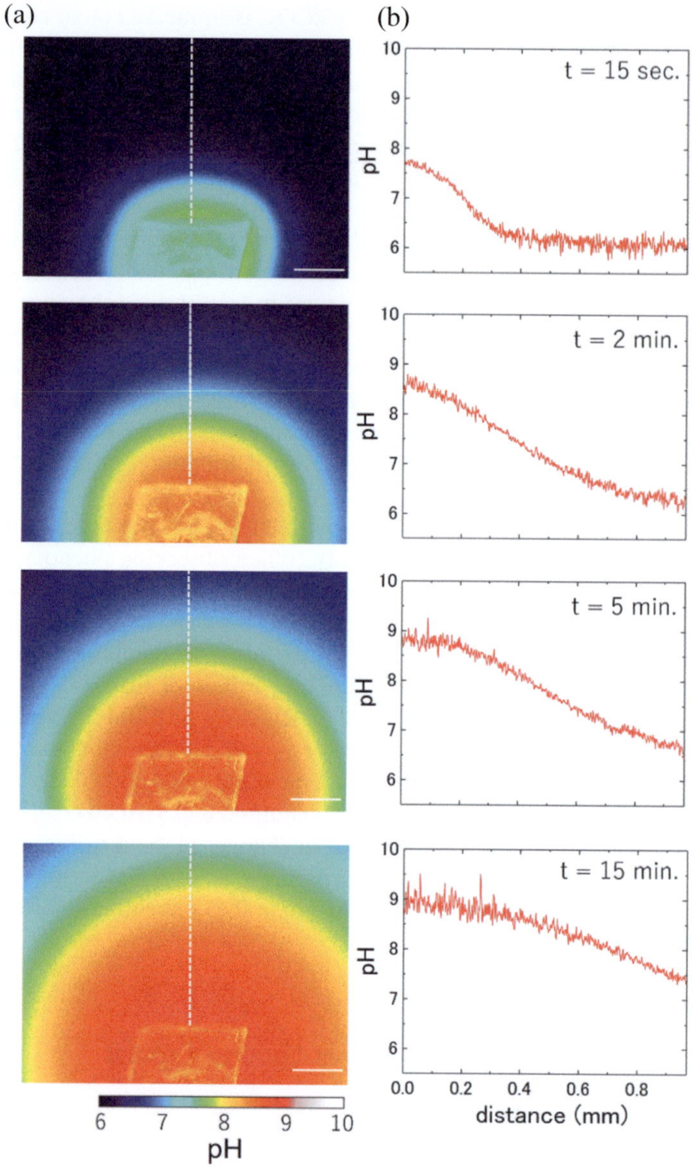


**Fig. 5.3** **a** Two-dimensional distributions of pH around dissolving calcite. **b** Change in pH versus the distance from the surface on the white dotted transect line in each figure. Scale bar is 200 mm. Adopted with permission from [3]. Copyright 2019, American Chemical Society

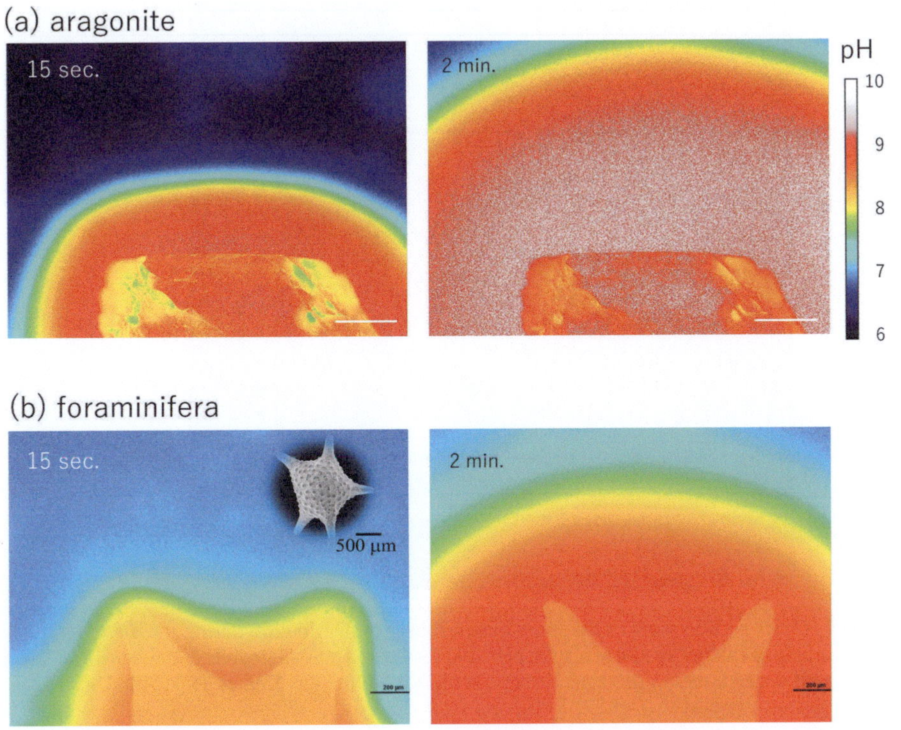


**Fig. 5.4** Time variation of pH distribution around the dissolution of a **a** single crystal of aragonite in water and **b** foraminifera shell in water. Scale bars in pH images are 200 μm. Figures (**a**) are adopted with permission from [3]. Copyright 2019, American Chemical Society

the location of crystal formation in the gel, although theoretical estimations based on the diffusion equation have been conducted to elucidate this [33]. The visualization method introduced in this chapter can be applied not only to aqueous solutions but also to gel-like materials and can determine the chemical environment of the location in the gel where the crystals are precipitating.

Genovese et al. first succeeded in visualizing the pH distribution of an entire gel where $CaCO_3$ formed using the double-diffusion method using a HPTS fluorescent probe [28]. We also observed pH distribution in gels during $CaCO_3$ formation under fluorescent microscopy [34] (Fig. 5.6). Both the agar gel and solutions to be diffused ($CaCl_2$ and $NaHCO_3$) contained HPTS to visualize the ph. Note that the thicker the gel, the narrower the applicable pH range. In this study, a 4-mm thick gel was used, and the applicable ph range was approximately 6–9. Figure 5.6 visualizes the pH distribution across the gel 6 h after the start of solution diffusion. The pH increases as the $HCO_3^-$ ions diffuse and decreases as crystallization begins. The polymorphism and morphology of the formed crystals differed depending on their location in the gel, suggesting that the different trends in pH change at each location may have an effect on the formation process. Furthermore, using confocal laser scanning microscopy, pH

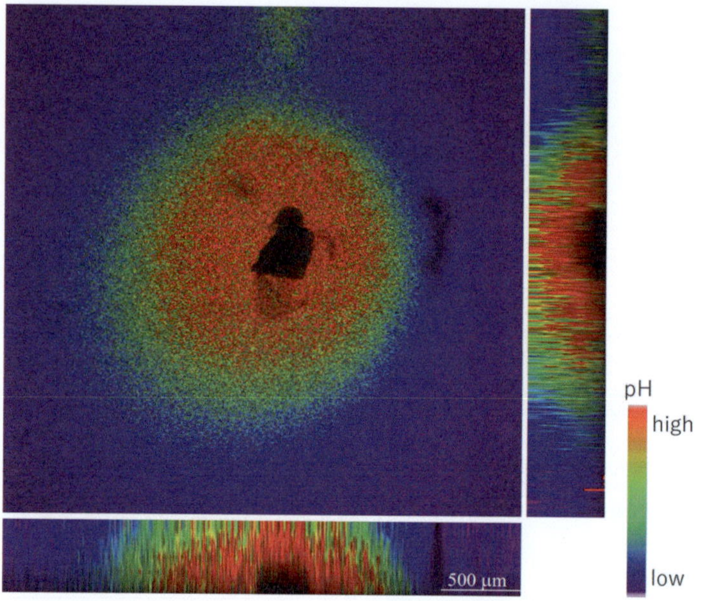


**Fig. 5.5** Cross-section view of the three-dimensional pH distribution during the dissolution of a small particle of calcite (shown in black). pH is expressed as a ratio of fluorescence intensities. Taken with a high-speed laser confocal microscope at the Nikon Imaging Center, Hokkaido University

changes in the vicinity of forming crystals can be captured in more detail; therefore, differences in polymorphism and morphology can be discussed based on differences in the local chemical environment around forming crystals.

## 5.5 Future Prospects

Although this visualization method was only recently applied to inorganic materials, it is now being applied to a wide variety of materials. This method can be applied to not only aqueous solutions but also reactions in gel-like materials, pseudobiological solutions, and polymer-containing solutions. Its sensitivity to temporal and spatial changes is very high, such that changes in pH as small as 0.1 can be detected, which is a major strength of this system. However, accurately quantifying absolute values of ionic concentrations across different experimental systems remains a challenging problem. The fluorescence intensity emitted by the probe is sensitive to the composition and thickness of the solution and the solution environment. Therefore, for accurate quantification, great care must be taken to ensure that the measurement results reproduce changes in fluorescence intensity under calibrated conditions. Currently, calibration curves must be prepared for each experimental system. Therefore, to apply this method to a wider range of phenomena, it is desirable to develop a

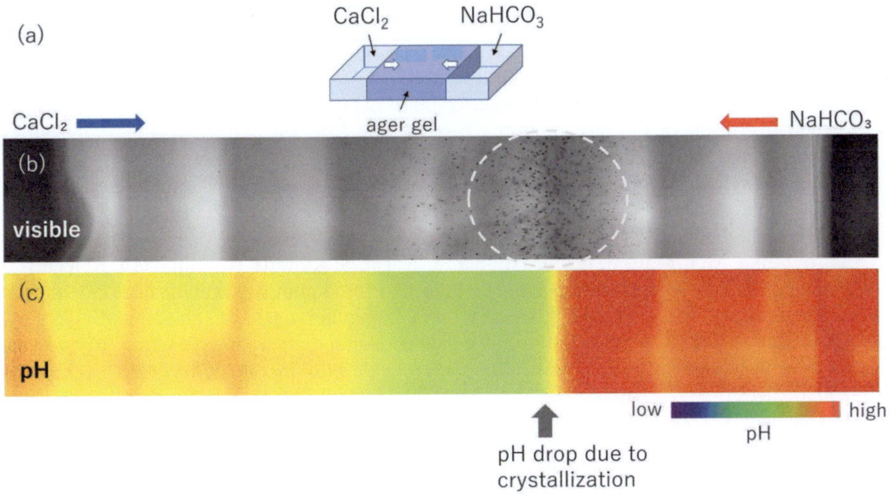


**Fig. 5.6** Snapshot of the pH distribution in a gel during $CaCO_3$ precipitation using the double-diffusion method. **a** Schematic drawing of the double-diffusion method, **b** transmission image, and **c** pH distribution. The width of the gel is approximately 2 cm. Multiple images were taken by moving the automated stage, which were then stitched together to form a single image. As $CaCO_3$ precipitation begins to be observed in the transmission image (inside the dotted line), the pH decreases

uniform and robust calibration method. Furthermore, high-resolution imaging using confocal laser scanning microscopy and an increase in the number of ion species to be monitored are expected to enable the application of this technique to a wider variety of crystallization phenomena. This leads to a discussion of the mechanisms of various crystal growth or dissolution reactions based on the local behavior of ions at or near the crystal–solution interfaces.

**Acknowledgements** We thank S. Matsumoto, K. Miki, Y. Nagai, and A. Ueda for their help in the experiments. The authors are grateful to T. Nagai, H. Teng, and Y. Kimura for their valuable comments and support. Observations using a laser confocal microscope were performed at the Nikon Imaging Center at Hokkaido University with the support of K. Kobayashi. This work was partly supported by JSPS KAKENHI Grant Numbers 21K18798, 20H05657, JP16H06280 and JP22H04926.

## References

1. K. Miyata, H. Asakawa, T. Fukuma, Real-time atomic-resolution imaging of **crystal growth** process in water by phase modulation atomic force microscopy at one frame per second. Appl. Phys. Lett. **103**, 203104 (2013). https://doi.org/10.1063/1.4830048
2. K. Miyata, J. Tracey, K. Miyazawa, V. Haapasilta, P. Spijker, Y. Kawagoe, A.S. Foster, K. Tsukamoto, T. Fukuma, Dissolution processes at step edges of calcite in water investigated

by high-speed frequency modulation atomic force microscopy and simulation. Nano Lett. **17**, 4083–4089 (2017). https://doi.org/10.1021/acs.nanolett.7b00757
3. J. Kawano, T. Toyofuku, K. Nishimura, A. Ueda, Y. Nagai, S. Kawada, H. Teng, T. Nagai, Direct two-dimensional time series observation of pH distribution around dissolving calcium carbonate crystals in aqueous solution. Cryst. Growth Des. **19**(8), 4212–4217 (2019). https://doi.org/10.1021/acs.cgd.9b00045
4. L.J. de Nooijer, T. Toyofuku, H. Kitazato, Foraminifera promote calcification by elevating their intracellular ph. Proc. Natl. Acad. Sci. U. S. A. **106**, 15374–15378 (2009). https://doi.org/10.1073/pnas.0904306106
5. T. Toyofuku, M.Y. Matsuo, L.J. de Nooijer, Y. Nagai, S. Kawada, K. Fujita, G.J. Reichart, H. Nomaki, M. Tsuchiya, H. Sakaguchi, H. Kitazato, Proton pumping accompanies calcification in foraminifera. Nat. Commun. **8**, 14145 (2017). https://doi.org/10.1038/ncomms14145
6. L. Addadi, S. Raz, S. Weiner, Taking advantage of disorder: amorphous calcium carbonate and its roles in biomineralization. Adv. Mater. **15**, 959–970 (2003).https://doi.org/10.1002/adma.200300381
7. D. Gebauer, A. Völkel, H. Cölfen, A. Volkel, H. Colfen, Stable prenucleation calcium carbonate clusters. Science **322**, 1819–1822 (2008). https://doi.org/10.1126/science.1164271
8. R. Demichelis, P. Raiteri, J.D. Gale, D. Quigley, D. Gebauer, Stable prenucleation mineral clusters are liquid-like ionic polymers. Nat. Commun. **2**, 590 (2011). https://doi.org/10.1038/ncomms1604
9. J.H.E. Cartwright, A.G. Checa, J.D. Gale, D. Gebauer, C.I. Sainz-Díaz, Calcium carbonate polyamorphism and its role in biomineralization: How many amorphous calcium carbonates are there? Angew. Chemie-Int. Ed. **51**, 11960–11970 (2012). https://doi.org/10.1002/anie.201203125
10. J. Kawano, N. Shimobayashi, A. Miyake, M. Kitamura, Precipitation diagram of calcium carbonate polymorphs: its construction and significance. J. Phys. Condens. Matter. **21**, 425102 (2009). http://iopscience.iop.org/0953-8984/21/42/425102
11. J. Kawano, S. Maeda, T. Nagai, The effect of Mg2+ incorporation on the structure of calcium carbonate clusters: investigation by the an harmonic downward distortion following method. Phys. Chem. Chem. Phys. **18**, 2690–2698 (2016). https://doi.org/10.1039/C5CP05139H
12. M.H. Nielsen, S. Aloni, J.J. De Yoreo, In situ TEM imaging of $CaCO_3$ nucleation reveals coexistence of direct and indirect pathways. Science **345**, 1158–1162 (2014). https://doi.org/10.1126/science.1254051
13. H.H. Teng, P.M. Dove, C.A. Orme, J.J. De Yoreo, Thermodynamics of calcite growth: baseline for understanding biomineral formation. Science **282,** 724–727 (1998). https://doi.org/10.1126/science.282.5389.724
14. C.A. Orme, A. Noy, A. Wierzbicki, M.T. McBride, M. Grantham, H.H. Teng, P.M. Dove, J.J. DeYoreo, Formation of chiral morphologies through selective binding of amino acids to calcite surface steps. Nature **411**, 775–779 (2001). https://www.nature.com/articles/35081034
15. H.H. Teng, P.M. Dove, J.J. De Yoreo, Kinetics of calcite growth: surface processes and relationships to macroscopic rate laws. Geochim. Cosmochim. Acta **64**, 2255–2266 (2000). https://doi.org/10.1016/S0016-7037(00)00341-0
16. P. Fenter, P. Geissbühler, E. DiMasi, G. Srajer, L.B. Sorensen, N.C. Sturchiom, Surface speciation of calcite observed in situ by high-resolution X-ray reflectivity. Geochim. Cosmochim. Acta **64,** 1221–1228 (2000). https://doi.org/10.1016/S0016-7037(99)00403-2
17. P. Fenter, S. Kerisit, P. Raiteri, J.D. Gale, Is the calcite-water interface understood? Direct comparisons of molecular dynamics simulations with specular X-ray reflectivity data. J. Phys. Chem. C **117**, 5028–5042 (2013). https://doi.org/10.1021/jp310943s
18. N.H. de Leeuw, S.C. Parker, Surface structure and morphology of calcium carbonate polymorphs calcite, aragonite, and vaterite: an atomistic approach. J. Phys. Chem. B **102**, 2914–2922 (1998). https://doi.org/10.1021/jp973210f
19. N.H. de Leeuw, S.C. Parker, C.R.A. Catlow, G.D. Price, Modelling the effect of water on the surface structure and stability of forsterite. Phys. Chem. Miner. **27**, 332–341 (2000). https://doi.org/10.1007/s002690050262

20. T. Ogino, T. Suzuki, K. Sawada, The formation and transformation mechanism of calcium carbonate in water. Geochim. Cosmochim. Acta **51**, 2757–2767 (1987). https://doi.org/10.1016/0016-7037(87)90155-4
21. pH indicators, in *The Molecular Probes Handbook: A Guide to Fluorescent Probes and Labeling Technologies*, 11th edn, ed. by I. Johnson, M.T.Z. Spence (Life Technologies, 2010), p. 35
22. Indicators for $Ca^{2+}$, $Mg^{2+}$, $Zn^{2+}$ and other metal ions, in *The molecular Probes Handbook: A Guide to Fluorescent Probes and Labeling Technologies* ,11th edn, ed. by I. Johnson, M.T.Z. Spence (Life Technologies, 2010)
23. R.M. Paredes, J.C. Etzler, L.T. Watts, W. Zheng, J.D. Lechleiter, Chemical calcium indicators. Methods **46**, 143–151 (2008). https://doi.org/10.1016/j.ymeth.2008.09.025
24. J. Kawano, M. Yazaki, Y. Konno, T. Toyofuku, Y. Nagai, T. Nagai: Visualization of two-dimensional distributions of pH and $Ca^{2+}$ concentration around dissolving/forming $CaCO_3$ minerals. Goldschmidt2020 Abstr. 1264 (2020). https://doi.org/10.46427/gold2020.1264
25. H.K. Henisch, J. Dennis, J.I. Hanoka, Crystal growth in gels. J. PHys. Chem. Solids **26**, 493–496 (1965). https://doi.org/10.1016/0022-3697(65)90123-X
26. H.J. Nickl, H.K. Henisch, Growth of calcite crystals in gels. J Electrochem. Soc. **116**, 1258 (1969). https://doi.org/10.1149/1.2412292
27. Y. Oaki, S. Hayashi, H. Imai, A hierarchical self-similar structure of oriented calcite with association of an agar gel matrix: inheritance of crystal habit from nanoscale. Chem. Commun. **1**, 2841 (2007). https://doi.org/10.1039/B704129B
28. D. Genovese, M. Montalti, F. Otálora, J. Gómez-Morales, M. Sancho-Tomás, G. Falini, J.M. García-Ruiz, Role of $CaCO_3{}^{\circ}$ neutral pair in calcium carbonate crystallization. Cryst. Growth Des. **16**, 4173–4177 (2016). https://doi.org/10.1021/acs.cgd.6b00276
29. M. Sancho-Tomás, S. Fermani, S. Goffredo, Z. Dubinsky, J.M. García-Ruiz, J. Gómez-Morales, G. Falini, Exploring coral biomineralization in gelling environments by means of a counter diffusion system. CrystEngComm **16**, 1257 (2014). https://doi.org/10.1039/C3CE41894D
30. J.M. García-Ruiz, E. Nakouzi, E. Kotopoulou, L. Tamborrino, O. Steinbock, Biomimetic mineral self-organization from silica-rich spring waters. Sci. Adv. **3,** 1–7 (2017). https://doi.org/10.1126/sciadv.1602285
31. H. Li, L.A. Estroff, Calcite growth in hydrogels: assessing the mechanism of polymer-network incorporation into single crystals. Adv. Mater. **21**, 470–473 (2009). https://doi.org/10.1002/adma.200801498
32. E. Asenath-Smith, H. Li, E.C. Keene, Z.W. Seh, L.A. Estroff, Crystal growth of calcium carbonate in hydrogels as a model of biomineralization. Adv. Funct. Mater. **22**, 2891–2894 (2012). https://doi.org/10.1002/adfm.201200300
33. H.K. Henisch, J.M. García-Ruiz, Crystal growth in gels and Liesegang ring formation I. diffusion relationships. J. Cryst. Growth **75**, 195–202 (1986). https://doi.org/10.1016/0022-0248(86)90028-X
34. J. Kawano, S. Matsumoto, K. Miki, T. Toyofuku, Y. Nagai, T. Nagai, Direct observation of pH distribution during the formation of calcium carbonate polymorphs in gel media. Goldschmidt2023 Abstr. 19407 (2023). https://doi.org/10.7185/gold2023.19407

# Chapter 6
# Crystal Structure Control Technology for Organic and Biomaterials Using Light

## Crystal Structure Control Technology Using Light

**M. Maruyama, Y. Tsuri, H. Y. Yoshikawa, K. Takano, M. Yoshimura, and Y. Mori**

**Abstract** This chapter explores the advanced developments in laser-induced crystal nucleation technology. The initial phase of this technology enabled the synthesis of previously inaccessible substances, while the current phase, addresses the separation of similar substances via laser irradiation, thereby facilitating precise control over crystal polymorphism. Considerable advancements have been achieved in the metastable crystallization of aspirin, a compound that has historically posed significant challenges for polymorphism control. Nevertheless, numerous materials continue to exhibit complexities in crystallization and polymorphism management. To broaden the applicability of this technology to a wider array of substances, it is imperative to conduct a more thorough investigation of the phenomena induced by laser irradiation. Future endeavors will focus on leveraging light to actively manipulate crystal phase transitions, thereby enabling precise molecular-level interventions. This approach aspires to develop materials that replicate the intricate polymorphism control mechanisms inherent in biological systems.

M. Maruyama (✉) · Y. Tsuri · H. Y. Yoshikawa · Y. Mori
Graduate School of Engineering, Osaka University, 2-1, Yamadaoka, Suita City, Osaka 565-0871, Japan
e-mail: maruyama@eei.eng.osaka-u.ac.jp

M. Maruyama · K. Takano
Graduate School of Life and Environmental Sciences, Kyoto Prefectural University, 1-5, Hangi-Cho, Shimogamo, Sakyo-Ku, Kyoto 606-8522, Kyoto, Japan

Y. Tsuri
Division of Materials Science, Nara Institute of Science and Technology, 8916-5 Takayama-Cho, Ikoma City, Nara 630-0192, Japan

M. Yoshimura
Institute of Laser Engineering, Osaka University, 2-6, Yamadaoka, Suita City, Osaka 565-0871, Japan

 
F. Ishikawa et al. (eds.), *Engineering Crystal Habit*, Topics in Applied Physics 152,
https://doi.org/10.1007/978-981-96-0266-7_6

## 6.1 Introduction

Crystalline materials are used in various fields, and can be expected to have various functions, depending on the crystal composition and structure. These examples include the areas of electronic materials, optical elements, pharmaceuticals, foods, cosmetics, etc. In addition, structural analysis of protein crystals is performed using X-rays and neutron beams, and there is a need for high-quality crystals to investigate the structures and functions of crystalline substances.

We have developed a technology by which crystal growth is controlled using light, targeting organic materials in particular. Organic materials have larger growth units than inorganic materials, and many consist of individual molecules with flexible structures, so crystal nucleation does not occur readily. In particular, one typical example in which nucleation is difficult is protein crystals. Proteins consist of amino acids linked in a chain, forming a complex three-dimensional structure. Depending upon the individual structures, they are high-molecular-weight molecules with functions essential for vital activities [1]. Among materials in which crystal nucleation is challenging to achieve solely by precise control of conditions such as solution concentration, temperature, viscosity, and additives, our research group achieved crystallization in numerous proteins using femtosecond laser irradiation [2–6]. After the successful induction of nucleation of membrane proteins in which crystallization is particularly difficult, the authors' aim changed to developing polymorphism control technologies for organic crystals, involving separation of the different crystal forms. Our aim was an advance in technology, from production of materials that could not previously be crystallized, to separation of structurally similar substances. Crystallization of various organic materials cannot be controlled simply by blindly repeating the evaluation of conditions. Thus, we have also focused on and achieved progress with research to elucidate the mechanism by which femtosecond lasers induce crystal nucleation.

The present report, in addition to providing a review of previous research, introduces the research on polymorphism control achieved by laser nucleation induction.

## 6.2 What is Crystal Polymorphism?

Crystal polymorphism refers to substances with the same chemical composition but different crystal structures. The most famous examples are graphite and diamond (also called allotropes, as they consist of a single element). Although both are carbon, they have different crystal structures, so their physical properties are different. Most low-molecular-weight organic compounds that are used as drugs show crystal polymorphism, with the different forms showing differences in properties that are highly important for use as medicine, including stability, solubility, and bioavailability (an index of the degree to which a formulation administered to a human exerts effects as a drug) [7]. In the drug development stage, it is essential to select the crystal structure most suitable for formulations, and research on polymorphism control is therefore crucial for low-molecular-weight organic compounds.

In the case of pharmaceuticals, crystal forms with low solubility in the constant-temperature and constant-pressure environments are the stable form and used for drugs, and other crystal forms, with high solubility, are considered to be metastable forms and not selected as drugs. Various techniques have been established for obtaining metastable forms. For example, there have been reports of the application of nonequilibrium states by means of the quenching method [8], multi-step cooling method [9], etc. Although the first form obtained under extreme conditions may be a metastable phase, it cannot be available if this subsequently undergoes phase transition to the stable form at constant temperature and pressure [10]. The requirement is for stability at least sufficient for evaluation of the physical properties of the metastable form, and, if the aim is a metastable formulation, even higher stability is needed [11]. Due to the molecule being the same, the crystallization conditions of polymorphs are often very similar [12]. This chapter will introduce the attempt to control polymorphism by irradiation with femtosecond lasers. We selected indomethacin (an anti-inflammatory agent) as a model substance.

## 6.3 Indomethacin Crystal Polymorphism Control by Laser Irradiation

Several research groups have also reported crystal nucleation by laser irradiation. The physicochemical phenomena that are focused on differ between groups; for details, please refer to the previous publications [2, 12–14, 13–31]. The laser technology developed by our research group involves the generation of cavitation bubbles by irradiating a solution with femtosecond lasers, thus inducing crystal nucleation (Fig. 6.1). The cavitation bubbles generated in the solution repeatedly expand and contract over periods of several tens μs, and localize solute concentration fluctuations, therefore, occur in the areas surrounding bubbles. If sites of localized high concentration formed, these possibly induce nucleation [15–17, 16–18]. However, in the case of polymorphism control, a highly stable metastable form cannot be achieved solely by inducing nucleation. The crucial points for metastable form crystallization are to take particular measures in the laser conditions and the positioning of laser irradiation to achieve rapid crystallization of the metastable form alone and to prevent this from being followed by crystallization of the stable form [18, 19].

Indomethacin is a compound with low aqueous solubility, and it shows crystal polymorphism, having α, β and γ forms [20]. Among these, the γ form is the most stable and is used for commercial formulation. The authors have attempted selective crystallization of the α form, which is metastable and has a high in vivo absorption rate [19]. Firstly, crystals of the γ form were dissolved in acetonitrile to prepare a supersaturated solution (i.e., a solution with the solute dissolved to at least the limit

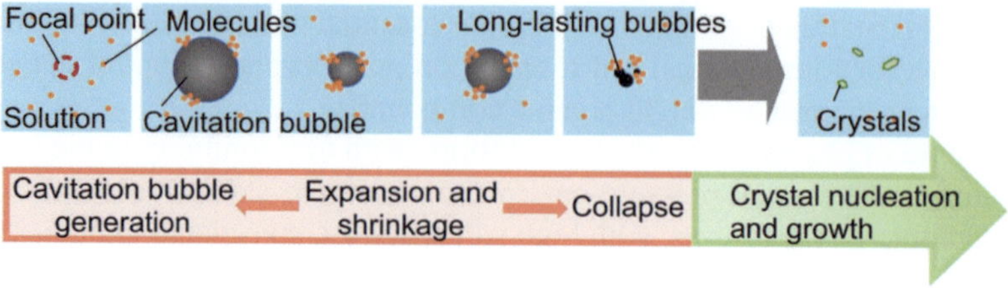


**Fig. 6.1** Schematic diagram of the crystallization mechanism by laser irradiation

of its solubility). Glass vials were then filled with this solution, and the solution was irradiated from the base of each vial with a femtosecond laser (pulse duration: approx. 200 fs; wavelength: 800 ± 10 nm; repetition frequency: 1 kHz; IFRIT-SA10, Cyber Laser, Inc.), focused using a 10x objective lens (NA = 0.4). The laser energy was 30 μJ/pulse, and the pulse number was approximately 60,000. One of the parameters in this study was laser focal position, and there were three conditions for this: the center of the glass vial, close to the glass surface, and at the gas–liquid interface. The schematic diagram in Fig. 6.2a provides information about these positions. For comparison, assessments were made under the same dissolution conditions, with vials left standing (controls), and with stirring stimulation using a magnetic stirrer. The crystallization probability under each condition is shown in Fig. 6.2a. No crystallization occurred in the control, whereas crystallization did occur under the condition of stirring the solution, and the condition of laser focusing. Under the solution stirring conditions, the probability of obtaining crystals was 100%, with the form immediately after crystallization being α (metastable form), but in all cases, this underwent subsequent phase transition to the γ form (stable form) within 24 h.

This is because the α and γ forms were mixed together during the crystallization stage. When even a small amount of the stable form is present in the same environment as the metastable form, solution-mediated phase transition causes all of the compound to transition to the stable form over time [10]. Interestingly, if the laser is focused on the central part of the vial, tabular, γ-form crystals (stable form) are obtained (Fig. 6.2b), whereas when the focal position is close to the vial wall or at the gas–liquid interface, the probability of formation of rod-shaped, α-form crystals (metastable form) is high (Fig. 6.2c). The authors consider this difference to be due to differences in crystallization speed resulting from laser irradiation [19]. As shown in Fig. 6.1, at the laser focal points, cavitation bubbles are generated, and repeated expansion and contraction of the bubbles occur. At that time, concentration fluctuations in the marginal regions promote crystal nucleation, but the cavitation bubbles are, in fact, finally crushed by pressure, after which numerous micro-bubbles (hereafter referred to as residual bubbles) are generated and dispersed widely. Careful observation of indomethacin crystallization showed that with irradiation close to the vial wall surface (B) or at the gas–liquid interface (C), the residual bubbles remained at the vial wall surface or solution meniscus area, respectively, for a considerable time, after which gradual nucleation of α-form crystals occurred. On the other hand, with irradiation in the center of the solution (A), the residual bubbles disappeared rapidly, so the

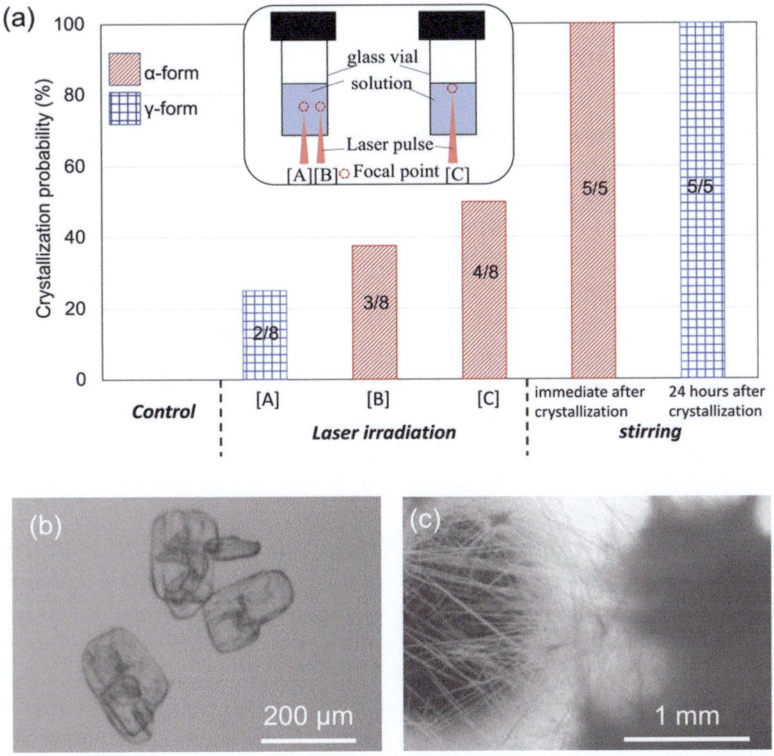


**Fig. 6.2** **a** Probability of crystallization for each case: without forcible nucleation techniques (control), with laser irradiation (at the center of the solution [A], near the side wall of the glass vial [B], and the air–liquid interface [C]), and with stirring (immediately after crystallization and about 24 h after). **a**, **b** and **c** correspond to the positions indicated in the inserted image. **b** Typical g form (stable form) of indomethacin. **c** A typical a form (metastable form) of indomethacin. (Reprinted with permission from the Japan Society of Applied Physics, Reference [19])

crystallization speed was lower than with B and C. On the basis of these differences in speed, it is considered that γ-form crystal nucleation followed, so that there was a mixture of two crystal forms, as a result of which phase transition took place, leading to complete replacement with the γ form. In the example of indomethacin, crystal nucleation due to laser irradiation was clearly shown to occur not only at the contraction–expansion interface of cavitation bubbles, but also at the interfaces of residual bubbles after disintegration. In addition, the importance of kinetics for crystal polymorphism control was demonstrated, that is, how quickly nucleation of the target crystal form occurs; and this has become an important viewpoint for research on polymorphism control [12, 21].

## 6.4 Attempts to Elucidate the Mechanism of Laser Nucleation Induction

The authors have previously attempted to elucidate the mechanism of induction of crystal nucleation by laser irradiation. By observation of concentration fields, using the red-colored protein cytochrome C [17], observation of concentration changes based on fluorescence intensity using fluorescent-labeled lysozyme which is a protein from egg white [16], and observation of the nucleation of urea that clearly shows crystallization from the laser focal point when conditions are aligned appropriately [22]. These results proposed the hypothesis presented in Fig. 6.1 [18]. However, in order to apply these techniques to various types of organic material, there is a need for more detailed understanding, such as indices for the laser irradiation conditions (energy, pulse duration, wavelength, etc.) that are appropriate to the material. In this context, the authors investigated the changes in crystallization probability and time needed for crystal nucleation, taking urea as the model substance, and laser pulse duration and energy as the parameters [23].

Figure 6.3 shows the mode of crystallization when irradiating a supersaturated urea solution with a laser with pulse duration ($t_p$) of 300 fs. After irradiation for a certain time, the crystals grow rapidly in a radial pattern from the laser focal point. For reliable observation of when and where crystal nucleation occurred, urea was the model substance for elucidation of the mechanism. The authors changed the pulse duration ($t_p$) over the 300 to 1200 fs range, and measured the induction time, that is, the time needed to wait until the onset of crystallization. Although the induction time showed considerable scatter, the time required for crystallization tended to decrease with increasing pulse duration. Next, under three $t_p$ conditions, 300, 700 and 1000 fs, the pulse energy dependence of the crystallization probability ($P_{crystal}$) was evaluated (Fig. 6.4). The energy threshold for crystal nucleation varied with $t_p$, so when $t_p$ was 300 fs, crystal nucleation were achieved at 100 μJ/pulse or higher. In contrast, when $t_p$ was 700 or 1000 fs, it was achieved at 70 μJ/pulse or higher. When the threshold value was exceeded, $P_{crystal}$ increased dependently on pulse energy at all conditions, that is, $t_p$ of 300, 700 or 1000 fs.

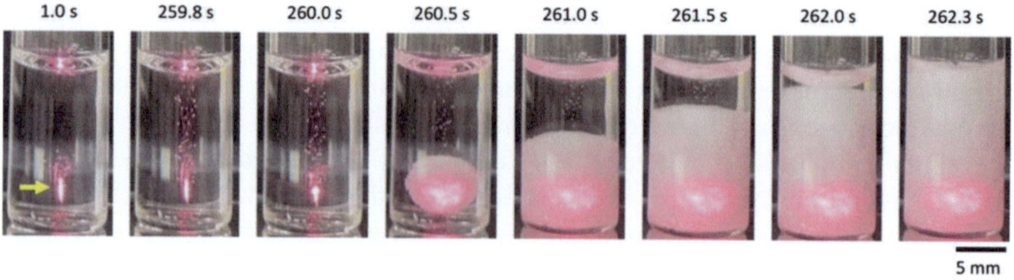


**Fig. 6.3** Sequential photographs of urea crystallization induced by laser irradiation with $t_p = 300$ fs at 260 μJ/pulse. The initiation of laser irradiation was defined as 0 s. The arrows show the laser focal point. (Reprinted with permission from AIP publishing, Reference [24])

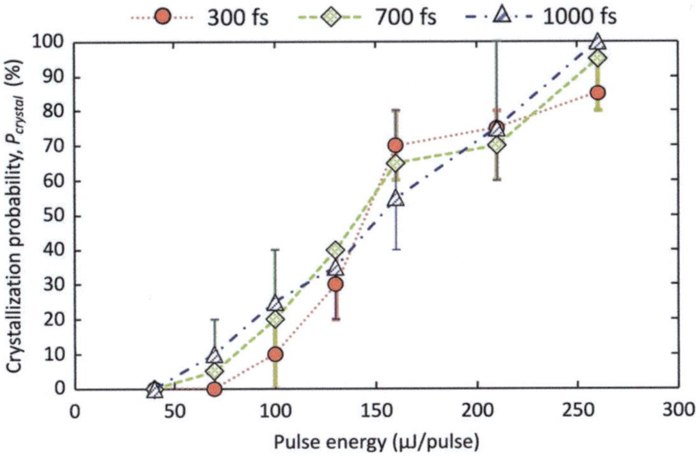


**Fig. 6.4** Crystallization probability as a function of pulse energy with $t_p = 300$, 700, and 1000 fs. (Reprinted with permission from AIP publishing, Reference [24])

The authors have focused on the cavitation bubble generation by laser irradiation as an essential physical phenomenon for the induction of crystal nucleation [18]. In this context, laser irradiation was performed at a $t_p$ of 300–1200 fs, and a bubble energy of 3–260 μJ/pulse, the behavior of the cavitation bubbles formed was observed using a high-speed camera (frame speed: 500 kfps), and parameters such as the maximum bubble diameter ($R_{max}$) were evaluated. Figure 6.5a and b show the expansion and contraction of cavitation bubbles at $t_p$ of 300 and 1000 fs, respectively. Comparison of these conditions by continuous photography showed that when $t_p$ was 1000 fs, $R_{max}$ was greater, and the expansion–contraction cycle was longer. Similar observations were performed with $t_p$ and pulse energy changed, and the relationships of $R_{max}$ and pulse energy is shown in Fig. 6.5c. As in previous reports, the relationship between $R_{max}$ and pulse energy was highly consistent with a logarithmic function [24]. Over the energy range at which crystallization occurs, 70 μJ/pulse or higher, it was found that $R_{max}$ tended to increase with increasing $t_p$.

The energy ($E_B$) of cavitation bubbles can be obtained by Eq. (6.1), from the hydrostatic pressure ($P_o$) and intra-bubble pressure ($P_v$) [25, 26].

$$E_B = \frac{4}{3}\pi\left(\frac{R_{max}}{2}\right)^3 (P_0 - P_v) \tag{6.1}$$

From Eq. (6.1), $E_B$ and ${R_{max}}^3$ can be seen to be in a proportionate relationship. In this context, assuming that cavitation bubbles provide the energy needed for crystal nucleation, with respect to crystal nucleation probability ($P_{crystal}$) and cavitation bubble maximum diameter ($R_{max}$), at $t_p$ of 300, 700 and 1000 fs, the relationship between ln ($P_{crystal}$) and ${R_{max}}^3$ is plotted in Fig. 6.6, and a linear relationship was established. In addition, the gradient of that line decreased with increasing $t_p$. This

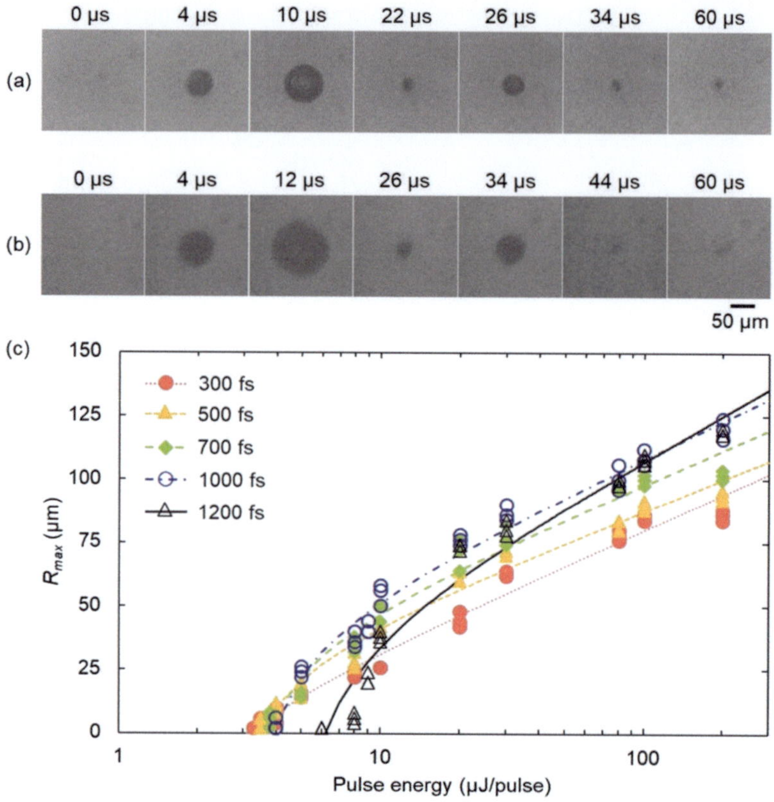


**Fig. 6.5** Bright-field images of cavitation bubbles generated by laser irradiation with $t_p = 300$ fs (**a**), and 1000 fs (**b**) at 80 mJ/pulse in aqueous urea solutions. **c** Laser pulse energy dependence of maximum radius of cavitation bubbles, $R_{\max}$, generated by laser irradiation with $t_p = 300$–1200 fs. $R_{\max}$ was determined from the obtained bright field images. (Reprinted with permission from AIP publishing, Reference [24])

shows that as $t_p$ increases, the dependence of $P_{crystal}$ on $R_{max}$ decreases. In other words, if there is a long pulse duration, large cavitation bubbles are required for the induction of crystallization. The authors consider this difference to be caused by changes in the solution temperature due to laser irradiation. In fact, with pulse energy at 260 μJ/pulse and continuous irradiation of the urea solution for 600 s, when $t_p$ was 300 fs, the temperature of the entire solution increased by less than 1 °C. In contrast, when $t_p$ was 1200 fs the temperature of the whole solution increased by 2 °C, the thermal effects thus being increased approximately two-fold [23]. In the case of substances such as urea, with temperature-dependent solubility, the level of supersaturation decreases as the temperature increases, and it is valid to consider that this leads to decreased $P_{crystal}$.

On the basis of the above results, the authors have proposed indices for laser irradiation conditions depending upon the substance used (Fig. 6.7). If the thermal

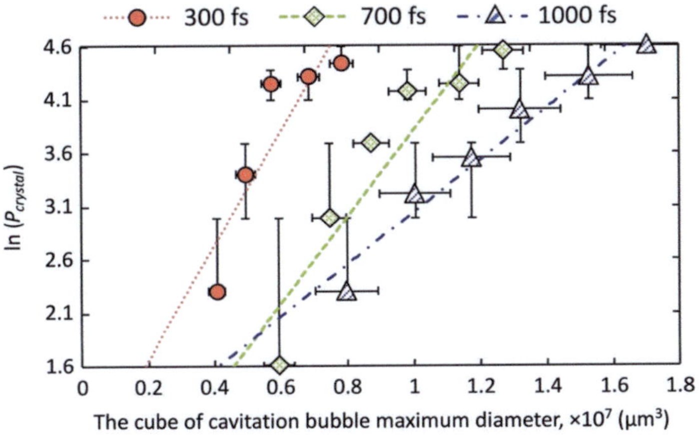


**Fig. 6.6** Plotting the logarithm of the crystallization probability ($P_{crystal}$) against the cube of the maximum diameter ($R_{max}$) of the cavitation bubble. $(R_{max})^3$ and $\ln(P_{crystal})$ are linearly related. (Reprinted with permission from AIP publishing, Reference [24])

resistance of a compound or solution used for crystallization is low, in order to suppress the thermal effects of laser irradiation, it is preferable for $t_p$ to be low. On the other hand, if the thermal resistance is high, evaluation under long-$t_p$ conditions is also feasible. However, even if the thermal resistance of a compound or solution is insufficient, if the solubility of a compound is strongly dependent upon temperature, an increase in temperature under long-$t_p$ conditions results in a gradual decrease in solution supersaturation, making crystal nuclei difficult. Therefore, a decision can be made as to the strategy employed: laser irradiation under long-$t_p$ conditions and at high pulse energy, without concern about thermal effects, with the aim of inducing rapid crystal nucleation; or careful laser irradiation over a considerable time, under short-$t_p$ conditions, to avoid degradation of the substance.

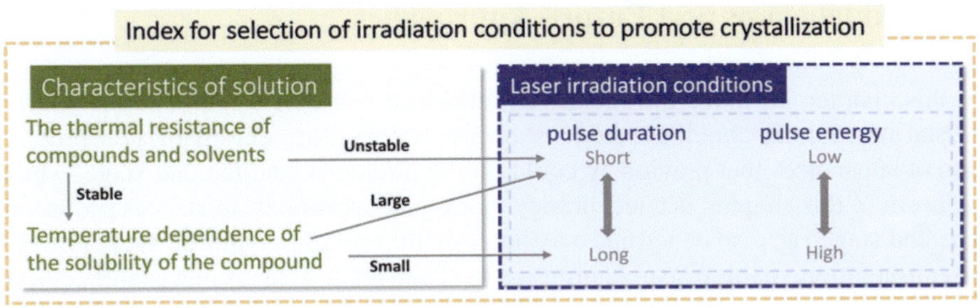


**Fig. 6.7** Indicators for selecting irradiation conditions for promoting crystallization with a pulsed laser

## 6.5 Strategies for Most Difficult Polymorphism Control: Aspirin

For testing the maturity of technologies, the preferred strategy is to apply them to materials that are difficult to crystallize. Therefore, the authors challenged crystallization of the metastable crystal form of aspirin, an antipyretic analgesic. Aspirin exists in forms I and II, which are stable and metastable, respectively (other addition forms have been reported, but these only occur under special environmental conditions such as high pressure, so only forms I and II are usually considered). The crystal structures of forms I and II are very similar, showing almost no difference in solubility. In addition, while observing the initial stage of crystallization in detail, it was seen that the crystal surface of form II becomes covered with form I, which continues to grow. In other words, even if form II is obtained, this will, without dissolving, be replaced by form I after a certain time [27]. Aspirin has extremely difficult to achieve polymorphism control compared to indomethacin and acetaminophen. Thus, no research group previously has achieved the control. The authors considered short-pulse conditions (300 fs) to be appropriate for preventing thermal degeneration of aspirin, on the basis of data obtained previously about laser irradiation. When pulse energy was evaluated over the range of 1.5–30 μJ/pulse, it was shown that the total number of crystal nuclei decreased with decreasing pulse energy. In addition, in additional evaluations, taking laser irradiation time as a new parameter, it was found that the number of crystal nuclei tends to decrease with decreasing laser irradiation time. Finally, with pulse energy at 1.5 μJ/pulse, and the short laser irradiation time of 0.5 s, aspirin single crystal containing only form II were obtained successfully (Fig. 6.8). These form II single crystal was found to be highly stable, remaining unchanged until subjected to external stimulation [28]. After adding form I crystals as seeds intentionally, form II dissolution and form I growth co-occurred, and finally, the crystal surface was covered with form I.

## 6.6 Conclusions and Future Approaches

In this chapter, we have introduced an advanced research stage of laser-induced crystal nucleation technology; the 1st stage was a technology that enables the production of substances that previously could not be produced, and the 2nd stage, which is shown in this chapter, is a technology that separates similar substances produced. The 2nd stage can also be termed a technology for crystal polymorphism control by laser irradiation. In this research, progress has been made as far as metastable crystallization of aspirin, in which polymorphism control has previously been challenging, but there remain numerous substances in which crystallization and polymorphism control are complex. In order to extend this technology even to those substances, it will be necessary to evaluate in more detail the types of phenomena that occur as a result of laser irradiation. In addition, although this issue is not covered in the present

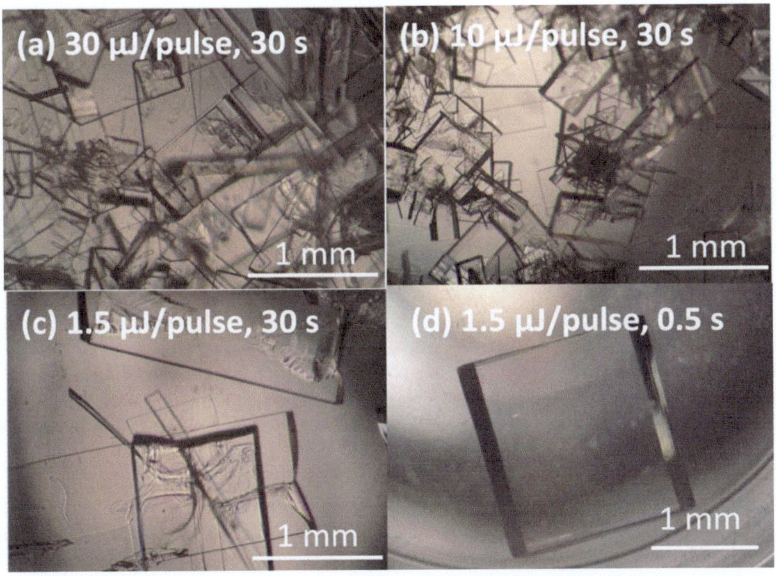


**Fig. 6.8** State of aspirin nucleation according to laser irradiation conditions. The number of crystal nuclei decreases as the pulse energy is reduced and laser irradiation time becomes shorter. Under the irradiation condition (**d**) of 1.5 mJ/pulse, 0.5 s, we obtained a single form of aspirin II. (Reprinted with permission from the Japan Society of Applied Physics, Reference [29])

report, the authors have succeeded in developing technologies that not only control crystal nucleation but also locally and actively control crystal growth mechanisms by laser irradiation [29–31]. Using light, this technology can change the crystals' shape and improve crystal quality. Therefore, if the technology is extended to various organic materials, it is possible to control the crystal shape as it can be those preferred, as appropriate to the aims of use.

The authors' next objective is actively controlling crystal phase transition using light. We want to enhance a transition from the initial phase to a different phase by laser irradiation for crystals with specified structures. Biominerals, such as bone, teeth, and urinary calculi, crystallize their stable form and unstable at the same time, as markedly controlling the speed of phase transition. The mechanism generates flexible and tough structures resulting from the co-existence of both stable and metastable crystal forms. Using light as a tool, our aims are to enable more precise direct involvement in molecular behavior at the nanometer scale, and thus to develop a technology that readily produces materials imitating biological polymorphism control strategies.

**Acknowledgements** This work was supported by Grant-in-Aid for JSPS Fellows (Nos. 20J10452 and 22J00529 for Y. T) and JSPS KAKENHI Grants (Nos. 19H02613, 20K21117, 22H00302, and 22H05423 for H. Y. Y and 19K229659 for M.M.). This work was also supported by grants from AMADA Foundation, the Uehara Memorial Foundation, Global Center for Medical Engineering and Informatics of Osaka University (MEI Grant 2021), Iketani Science and Technology Foundation (No. 0331016-A), and 2022 1st Kao Crescent Award.

**Conflict of Interest** No potential conflicts of interest exist for authors to report.

**Ethical Approval** No ethical approval to report as it does not include experiments on humans and animals.

## References


1. S. Sugiyama, *Frontiers of protein crystals* (CMC Books, Tokyo, Japan, 2013)
2. H. Adachi, K. Takano, Y. Hosokawa, T. Inoue, Y. Mori, H. Matsumura, M. Yoshimura, Y. Tsunaka, M. Morikawa, S. Kanaya, H. Masuhara, Y. Kai, T. Sasaki, Laser irradiated growth of protein crystal. Jpn. J. Appl. Phys., **42** (2, No. 7B), L798–L800 (2003)
3. H. Adachi, S. Murakami, A. Niino, H. Matsumura, K. Takano, T. Inoue, Y. Mori, A. Yamaguchi, T. Sasaki, Membrane protein crystallization using laser irradiation. Jpn. J. Appl. Phys. **43**(10B), L1376–L1378 (2004). https://doi.org/10.1143/JJAP.43.L1376
4. T. Numata, Y. Ikeuchi, S. Fukai, H. Adachi, H. Matsumura, K. Takano, S. Murakami, T. Inoue, Y. Mori, T. Sasaki, T. Suzuki, O. Nureki, Crystallization and preliminary X-ray analysis of the tRNA thiolation enzyme MnmA from Escherichia coli complexed with tRNA(Glu). Acta Crystallographica. Sect. F, Struct. Biol. & Cryst. Commun., **62**(4), 368–371 (2006). https://doi.org/10.1107/S174430910600738X
5. T. Tsukazaki, H. Mori, S. Fukai, T. Numata, A. Perederina, H. Adachi, H. Matsumura, K. Takano, S. Murakami, T. Inoue, Y. Mori, T. Sasaki, D.G. Vassylyev, O. Nureki, K. Ito, Purification, crystallization and preliminary X-ray diffraction of SecDF, a translocon-associated membrane protein, from Thermus thermophilus. Acta Crystallographica. Sect. F, Struct. Biol. & Cryst. Commun., **62**(4), 376–380 (2006). https://doi.org/10.1107/S1744309106007779
6. H. Adachi, J. Jpn. Assoc. Cryst. Growth, **29**, 445–449 (2002)
7. E.H. Lee, A practical guide to pharmaceutical polymorph screening & selection. Asian J. Pharm. Sci. **9**(4), 163–175 (2014). https://doi.org/10.1016/j.ajps.2014.05.002
8. C. Sudha, K. Srinivasan, Nucleation control and separation of paracetamol polymorphs through swift cooling crystallization process. J. Cryst. Growth **401**, 248–251 (2014). https://doi.org/10.1016/j.jcrysgro.2013.10.025
9. M.A. Mikhailenko, Growth of large single crystals of the orthorhombic paracetamol. J. Cryst. Growth **265**(3–4), 616–618 (2004). https://doi.org/10.1016/j.jcrysgro.2004.02.091
10. M. Maruyama, H.Y. Yoshikawa, K. Takano, M. Yoshimura, Y. Mori, Solution-mediated phase transition of pharmaceutical compounds: case studies of acetaminophen and aspirin. J. Cryst. Growth **602**, 126990 (2023). https://doi.org/10.1016/j.jcrysgro.2022.126990
11. M. Yoshimura, J Jpn Assoc Cryst Growth, 45(4), 1–8 (2018)
12. K. Kawakami, *Cutting edge technologies and practical view for crystallization and physicochemical characterization of drug compounds* (CMC Books, Tokyo, Japan, 2019)
13. B.A. Garetz, J.E. Aber, N.L. Goddard, R.G. Young, A.S. Myerson, Nonphotochemical, polarization-dependent, laser-induced nucleation in supersaturated aqueous urea solutions. Phys. Rev. Lett. **77**(16), 3475–3476 (1996). https://doi.org/10.1103/PhysRevLett.77.3475
14. Y. Hosokawa, H. Adachi, M. Yoshimura, Y. Mori, T. Sasaki, H. Masuhara, Femtosecond laser-induced crystallization of 4-(dimethylamino)- N -methyl-4-stilbazolium tosylate. Cryst. Growth Des. **5**(3), 861–863 (2005). https://doi.org/10.1021/cg049709y
15. T. Okutsu, K. Furuta, M. Terao, H. Hiratsuka, A. Yamano, N. Ferté, S. Veesler, Light-induced nucleation of metastable hen egg-white lysozyme solutions. Cryst. Growth Des. **5**(4), 1393–1398 (2005). https://doi.org/10.1021/cg049604v
16. R. Murai, H.Y. Yoshikawa, Y. Takahashi, M. Maruyama, S. Sugiyama, G. Sazaki, H. Adachi, K. Takano, H. Matsumura, S. Murakami, T. Inoue, Y. Mori, Enhancement of femtosecond laser-induced nucleation of protein in a gel solution. Appl. Phys. Lett. **96**(4), 043702 (2010). https://doi.org/10.1063/1.3294622

17. N. Iefuji, R. Murai, M. Maruyama, Y. Takahashi, S. Sugiyama, H. Adachi, H. Matsumura, S. Murakami, T. Inoue, Y. Mori, Y. Koga, K. Takano, S. Kanaya, Laser-induced nucleation in protein crystallization: local increase in protein concentration induced by femtosecond laser irradiation. J. Cryst. Growth **318**(1), 741–744 (2011). https://doi.org/10.1016/j.jcrysgro.2010.10.068
18. H.Y. Yoshikawa, R. Murai, H. Adachi, S. Sugiyama, M. Maruyama, Y. Takahashi, K. Takano, H. Matsumura, T. Inoue, S. Murakami, H. Masuhara, Y. Mori, Laser ablation for protein crystal nucleation and seeding. Chem. Soc. Rev. **43**(7), 2147–2158 (2014). https://doi.org/10.1039/c3cs60226e
19. K. Ikeda, M. Maruyama, Y. Takahashi, Y. Mori, H.Y. Yoshikawa, S. Okada, H. Adachi, S. Sugiyama, K. Takano, S. Murakami, H. Matsumura, T. Inoue, M. Yoshimura, Y. Mori, Selective crystallization of the metastable phase of indomethacin at the interface of liquid/air bubble induced by femtosecond laser irradiation. Appl. Phys. Express **8**(4), 045501 (2015). https://doi.org/10.7567/APEX.8.045501
20. H. Yamamoto, 1-acyl-indoles. II. A new syntheses of 1-(p-chlorobenzoyl)-5-methyoxy-3-indolyacetic acid and its polymorphism. Chem & Pharm Bull, **16**(1), 17–19 (1968). https://doi.org/10.1248/cpb.16.17
21. M. Maruyama, M. Yoshimura, K. Takano, Chapter 13. Development of polymorphic control technology for pharmaceutical compounds, in *Kinetic Control in Synthesis and Self-Assembly*. ed. by M. Numata, S. Yagai, T. Hamura (Academic Press, USA, 2019), pp.269–291
22. H.Y. Yoshikawa, Y. Hosokawa, H. Masuhara, Spatial control of urea crystal growth by focused femtosecond laser irradiation. Cryst. Growth Des. **6**(1), 302–305 (2006). https://doi.org/10.1021/cg050190v
23. Y. Tsuri, M. Maruyama, K. Tsukamoto, H. Adachi, K. Takano, S. Usami, M. Imanishi, M. Yoshimura, H.Y. Yoshikawa, Y. Mori, Effects of pulse duration on laser-induced crystallization of urea from 300 to 1200 fs: impact of cavitation bubbles on crystal nucleation. Appl. Phys. A, **128** (9), (2022). https://doi.org/10.1007/s00339-022-05909-y
24. H.Y. Yoshikawa, R. Murai, S. Maki, T. Kitatani, S. Sugiyama, G. Sazaki, H. Adachi, T. Inoue, H. Matsumura, K. Takano, S. Murakami, T. Sasaki, Y. Mori, Laser energy dependence on femtosecond laser-induced nucleation of protein. Appl. Phys. A **93**(4), 911–915 (2008). https://doi.org/10.1007/s00339-008-4790-x
25. A. Vogel, S. Busch, U. Parlitz, Shock wave emission and cavitation bubble generation by picosecond and nanosecond optical breakdown in water. J. Acoust. Soc. Am. **100**(1), 148–165 (1996). https://doi.org/10.1121/1.415878
26. A. Vogel, J. Noack, K. Nahen, D. Theisen, S. Busch, U. Parlitz, D.X. Hammer, G.D. Noojin, B.A. Rockwell, R. Birngruber, Energy balance of optical breakdown in water at nanosecond to femtosecond time scales. Appl. Phys. B **68**(2), 271–280 (1999). https://doi.org/10.1007/s003400050617
27. Y. Tsuri, M. Maruyama, H.Y. Yoshikawa, S. Okada, H. Adachi, K. Takano, K. Tsukamoto, M. Imanishi, M. Yoshimura, Y. Mori, Intergrowth of two aspirin polymorphism observed with Raman spectroscopy. J. Cryst. Growth **532**, 125430 (2020). https://doi.org/10.1016/j.jcrysgro.2019.125430
28. Y. Tsuri, M. Maruyama, R. Fujimoto, S. Okada, H. Adachi, H.Y. Yoshikawa, K. Takano, S. Murakami, H. Matsumura, T. Inoue, K. Tsukamoto, M. Imanishi, M. Yoshimura, Y. Mori, Crystallization of aspirin form II by femtosecond laser irradiation. Appl. Phys. Express **12**(1), 015507 (2019). https://doi.org/10.7567/1882-0786/aaf419
29. Y. Tominaga, M. Maruyama, M. Yoshimura, H. Koizumi, M. Tachibana, S. Sugiyama, H. Adachi, K. Tsukamoto, H. Matsumura, K. Takano, S. Murakami, T. Inoue, H.Y. Yoshikawa, Y. Mori, Promotion of protein crystal growth by actively switching crystal growth mode via femtosecond laser ablation. Nat. Photonics **10**(11), 723–726 (2016). https://doi.org/10.1038/nphoton.2016.20231

30. H. Takahashi, M. Yamaji, J. Ikeyama, Growth enhancement of organic nonlinear optical crystals by femtosecond laser ablation. J. Phys. Chem. **125**(15), 8391–8397 (2021). https://doi.org/10.1021/acs.jpcc.0c10636
31. M. Maruyama, H. Yoshikawa, J. Photochem. Photobiol. **49**, 71–77 (2018)

# Part III
# Applied Microtexture and Polymorphism

# Chapter 7
# Microtexture and Polymorphism Observed During the Molecular-Beam Epitaxial Growth of Group III–V Semiconductor Nanostructures

**Fumitaro Ishikawa, Kazuki Nagashima, Takeshi Yanagida, Robert D. Richards, and Irina A. Buyanova**

This study explores experimental research on the molecular-beam epitaxial growth of compound semiconductor nanowires (NWs). Unique crystals, such as crystal whiskers, can be grown based on the understanding of the growth mechanism of III–V semiconductor thin films. This mechanism has been recently developed to produce NWs. These crystal whiskers can be considered as NWs with characteristic microtextures and polymorphism. Several studies have addressed the reason for the generation of these crystals, especially the vapor–liquid–solid (VLS) growth mechanism. This growth mechanism can be interpreted based on a well-established understanding of crystal growth. Particularly, an ultrahigh vacuum environment of molecular-beam epitaxy (MBE) will be discussed. Moreover, the status of our work on crystal whiskers, which is referred to as NWs, is reported.

The highly developed MBE technique utilizes a vacuum comparable to that in the upper atmosphere, which allows the generation of ultra-high-purity crystals. Furthermore, the growth of group III–V compound semiconductor heterostructures, which

F. Ishikawa (✉)
Research Center for Integrated Quantum Electronics, Hokkaido University, Sapporo 060-0813, Japan
e-mail: Ishikawa.fumitaro@rciqe.hokudai.ac.jp

K. Nagashima
Research Institute for Electronic Science, Hokkaido University, Sapporo 001-0021, Japan

T. Yanagida
Graduate School of Engineering, The University of Tokyo, Tokyo 113-8656, Japan

R. D. Richards
Department of Electronic and Electrical Engineering, University of Sheffield, Sheffield, UK

I. A. Buyanova
Department of Physics, Chemistry and Biology, Linköping University, 58183 Linkoping, Sweden

 
F. Ishikawa et al. (eds.), *Engineering Crystal Habit*, Topics in Applied Physics 152,
https://doi.org/10.1007/978-981-96-0266-7_7

are structurally controlled at the single-atomic-layer level, can be used in highly functional optoelectronic devices, which was conducted by precisely controlling the source material vapor pressures and by using shutter controls. In addition, this article discusses GaAs NWs synthesized via VLS growth, in which Ga acts as a self-catalyst, the growth of related materials, and various microstructures and polymorphs of the crystals produced by the formation of laminated structures. This study confirms that a detailed understanding of these species can reveal the formation of a promising quantum structure and physical properties unique to fine structures, which would be useful for several device applications. This suggests that new crystals can be grown and new application will be prospected.

## 7.1 Molecular-Beam Epitaxy of Compound Semiconductors

MBE is used under ultrahigh vacuum to grow compound semiconductor crystals [1, 2]. Compound semiconductors are materials used in developed semiconductor lasers and detectors. Figure 7.1a shows an MBE system. A growth chamber equipped with a cooled shroud filled with liquid N is used to achieve an ultrahigh vacuum with a background pressure of the order of $10^{-10}$ Torr. The vacuum level used in this study, i.e., ~$10^{-10}$ Torr (Fig. 7.1b), shown on the ion-gage monitor is broadly equivalent to the space environment at an altitude of 1,000 km above a geostationary satellite orbit where the particle density in the atmosphere is extremely suppressed to a level below $10^8$ cm$^{-3}$. This vacuum level provides a particle mean free path of the order of a few km and is expected to be suitable for the synthesis of ultrapure crystals [3]. The interior of the experimental apparatus is under a space-level ultrahigh vacuum and the outside of the device is under standard atmospheric pressure; thus, the environmental difference between the inside and outside of the chamber resembles that of an actual spacecraft. The environment of the chamber is rather the direct opposite of that of a spacecraft, of which the inside contains a normal atmosphere and the outside is the vacuum of space.

As mentioned above, crystal growth is conducted using MBE. Epitaxy can be described as "the deposition of a crystalline overlayer on a crystalline substrate" [4]. In epitaxy, elements accumulate onto a substrate to form a structure similar to that of the initial substrate. The growth mechanism is discussed as following. Elements supplied into the chamber (Fig. 7.1c) are heated and evaporated, and the ultrahigh vacuum environment then allows the evaporated materials to travel unimpeded from the point of evaporation to the substrate in the form of a beam. When an element is supplied in this way, it initially contacts the substrate surface, bonds loosely, and migrates to find a suitable lattice site before incorporating, resulting in crystal growth. The source material cell used for this purpose incorporates a shutter that can be rapidly actuated to block or pass the molecular beam. Using this technique, fine crystal growth is possible because the structure can essentially be controlled at the

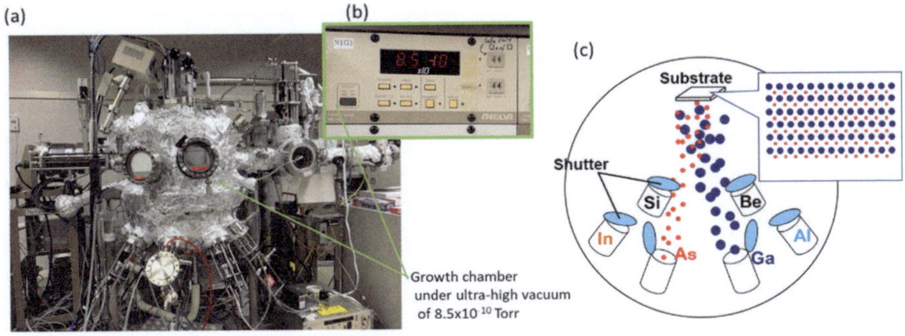


**Fig. 7.1** **a** Molecular-beam epitaxy (MBE) system and **b** its vacuum-level monitor, showing a Torr range. **c** Schematically illustrated crystal growth of the GaAs compound semiconductor using MBE conducted by supplying an atomic flux in the ultrahigh vacuum chamber where the directed beam has a mean free path of the elements of over 100 m at a pressure of $< 10^{-7}$ Torr. Elements can be supplied by heating a crucible to control precisely its vapor pressure. In principle, the epitaxial crystal can be formed layer by layer on the substrate, reflecting the crystal structure of the substrate

atomic-layer level, i.e., at the sub-nanometer level because usually an atomic step corresponds to approximately 0.2–0.3 nm, enabling highly detailed and structurally controlled crystal growth.

## 7.2 III–V Compound Semiconductors

In this section, the materials used in this study are described. GaAs crystals, which are compound semiconductors of group III and group V elements, i.e., III–V semiconductors, are grown. This crystal exhibits superior electron mobility and photoelectric conversion efficiency because of its direct bandgap structure [5, 6]. This group III–V semiconductor, i.e., GaAs, has an ideal photoelectric conversion efficiency and high electron mobility; thus, it is used in transistors, lasers, solar cells, and other extremely high-performance devices. Moreover, the growth of multi-elemental compounds consisting of group III and V elements, such as AlGaAs, InGaAs, and InGaAsP, can be achieved through bandgap engineering and the growth of heterostructures to construct various sophisticated semiconductor devices and quantum structures [7–9].

The supplied elemental-beam flux is monitored by observing its equivalent pressure using a retractable ion gauge. This can be precisely controlled by simply varying the temperature of the crucible where the source elements, such as Ga, Al, As, or Si, are introduced in the chamber. The equivalent pressure itself is linearly proportional to the flux provided onto the substrate surface of the crystal. A supply rate of approximately one atomic layer per second can be achieved for almost any element. The shutter that cuts off the molecular beam can be controlled at 0.1-s or shorter intervals, allowing the crystal growth in thin films with a precision at or below the atomic level. Therefore, extremely flat layered structures (flatness = 0.5 nm), which correspond

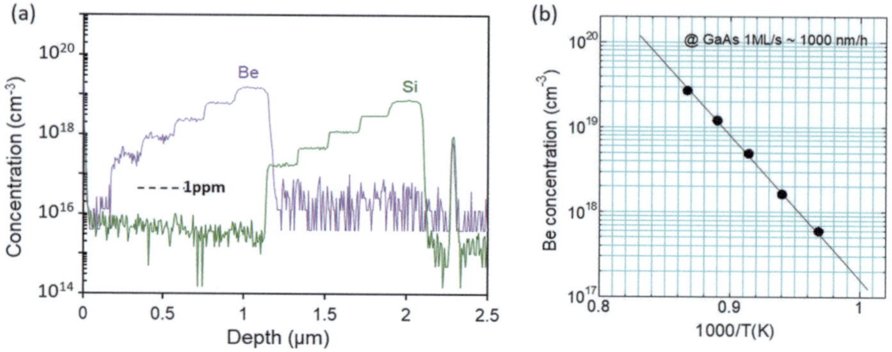


**Fig. 7.2** **a** Secondary ion mass spectrometry depth profiles of the Be and Si dopants in a GaAs epitaxial structure grown by MBE, resulting in p-type and n-type electrical conductivity, respectively, and **b** Arrhenius plot of the Be concentration as a function of the effusion cell temperature

to the size of a double atomic layer step, can be fabricated [10]. Using this technique, introduction of an atomic layer of a different element is possible by switching the shutter at the correct time. This technique can generate a typical quantum well structure with a width of several nanometers, corresponding to approximately 10 atomic layers. This several-nanometer thin layer quantum well enables the formation of a light-emitting active layer, in which electrons and holes can be confined [9].

Figure 7.2a shows an impurity profile of a GaAs-based epitaxial structure obtained using secondary ion mass spectrometry. The conductivity of the matrix can be controlled to be p-type and n-type by introducing doping elements, i.e., Be and Si, respectively. During doping, the group III Ga atom is substituted with these elements. The doping concentration can be controlled with a precision approaching 1 ppm (Fig. 7.2a), with absolute values of the order of $10^{17}$–$10^{19}$ $cm^{-3}$. The profiles of Be and Si significantly change, showing clear step-like profiles, which are realized by controlling the shutter and the vapor pressure control as described above. As shown in Fig. 7.2b, the dopant concentration varies with source temperature in a way that neatly fits an Arrhenius plot constructed under vapor pressure. The electronic conductivity of the obtained structures is precisely proportional to the dopant concentrations; thus, highly precise electronically controlled thin epitaxial layers can be obtained for sophisticated semiconductor device applications [2].

## 7.3 Growth of Nanostructures and Nanowires

In this study, a crystal growth method is used to form highly pure thin, flat films with dopant concentrations of < 1 ppm. This method is used to form NWs that can be used in numerous applications. This section discusses the crystal growth of NWs based on the study of crystal whisker growth. Thin film production has significantly progressed over the last few decades in terms of applying different crystal growth modes and developing the microscopy techniques to characterize the resulting materials. Even at the level of university laboratory research, previously unusable atoms can now be detected, observed, and utilized.

A historical example of a growth mode being developed is the VLS growth discovered by Wagner and Ellis in 1964 [11]. This is the mechanism of crystal whisker formation by crystal nucleation and growth through the transitions between the vapor, liquid, and solid (VLS) phases. In this process, a catalyzing droplet of liquid source material is coalesced on the surface of the substrate. Then, this droplet dissolves the source materials from the flux and deposits them at the base of the droplet. As the material is deposited, the droplet rises on the newly formed material. Under appropriate conditions, this will form a freestanding pillar (whisker) of epitaxial material, with the droplet at its tip. Hiruma and coworkers developed a VLS growth method for the growth of GaAs and InAs nanoscale whiskers [12–16]. Even at this early period (1992), researchers discovered that crystal whiskers contain a metastable wurtzite (WZ) phase within the dominant zinc blende (ZB) phase in the GaAs whiskers [17].

Several studies have used the word "tree," drawing imagery from nature, to describe these nanostructures. These nanoscale trees were actually generated in studies conducted after the year 2000 [18]. The generated crystal trees were a few hundred nanometers in trunk diameter and were a few microns long. Thus, trees with extremely complex structures can be formed, e.g., Eshelby twist [19]. The application of this technology allows for various highly advanced manipulations. Branches can be fused together at points along their length to form conductor wires, enabling the formation of superstructures, which were not previously achievable, opening the gate for the fabrication of nanoscale electronic devices [20].

Here, the current state of crystal whisker applications and the potential of controlling their crystal growth are discussed. In 2002, two studies by the Charles Lieber Group on the applications of these NWs resulted in a considerable change and progress in this field [21, 22]. In VLS growth, a certain material is used to switch the elements midway to produce a heterostructure. This enables the production of extremely advanced devices. In addition to the vertical shape, a lateral core with a material wrapped around it can be formed to create a core–shell structure. Manmade crystals with controlled complex three-dimensional structures can be grown. The idea that this type of control could be applied to device production was first proposed in 2002, promoting research that allows its use for various applications [23–25]. Lasers and electronic devices can be fabricated using these developed NWs [26, 27]. The NWs produced using this technique can change the current roadmap of electronic device production based on Si [28–31]. This technology could allow solar cells to overcome their current limitations [32, 33]. In addition, photoelectric energy conversion could achieve higher efficiency levels. Thus, this technology has a significant potential for numerous applications.

## 7.4 Growth of Core–Shell Nanowires by Molecular-Beam Epitaxy Using Precisely Controlled Shell Structures

The introduction of heterostructures allows the manipulation of the semiconductor system through band engineering, which results in unique quantum structures and devices with superior functions as described in the previous section. Figure 7.3 shows the formation mechanism of the core–shell type NWs investigated in this study using MBE. Figure 7.3 shows a representative standard material system of GaAs/AlGaAs core–shell type NWs [34–37]. The NW growth is initiated by the formation of a Ga droplet on the (111) Si substrate. A thin native oxide (thickness = 2–3 nm) is formed on the substrate surface. A number of pinholes are observed on the surface of this native oxide. These pinholes can act as nucleation sites for the NWs as a result of the formation of nanoscale craters where Ga droplets accumulate [38, 39]. The density of NWs in an area, thus, depends on the conditions of the native oxide, which can be controlled using pregrowth treatments such as annealing [40, 41]. The wire predominantly starts to grow longitudinally under optimized growth conditions, where the growth proceeds under VLS mode with a Ga catalyst at the tip of the NW; furthermore, an As supersaturation is achieved on the growth front. Next, the growth is interrupted by terminating the Ga flux to switch the growth mode from a vertical VLS mode to a lateral vapor–solid mode. During the growth interruption, the Ga catalyst is consumed and crystallized. The subsequent supply of Al and Ga mostly results in the lateral growth of the AlGaAs shell layer, producing core–shell NWs. This procedure enables a precise control of the NW diameters and compositions of the shell layers based on the concepts of planar MBE thin film growth in the shell layers. In this work, this results in GaAs/AlGaAs NWs with various Al-concentrations, a GaAs/GaAsN core–shell system, and a GaAs/InGaAs core–shell system [34–39].

This study aims to explore the necessary steps to produce the crystal whiskers and to develop the whisker materials required for the practical application of the above-mentioned devices. At present, our level of understanding of the growth mechanism is confined to that demonstrated in Fig. 7.3. However, even this level is sufficient to fabricate various devices.

To precisely control the NW structure, the effects of the growth time and Ga flux on the NW diameter are investigated. Using the growth procedure illustrated in Fig. 7.3, these parameters were changed during the lateral (shell) growth period. Figure 7.4a and b show the development of the NW diameter after the growth interruption and the crystallization of the Ga catalyst [35]. As shown in the Fig. 7.4, the NW diameter linearly increases as a function of the growth time and Ga flux. The wire diameter can be finely controlled by adjusting these parameters, which is similar to the case of the planar epitaxial growth of conventional group III–V materials, even though the formation rate of the shell layer is approximately one fifth of the thin film growth rate [2]. These results also indicate that the composition of the NW shell layers can be controlled during the lateral growth similar to MBE thin film manipulation. Therefore, core–shell heterostructure NW can be precisely grown.

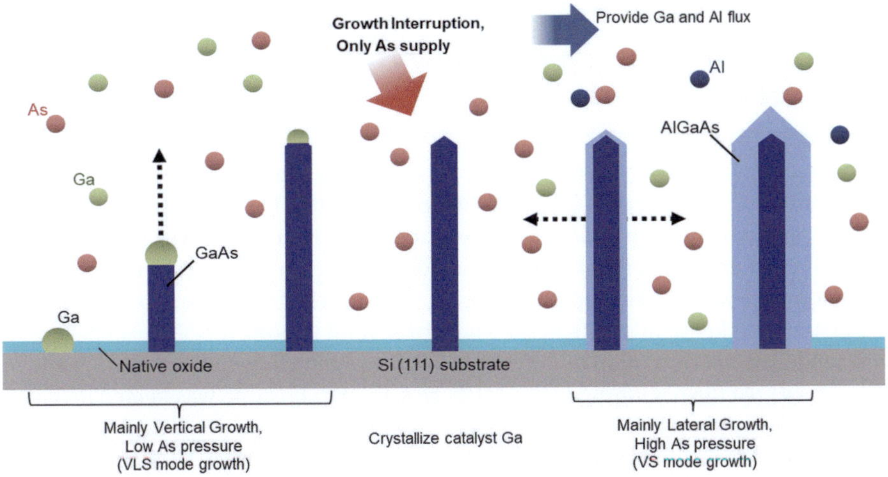


**Fig. 7.3** Schematic representation of a typical GaAs-based core–shell type nanowire growth mechanism. The growth of GaAs core is initiated under the VLS mode in the presence of a metallic droplet of a Ga catalyst. The Ga catalyst is consumed when the growth is interrupted by the termination of the Ga flux under continuing As irradiation. As flux was maintained throughout the growth. Next, the lateral growth becomes dominant and the Ga and Al supplies induce AlGaAs shell formation. The concept of MBE thin film growth is mostly applicable for the growth of the shell layer; thus, the growth rate and the resulting layer compositions can be precisely controlled as shown in Fig. 7.4

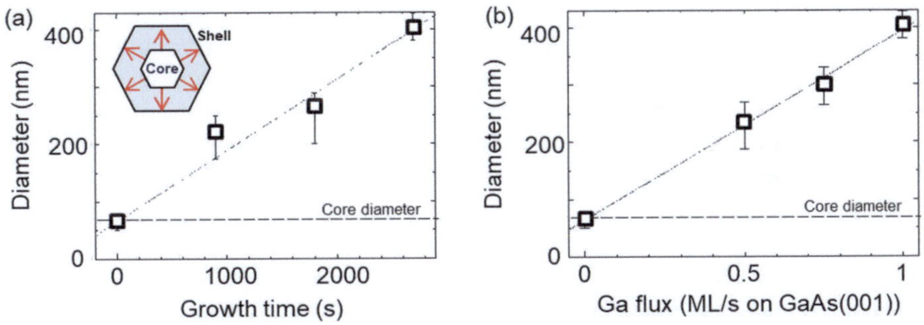


**Fig. 7.4** **a** Nanowire diameter as a function of growth time at a fixed total Ga flux of 1.0 monolayer/s on a (001) GaAs substrate. **b** Nanowire diameter as a function of the Ga flux at a fixed growth time of 45 min. Reproduced from Ref. [35] with a permission from the Institute of Physics

## 7.5 MBE Growth of Nanowires with Well-Ordered Microstructures: A Case Study of GaAs/GaNAs Core–Shell Nanowires

In this section, an MBE growth example of NWs with sharp-facets and well-ordered microstructures is discussed. The NWs were grown on (111) Si substrates. Ga and In were supplied using conventional solid-source effusion cells. As was supplied using a valved cracker cell operated in the $As_4$ mode at a cracking temperature of 550 °C. A plasma source was used to supply N [34, 35, 42, 43]. The Si surface was covered with a thin native oxide with a typical thickness of several nanometers. The nucleation of NWs requires the formation of Ga droplets at the pinholes on the surface oxide layer [35, 37, 44]. At the beginning of the MBE growth, the substrate was heated to 620 °C for 10 min to thermally clean the substrate. Ga and As flows were then supplied to the Si substrate to initiate the GaAs NW core growth through VLS growth using Ga as a self-catalyst [34, 35, 37, 45]. Next, the GaAs and GaAs/GaNAs core–shell structures were grown. The beam equivalent pressure (BEP) of As was maintained at $3 \times 10^{-5}$ Torr throughout the growth process. Before the growth, the Ga flux was set to be equivalent to that of a planar growth rate on (001) GaAs of 1.0 ML/s). The $As_4$/Ga BEP ratio was 37. The GaAs core was grown for 30 min at a substrate temperature of 580 °C. Then, the Ga catalyst was crystallized by terminating the Ga supply under the same $As_4$ beam irradiation. During the growth interruption, the Ga flux was reduced to 0.5 ML/s. A shell layer was then formed by the dominant lateral growth shown in Figs. 7.3 and 7.4. Next, a thin GaAs shell was grown for 20 min, and the growth was again interrupted to reduce the growth temperature to 500 °C and ignite the N plasma [35, 42, 43]. Subsequently, the GaAs shell was grown for 5 min, and the GaNAs layer was grown by opening the shutter of the plasma source for 30 min. NWs with a GaAs/GaNAs core–shell structure were then formed [34, 35]. Under these plasma conditions, the NW's shell is expected to contain 2% N.

Figure 7.5a–f show the scanning electron microscopy (SEM) images of the obtained GaAs/GaNAs core–shell NWs. The formation of NWs is clearly observed in the images. The density of NWs is approximately $2 \times 10^8$ cm$^{-2}$ [34, 35, 37, 44, 45]. The length and average diameter of the NWs are 4–6 μm and approximately 350 nm, respectively. Figure 7.5c and d show a plan view of the obtained NWs, which clearly illustrate that the wires have a hexagonal cross section [46, 47]. Figure 7.5e shows an enlarged SEM image of a single NW. The wire sidewall is sharp and straight and is originated from the facetted sidewall of the (110) plane. Figure 7.5f shows a cross-sectional bright-field scanning transmission electron microscopy (BF-STEM) image of a NW. A dark line is observed at the interface between the GaAs and GaNAs shell. The directions of the NW vertex are <112> [34, 35, 44–47]. These results indicate the clear formation of the GaAs/GaNAs core–shell structure in the wire (length = 5 μm and diameter = 350 nm). Thus, highly controlled NW multilayer structures can be achieved by combining the high-precision MBE growth technique of group III–V thin film multilayer structures with the VLS-type NW growth mechanism. These

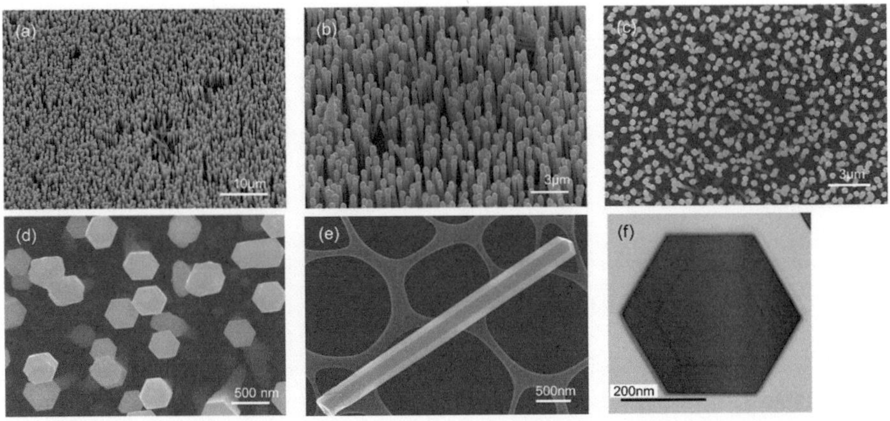


**Fig. 7.5** **a** 30°-tilted scanning electron microscopy (SEM) images of GaAs/GaNAs core–shell nanowires (NWs) nominally containing 2% N in the GaNAs shell and **b** their enlarged image. SEM images showing **c** the plan view and **d** the enlarged view of the NWs. **e** SEM image of a NW, in which the tip of the NW is at the top right of the image. The background mesh is a porous carbon film used for imaging. **f** Axial cross-sectional bright-field scanning transmission electron microscopy image of the GaNAs NW nominally containing 2% N

structures can be used as electrically and optically functional GaAs-based materials for semiconductor devices [21–25, 48, 49].

## 7.6 Microtextures Observed in the Epitaxial Nanostructure and Introduction of Bi During the Growth of GaAs Nanowires

Here, the characteristics of Bi-containing GaAs NWs are discussed. As described in the previous sections, the developed NWs could be used in functional materials and devices because of their one-dimensional self-assembled nanostructures [23–25]. When Bi is introduced to GaAs, the Bi oversaturation substantially differs from that of other elements and various structures are observed (Fig. 7.6a–k). Bi easily forms droplets; thus, GaAs NWs containing a low level of Bi can form a highly characteristic complex structure [50]. Therefore, the complex structures in Fig. 7.6 were investigated to discover an effective function. The formation of this complex structure can be attributed to the ease of droplet formation and diffusion on the growth surface, causing a spontaneous formation of branches, which generates extremely complex structures.

In the past, the available technology only allowed the production of thin, flat, featureless films. Currently, minute and unique crystals can be observed during the production of thin films. These crystals were not detectable during film production twenty years ago. However, the current technology enables the observation

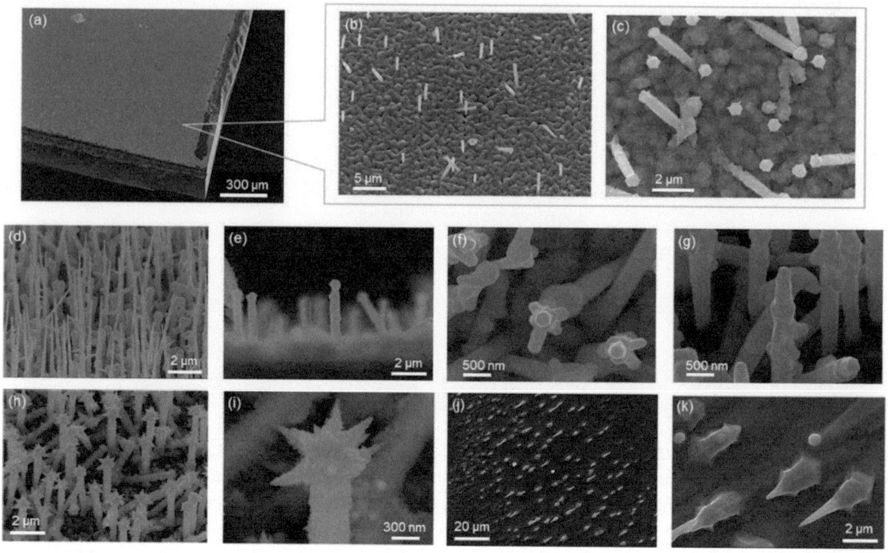


**Fig. 7.6** Scanning electron microscopy images of the microtextures in GaAs/GaAsBi heterostructure core–shell nanowires, which were unintentionally obtained by varying the growth conditions of the GaAsBi shell, i.e., growth time, Bi flux, or temperature: **a** a close up of the (111) Si substrate edge. Enlarged images **b**, **c** at different magnifications and **d**–**k** of various structures

and the intentional production of very small crystals. Several crystal variations are not produced intentionally. This study aimed to produce uniform and perpendicular columnar crystals. However, by slightly changing the manufacturing conditions, even if unintentional, numerous structural variations can be obtained. Knowledge of thin films is insufficient to overcome this problem. In this case, a deep understanding of natural crystalline ores and the growth mechanisms of macro crystals is required to achieve an interpretation of these variations. Thus, an interdisciplinary approach may enable a better understanding of various phenomena during the fabrication of these thin films and can result in additional novel achievements.

As mentioned, the formation of complex structures can be attributed to the ease of droplet formation and the large diffusion length of the droplets on the growth surface, causing a spontaneous nucleation of microcrystals, which results in extremely complex structures [50]. Next, the randomness of this growth, i.e., if it is random, was examined. When a horizontal cross section is cut in the developed NW structure, a highly uniform rod shape was observed. This reflects a hexagonal crystalline structure, resulting in the formation of an unusual structure, in which hexagonal branches form on the rod tips. This structure, i.e., the unique shape (wires with branches growing from them) revealed by the vertical cross section of the developed material, is important from several perspectives. Within the developed material, an abrupt interface with a change in one of the atomic layer was observed. A transition between a section that contains a large amount of Bi and a section that contains less Bi was observed within the GaAs matrix. These sections were not created intentionally.

## 7.7 Microstructure in GaAs/GaAsBi Core–Shell NWs

In this section, a special nanostructure that could be grown by adding Bi is discussed [50, 51]. The characteristics and growth mechanisms of GaAs/GaAsBi core–shell multilayered NWs grown on (111) Si substrates are illustrated, focusing on the structural deformation induced by Bi.

The NWs were grown by molecular-beam epitaxy using Ga-induced VLS growth on phosphorous-doped n-type (111) Si substrates (more details on its growth can be found in Ref. 50). First, the samples were heated to 580 °C under an $As_4$ BEP of $8 \times 10^{-6}$ Torr. The Ga supply was then set to obtain a planar growth rate of 1.0 ML/s on (001) GaAs to achieve a group-V As-rich growth condition. The core–shell structure was grown as shown in Figs. 7.3 and 7.4. The GaAs core NWs were grown, and the growth was then interrupted to crystalize the Ga catalyst. After the catalyst crystallization, the lateral growth became dominant and the formation of the core–shell type NWs was expected. A Ga flux was then initiated to form the GaAs shell, and the growth was again interrupted by further reducing the substrate temperature to 350 °C for the subsequent growth of GaAsBi. Next, Ga and Bi fluxes were initiated to form the GaAsBi shell. The widths of the GaAs core and GaAsBi shell were expected to be 100 and 80 nm, respectively. The BEP of $As_4$ was maintained at $8 \times 10^{-6}$ Torr throughout the growth process. The Bi BEP was adjusted to $4 \times 10^{-7}$ Torr for the growth of the GaAsBi shell. After the structure growth, the substrate temperature was reduced until it reached 300 °C under the $As_4$ pressure. Subsequently, the As pressure was terminated, and the samples were removed after the substrate temperature dropped below 100 °C. A GaAsBi thin film was simultaneously grown on GaAs (001) substrates placed side by side on the growth substrate holder. The Bi concentration of the GaAsBi thin films was determined using X-ray diffraction (XRD) to be 4.5% [52].

Figure 7.7a and b show the 45°-tilted SEM image of the GaAs/GaAsBi core–shell NWs, and Fig. 7.7c and d show their planar SEM image. The NWs exhibit a characteristic surface morphology, i.e., rough surfaces and branches. The tips of the wire branches are round hemispheres and are associated with the catalyst that induced the branch formation [53]. In contrast to the rough surface on the primary trunk, the branched NWs exhibit a smooth surface. The trunk wire length varies between 1 and 6 μm, and the rough NW trunk width exhibits a wide statistical distribution with an average diameter of 380 nm and a standard deviation of 70 nm. The NWs shown in Fig. 7.7b show branches, the nucleation of which started from the sharp-edged wire sidewall. The planar SEM image of the NWs (Fig. 7.7d) shows a defined growth of the branches from the trunk crystal planes, resulting in a hexagonally radiated structure. The branches are oriented to the <112> directions, which is further discussed in the following sections. The formed wires originate from the apexes, i.e., the vertices between the adjacent {110} side planes [54].

Figure 7.7e–h show SEM images and energy dispersive spectroscopy (EDS) maps of a typical branched NW. As seen in Fig. 7.7f–h, the primary trunk shows large Ga and As contents with weak Bi intensity. The trunk is a NW consisting of GaAsBi with

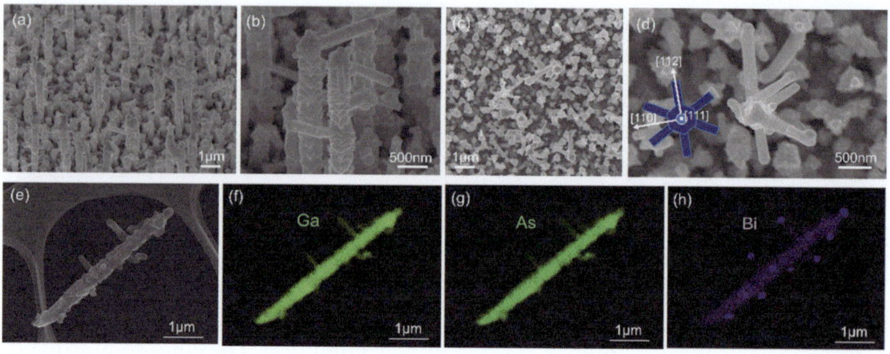


**Fig. 7.7** **a**, **b** 45°-tilted and **c**, **d** scanning electron microscopy (SEM) images of the plan view of the GaAs/GaAsBi core–shell NWs with a hexagonal guided structure in (**d**). The blue overlay schematic illustrates the arrangement and orientation of the trunk and branches, where the branches grow from the apexes. **e** SEM image and **f**–**h** energy dispersive spectroscopy (EDS) elemental mapping of Ga, As, and Bi, respectively, of a single branched GaAs/GaAsBi NW. Reproduced from Ref. [50] with a permission from the American Chemical Society

a Bi content of 2%–3% depending on the growth conditions [51]. The wire branches similarly exhibit high Ga and As contents and low Bi content, suggesting that the wire branches consist of GaAsBi. Considering the smooth surface of the branches and the rough surface of the trunk, the strain in the trunk should be significantly different from that in the branches. Both the primary trunk and branches exhibit a hemispherical tip morphology. As clearly seen in the Bi elemental mapping images (Fig. 7.7h), the hemisphere morphology can be attributed to Bi. In addition, no Ga is observed at the tip. In a previous report by Zha et al., a Ga droplet acted as the catalyst for the VLS growth of the branches [53]. Considering the Bi BEP at the growth of NWs with rough surfaces, over 70% of the irradiated Bi is not introduced into the GaAsBi layer, which generates excess metallic Bi on the growth front [50–52]. The cooling procedure, which includes an As overpressure after the growth process, could change the droplet characteristics on the surface [51]. The droplets should mainly comprise Bi because the presence of Bi droplets can possibly promote the formation of the NW branches.

We here show an example for the characteristics of the Bi droplets, e.g., on the (001) GaAs Substrate, mounted adjacent to the Si substrate during the growth of the sample with branched nanowires. The additional Ga and As overpressure and the presence of Bi droplets induce a characteristic supersaturation, resulting in the formation of needle-like nanocrystals. Figure 7.8a–f show SEM and EDS images of GaAsBi thin films on a (001) GaAs substrate placed side by side with the Si substrate during the growth of branched GaAs/GaAsBi core–shell NWs. The droplets formed on the substrate with a density of $1.3 \times 10^7$ cm$^{-2}$ and a diameter of 100 nm–2 μm. Figure 7.8a shows that some droplets were observed on the surface, and the higher magnification plan view and cross-sectional images (Fig. 7.8b and c), respectively) show the presence of a needle-like structure grown from a droplet. This

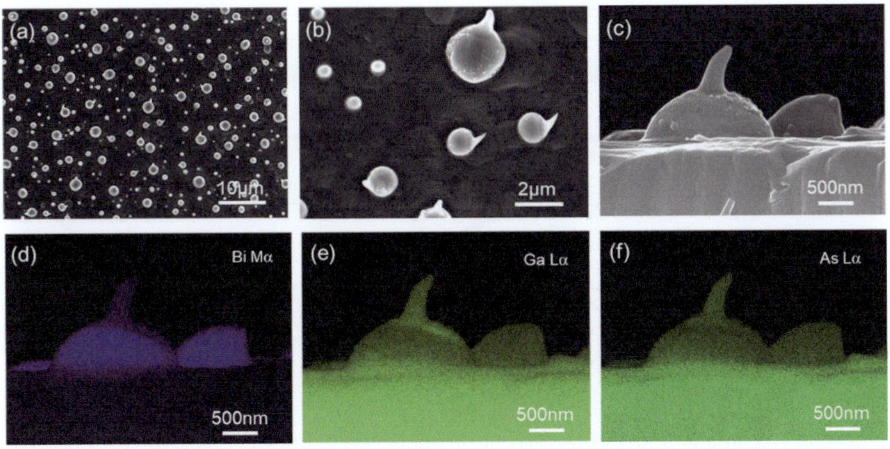


**Fig. 7.8** **a**, **b** Plan view and **c** cross-sectional scanning electron microscopy (SEM) image of a sample surface of the (001) GaAs substrate, which is grown simultaneously with the GaAs/GaAsBi core–shell NW sample shown in Fig. 7.7. Energy dispersive spectroscopy (EDS) elemental mapping of **d** Bi, **e** Ga, and **f** As for the corresponding enlarged cross-sectional SEM images shown in (**c**). Reproduced from Ref. [50] with a permission from the American Chemical Society

needle comprises GaAsBi, of which the composition is dominated by GaAs. This is probably a typical characteristic of Bi droplets, which exhibit supersaturation when they coexist with Ga on the surface under an As overpressure [55]. This induces the growth of needle-like GaAs nanocrystals containing small amounts of Bi. This phenomenon might be the reason for the rough surface and the complicated structures of the branched NWs shown in Figs. 7.6 and 7.7.

Figure 7.9 illustrates the formation mechanism of branched NWs [50]. After the growth of the GaAsBi shell to form the trunk, the As overpressure is maintained for a certain period, which allows As to sufficiently interact with the residual Ga and Bi on the growth front. Even when the Ga BEP is terminated at the end of the GaAsBi shell growth on the trunk, residual Bi remains and widely diffuses on the surface. The presence of the Bi hemispheres on the tip of the GaAsBi trunk and branches suggests that Ga would have existed in the metallic droplet during the VLS formation of the branches containing Bi, which acts as a catalyst. Thus, it is suggested that branch formation is induced by this process, where the residual Ga and Bi on the surface will eventually crystallize into NW branches in the presence of atmospheric As. It can be suggested that branch formation is induced by the residual Bi that forms into Bi droplets and acts as a catalyst for branch formation. After the trunk growth, the Ga is effectively transported to points around the NW in Bi droplets. These droplets then re-initiate VLS growth, consuming the Ga that remains in the droplets to form GaAs(Bi) NW branches with Ga-depleted Bi droplets at the tips.

Figure 7.10a–j show axial cross-sectional BF-STEM, high-angle annular dark field (HAADF)-STEM, EDS images, and fast Fourier transform (FFT) diffraction

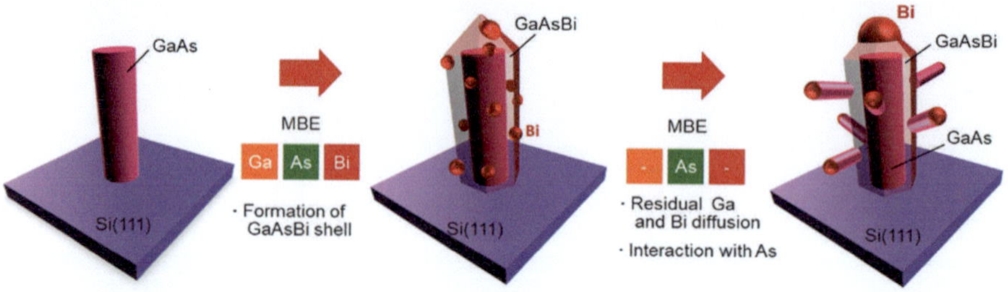


**Fig. 7.9** Schematic illustration of the growth mechanism of the branched GaAs/GaAsBi core–shell NWs with residual Bi on a nanowire growth front. Reproduced from Ref. [50] with a permission from the American Chemical Society

patterns of branched GaAs/GaAsBi core–shell NWs. Figure 7.10a illustrates a BF-STEM image of a branched NW, showing that the branch nucleation starts on the trunk of the NW. The FFT electron diffraction pattern shows identical crystal orientation of the trunk and the middle of the branch. Thus, the wire epitaxially grows from the trunk. The NW mainly consists of GaAs and a small amount of Bi in both the trunk and branch. As shown in Fig. 7.7, there is a Bi cluster at the tip of the NW. The HAADF image provides information about the elemental distribution based on its Z-contrast image, in which elements with larger atomic numbers exhibit bright contrast. The bright contrast in the HAADF image in Fig. 7.10d clearly corresponds to the Bi intensity in Fig. 7.10g. Figure 7.10h–j show high magnification BF-STEM and HAADF-STEM images as well as EDS elemental mapping for Bi in the trunk, respectively. The GaAs core in the NW trunk shows hexagonal vertexes. Based on the FFT diffraction patterns in Fig. 7.3a, the vertices are observed in the <112> directions between the adjacent {110} side planes. The lattice distances obtained from the FFT patterns are 0.194 and 0.201 nm for the $(2\,\overline{2}\,0)$ and $(0\,\overline{2}\,2)$ diffractions, respectively, at the center of the wire trunk, which are similar to the standard GaAs lattice constants. The same lattice constants measured in the middle of the branch are 0.213 and 0.204 nm for the $(2\,\overline{2}\,0)$ and $(0\,\overline{2}\,2)$ diffractions, respectively. This indicates that the GaAsBi at the middle of the wire branch is deformed by the introduction of Bi, which has a larger lattice constant than the GaAs in the NW core in the middle of the wire trunk. The directions of the branches are <112> , and they originate from a vertex of the GaAs core [54]. The structure of the outer shell has a characteristic hexagonal star structure that is probably induced by the presence of Bi at the vertex of the outermost shell. Moreover, the dark contrast in Fig. 7.10h suggests large concentration of stacking faults, which are likely induced by the accumulation of Bi. This is confirmed by the bright contrast in Fig. 7.10i and the Bi elemental mapping shown in Fig. 7.10h. According to the dark contrast of the BF-STEM image in Fig. 7.10h, several stacking faults are present at the outer shell of the vertexes where there are no branches. In contrast, the vertex with a branch exhibits much smaller stacking faults, which suggests that the branch formation suppresses the formation of crystalline defects. By comparing the BF-STEM and HAADF-STEM images as well as the Bi elemental distribution in Fig. 7.10h–j, the locations corresponding

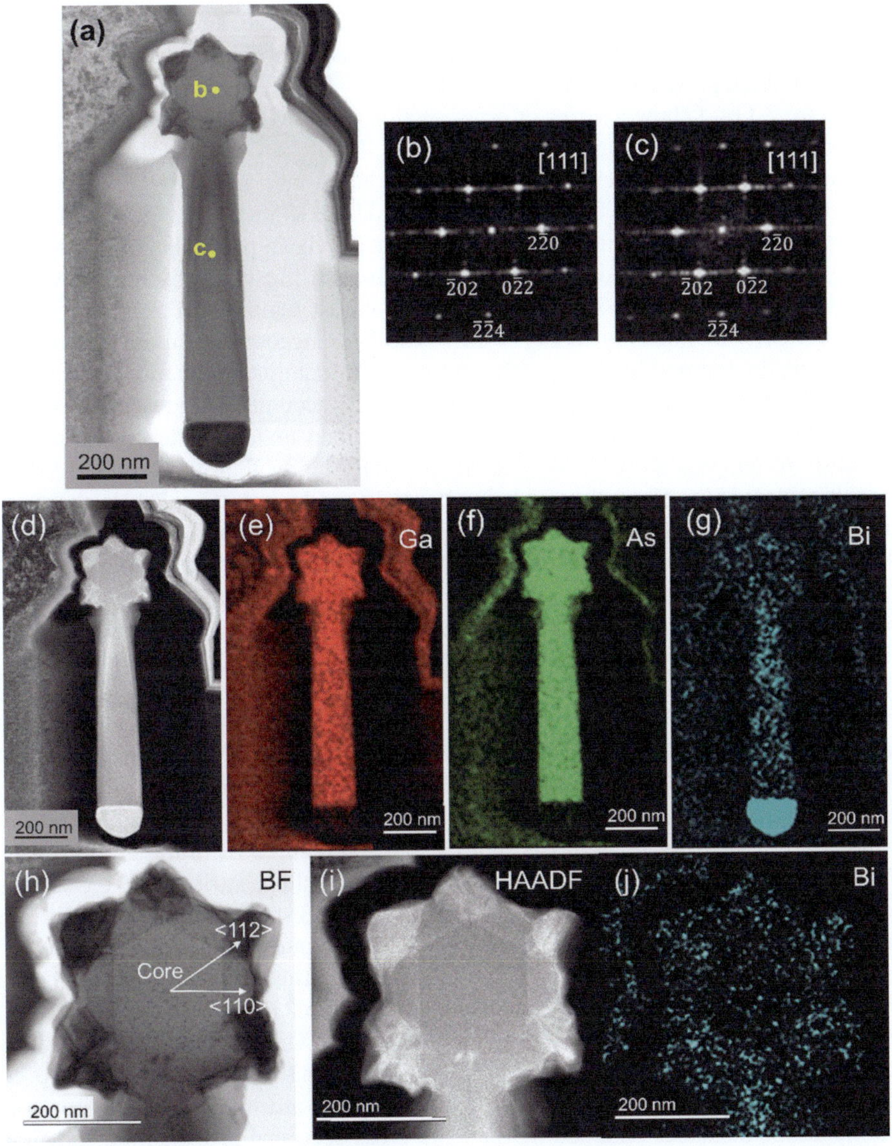


**Fig. 7.10** Axial cross-sectional scanning transmission electron microscopy (STEM) study of the branched GaAs/GaAsBi core–shell NWs, which are observed in the [111] direction. **a** Bright-field scanning transmission electron microscopy (BF-STEM) image with insets showing **b**, **c** fast Fourier transform (FFT) electron diffraction patterns obtained at the spots of the trunk and middle of the branch indicated in (**a**). **d–g** high-angle annular dark field–STEM and energy disruptive spectroscopy (EDS) elemental mapping of Ga, As, and Bi, respectively. **h** High magnification BF-STEM, **i** high-angle annular dark field (HAADF)-STEM images, and **j** EDS Bi elemental mapping of the cross section of the trunk in (**h**) and (**i**). Reproduced from Ref. [50] with permission from the American Chemical Society

to the stacking faults, bright contrast of the HAADF image, and the location of Bi are determined. These results suggest that there is a relationship between crystalline defects and the presence of Bi.

Figure 7.11a–f show a series of cross-sectional STEM images and EDS maps of the branched NW growing from the substrate. Figure 7.11 shows that the branch protrudes in the perpendicular direction to the NW trunk, i.e., they are oriented in the <112> directions. The bright contrast in the HAADF image in Fig. 7.11e clearly corresponds to the strong intensity of Bi in Fig. 7.11h. In particular, a horizontally striped switching can be observed within the branch illustrated by the contrast in the HAADF image (Fig. 7.11e). The bright contrast region is in good agreement with the EDS Bi intensity in Fig. 7.11h. The Bi concentration in the core of the trunk was determined based on EDS to be 0%. In contrast, the concentration was 7% within areas rich in group V elements in the Bi-accumulated area in the middle of the branches [50]. The Bi concentration obtained under As-rich conditions at a substrate temperature of 350 °C is significantly larger than the reported values for thin films [56]. The enlarged striped area in the branch (Fig. 7.11i) shows an interesting feature. Notably, the studied NWs dominantly comprised a ZB structure with some twins. A twin defect of a ZB GaAs structure is observed at the interface of the contrast switching. This is clearly observed in the filtered atomic HAADF image in Fig. 7.11j. The lower part of the image, which is labeled area B, has a brighter contrast corresponding to the Bi rich area of Fig. 7.11h. In this area, the present atoms can be determined based on the intensity of the observed atoms. As seen in the intensity profile superimposed in Fig. 7.11j, at area B, the upper left paired atoms exhibit higher intensity than the lower right ones, which suggests that they are As or Bi atoms. However, this clear intensity difference between the pair elements could not be observed in area A, suggesting that the atoms in this area are Ga and As, which have close atomic numbers, with a trace amount of Bi in the group V site. Thus, the observed higher Bi composition in area B could be attributed to the surface configuration at the growth front, where As-rich (112) B surfaces with higher surface energy are exposed, promoting the introduction of Bi into the layer [57–59].

The diffraction pattern in area A suggests that the lattice distance in the $(\overline{111})$ and (200) planes are 0.283 and 0.259 nm, respectively. In area B, the distance in the (111) and $(00\bar{2})$ planes are 0.288 and 0.267 nm, respectively. The standard lattice distances of the (111) and (200) phases of GaAs are 0.326 and 0.283 nm, respectively. However, the measured values of both orientations in this study are smaller. This is opposite to the lattice distances obtained by the axial cross-sectional investigation shown in Fig. 7.10a, which are longer (expanded) than the standard values. Thus, the lattice exhibits a three-dimensional distortion in the branch. This means that the lattice is extended in the axial and horizontal directions, but it is shrunken in the radial and vertical directions, especially for the {111} planes. The observed large strain (lowest value $< 5\%$) induces structural deformations, which result in surface undulations in GaAsBi thin films [60] and NWs [51]. This surface roughening causes various crystal orientations on the branch surface. Therefore, the (112)-oriented surface possibly acts as the preferential site for the nucleation of additional nanocrystals, inducing the formation of the observed branches [55]. Moreover, the twin defect interface

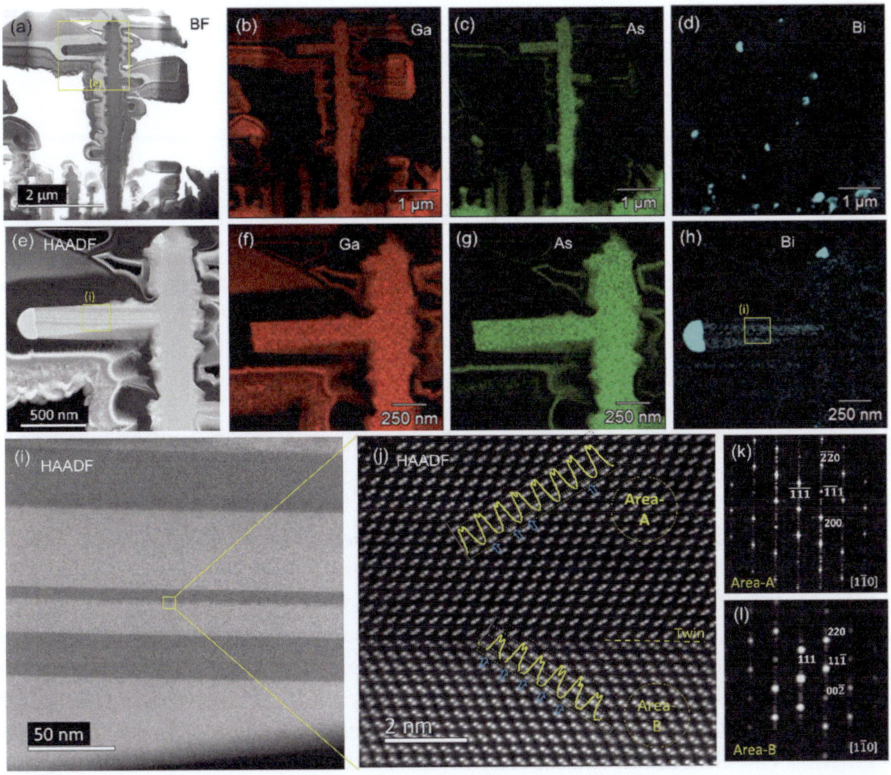


**Fig. 7.11** **a** Cross-sectional bright-field scanning transmission electron microscopy (BF-STEM) image of a branched wire growing out of the substrate and **b–d** the corresponding energy disruptive spectroscopy (EDS) elemental mappings of Ga, As, and Bi. **e** A high magnification high-angle annular dark field (HAADF)-STEM image of the area around the trunk tip delimited by the square in (**a**), and **f–h** the corresponding EDS elemental mappings of Ga, As, and Bi. **i** High magnification HAADF-STEM image of the marked area in (**e**). **j** Filtered HAADF image obtained at the interface between bright and dark contrast regions, and its high-resolution image of areas A and B, where brighter atoms correspond to group V and are indicated by arrows. **k, l** fast Fourier transform (FFT) patterns of areas A and B, respectively, as indicated by the circles in (**j**). Reproduced from Ref. [50] with a permission from the American Chemical Society

penetrates and is preserved from the trunk to the branch. The apparent contrast switch occurs at the branch [50]. Consequently, the twin defect is preserved at the epitaxial nucleation site of the wire branch.

The growth of the GaAs or GaAsBi occurs during the growth in the presence of the Bi catalyst at the tip of the NW. At this site, Bi is dominantly incorporated in the preferable crystal orientation that is switched by the twin, which is propagated from the GaAs trunk core. Particularly, the hetero-interface between GaAs and GaAsBi was defined by crystalline twin defects created through the characteristic growth mechanism attributed to Bi in this material system [61, 62]. Considering the smaller bandgap energy for the GaAsBi alloy, the structure can be considered a GaAs/GaAsBi/GaAs quantum confined structure.

### *7.7.1 Using Composition and Structural Modulations of Nanowires to Form Quantum Structures*

Figure 7.12a shows the 45°-tilted SEM image of a GaAs/GaAsBi/GaAs core–multi-shell NW. The growth process was conducted under growth conditions identical to that of the growth of branched NWs shown in Figs. 7.7, 7.10 and 7.11, except in this case, Ga is added for 15 min after the growth under an As flux. The wire does not show branches due to the suppression of the diffusion of the Bi droplet at the growth front by the formation of the outermost GaAs shell [50, 51]. The wire surface is corrugated not flat, which is induced by the introduction of Bi in the GaAsBi shell [51].

Figure 7.12b illustrates the monochromatic cathodoluminescence (CL) images of three detection energies superimposed on the corresponding SEM image of the NW. At the selected detection energies, which correspond to the high- and low-energy spectral positions, the emission originates from the lower part of the NWs because it exhibits a smaller twin defect density with dominant ZB segments compared to its upper part [39, 51]. Thus, the luminescence essentially originates from the lower part of the NWs, which exhibits a lower defect density and consists mainly of ZB segments. The CL obtained at 1.30 eV is observed at well-resolved NW segments. Carrier transfer occurs in the direction of the lower energy states; thus, the segment structure in the CL image is not visible in the images obtained at lower detection

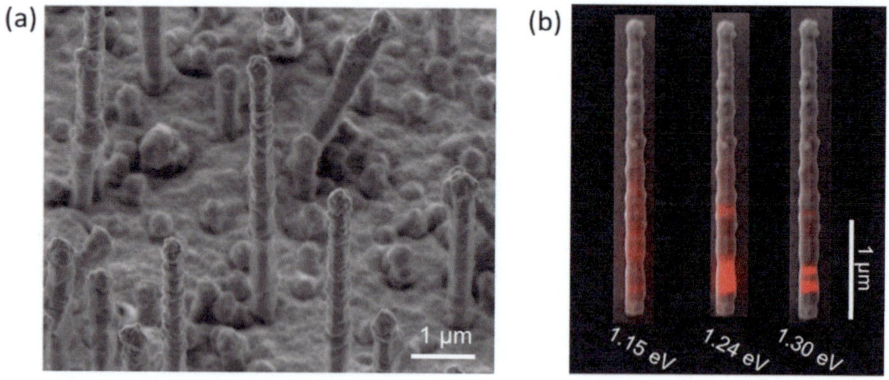


**Fig. 7.12** **a** Scanning electron microscopy (SEM) images of the 45°-tilted GaAs/GaAsBi nanowires. **b** and **c** Monochromatic cathodoluminescence (CL) maps superimposed on the respective SEM images for Wire 1 and Wire 2, respectively, obtained at the indicated CL detection energies. Both CL spectra and images have been acquired at 20 K. Reproduced from Ref. 51 with a permission from the American Chemical Society

energies(1.24 and 1.15 eV, Fig. 7.12b) [63]. These phenomena can be attributed to the presence of localized states due to the Bi introduction, which induces a morphological roughening and an inhomogeneous distribution of the ternary compositions. The localization of the carriers due to the structural and compositional inhomogeneities within the wire can be attributed to the coexistence of spatially modulated bandgap energies, defect states, and states at the GaAs/GaAsBi interface [64–66]. These localized light emissions, which could be attributed to a quantum scale carrier confinement, may indicate the presence of a quantum-structure formation.

## 7.8 Polymorphism in Nanowires

In recent years, the control of the polymorphism and stability of crystals has been significantly developed. Crystals inside wires grow through a very different mechanism from those in thin films; thus, their crystalline structures are also different. In case of GaAs, thin films can be grown in the stable ZB phase. However, in NWs, a mixture of metastable WZ structures can be observed. Recent studies have indicated that this phenomenon can be controlled and induced intentionally [67–70]. The WZ structure can increase the polarity of the crystal. Thus, the optical performance of materials (depending on the material) can be enhanced. This phenomenon has been further studied after the discovery of the WZ optically efficient direct bandgap (GaP) [71]. For example, Si and Ge are elements that have optically inefficient indirect bandgap; however, when they are combined and added to a hexagonal crystal, their bandgap changes into a direct one, enhancing their optical luminescent and absorptive efficiency [72–74].

### *7.8.1 Polytypism in GaAs/GaNAs Nanowires*

Examples of polytypism in the GaAs/GaNAs NWs are discussed here [75]. The structural characteristics of the NWs were investigated using SEM and XRD. Cross-sectional STEM and electron backscattered diffraction (EBSD) were conducted on sliced single NW samples. The samples were prepared by focused ion-beam processing. EBSD was conducted using a JSM-6500F TSL microscope from JEOL (Japan) operated at 30 kV with an injection current of 30 nA and equipped with a Digi View slow scan camera for orientation imaging microscopy (OIM). Images were analyzed using OIM data collection and OIM analysis software from TSL solutions (Japan).

Figure 7.13a shows the radial cross-sectional BF-STEM and enlarged filtered HAADF-STEM images of a GaAs/GaNAs core–shell NW nominally containing 2% N [75]. In the BF images, horizontal dark lines, which are assigned to twin defects, are observed. In the HAADF images, the coexistence of ZB and WZ polytypes was

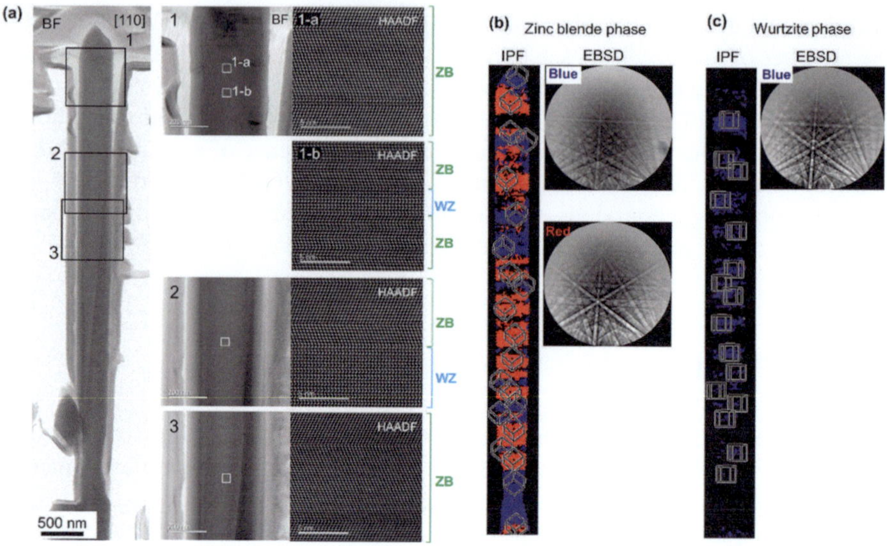


**Fig. 7.13** **a** Radial cross-sectional bright-field scanning transmission electron microscopy (BF-STEM) and enlarged filtered high-angle annular dark field (HAADF)-STEM images of the GaNAs nanowire nominally containing 2% N observed from the [011] direction. The areas shown in the individual images are numbered as indicated in the left-most whole wire image. The rectangle shown in the individual numbered BF images indicates the area magnified in the HAADF images. **b** Electron backscattered diffraction (EBSD) results of the GaAs/GaNAs/GaAs nanowire nominally containing 2% N. Phase map, pole figures of the respective orientations, and EBSD pattern of the **b** zinc blende and **c** wurtzite phases. The crystal structure at the local area is schematically illustrated for the inverse pole figure map. Reproduced from Ref. [75] with a permission from the Institute of Physics

observed. As shown in the Fig. 7.13, ZB and WZ polytypisms can be observed within the NW, and they randomly switch along the length of the wire [63, 76].

Figure 7.13b shows the EBSD results for the NW, i.e., the EBSD radial cross-sectional inverse pole Fig. 7.13 mapping and EBSD patterns of the ZB and WZ phases in the developed NW. As shown in Fig. 7.13a, the ZB and WZ polytypisms coexist within the NW. Based on these results, the fractions of the major ZB and minor WZ phases were determined to be 80% and 20%, respectively [75]. The ZB phase shows two components with opposite directions. This can be attributed to the presence of twin defects at the borders of the two ZB phases aligning in different directions. In contrast, the WZ phase does not include these different orientations; thus, it is a single phase.

### 7.8.2 Distributions of Stable ZB and Metastable WZ Phases

Figure 7.14a shows an image quality (IQ) map, phase map, and grain-size distributions of the ZB and WZ segments of a single pure GaAs NW (designated sample GaAs1) [77]. The phase map indicates that the sample is composed of 80% ZB and 17% WZ. The ZB is distributed throughout the wire, whereas the WZ is concentrated at the top of the wire. The ZB and WZ alternate along the wire length [68, 76, 78]. ZB and WZ segments are observed and their distribution in the horizontal directions is consistent with the wire diameter. Specifically, the segments mainly alternate in the vertical direction. In the horizontal direction, each ZB and WZ segment is distributed throughout the diameter.

Figure 7.14b shows an EBSD example of another GaAs NW (sample GaAs2). In contrast to GaAs1, sample GaAs2 has an inhomogeneous diameter. The effect of shadowing during the wire growth is considered because the density of the NWs is high ($2 \times 10^8$ cm$^{-2}$), which results in the observed narrowing of the wire diameter at the bottom side [79]. The large green fraction in the phase map (Fig. 7.14b) indicates that GaAs2 contains a larger WZ proportion than GaAs1 (Fig. 7.14a). In this NW, the proportions of ZB and WZ are 53% and 47%, respectively. Twin defects formed in the ZB segments of GaAs2. Even though the wire diameter is not homogeneous like

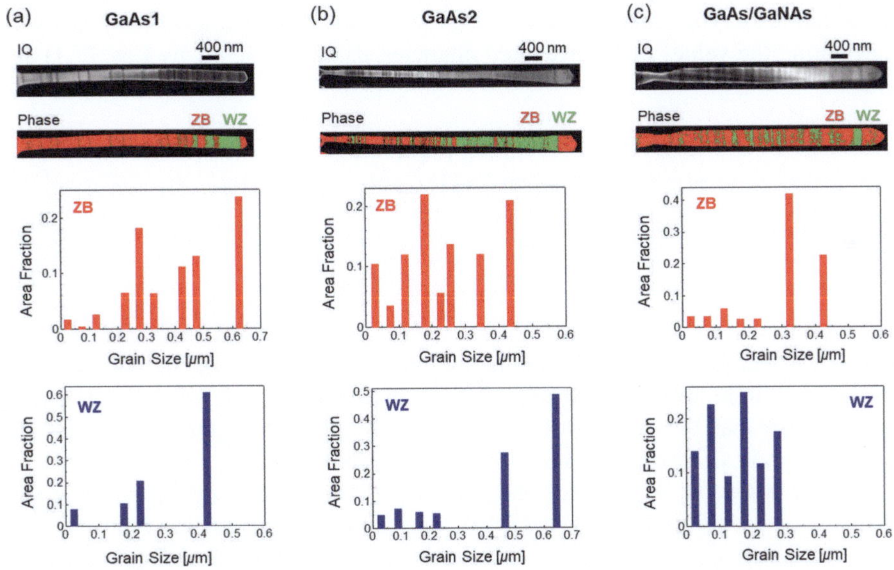


**Fig. 7.14** Electron backscattered diffraction of sample GaAs1. **a** Image quality (IQ) map and **b** phase map of zinc blende (ZB) and wurtzite (WZ). The red area is ZB and the green area is WZ. Grain-size distribution of the ZB and WZ segments of **a** and **b** GaAs1, **c** and **d** GaAs2, and **e** and **f** a GaAs/GaNAs/GaAs NWs. Reproduced from Ref. [75] with a permission from the Institute of Physics

that of GaAs1, both the ZB and WZ segments of GaAs2 are distributed throughout the diameter in the horizontal direction.

Figure 7.14c shows the EBSD results of a GaAs/GaNAs/GaAs core–multishell NW containing 2% N in the GaNAs shell [46, 75]. The diameter of the NW ($\approx$370 nm) is homogeneous throughout its length. The phase map in Fig. 7.14c shows that this NW is 80% ZB and 20% WZ. The phase map in the Fig. 7.14c shows that the dispersion of the WZ segments differs from those in the GaAs1 and GaAs2 NWs (Fig. 7.14a and b). The WZ segments in the GaAs NWs are horizontally distributed throughout the NWs, and their horizontal widths are almost equivalent to the NW diameters. In contrast, the width of the WZ segments in the GaAs/GaNAs/GaAs NWs (green areas in the phase map, Fig. 7.14c) is mostly smaller than the NW diameter. These characteristics suggest that the outermost surface of the NWs is mainly ZB and that the WZ segments are covered by ZB segments. Moreover, these results indicate that the polytypism of the NWs with a GaNAs shell differs from that of pure GaAs NWs. The ZB and WZ segments were roughly dispersed in the GaNAs NWs with no horizontal switching, resulting in an irregular distribution in the radial directions unlike that observed in the pure GaAs NWs.

The bottom parts of Fig. 7.14a–c summarizes the grain-size distribution of the investigated NWs, which was analyzed by the OIM analysis software [75]. The grain-size distribution is a histogram of each crystal grain, which is identified according to the grain maps in Fig. 7.14a–c. This histogram is used to calculate the equivalent-circle diameter. The grain-size distribution was obtained by computational analysis, calculating the values of the averaged diameter, $\overline{d_n}$, and area, $\overline{d_a}$, by Eqs. (7.1) and (7.2).

$$\overline{d_n} = \frac{1}{N}\sum_{i=1}^{N} d_i \tag{7.1}$$

$$\overline{d_a} = \frac{1}{N}\sum_{i=1}^{N} A_i \times d_i \tag{7.2}$$

where $N$, $d_{\mathrm{i}}$, and $A_{\mathrm{i}}$ are the total number of grains, diameters of the individual grains treated as a circle structure, and area ratio of individual grains, respectively. If a grain size is larger than the diameter, the grain size strongly reflects the NW length. The extracted grain size then becomes larger than the diameter.

In case of GaAs1 (Fig. 7.14a), the grain-size distribution of ZB is larger than the diameter of the NW (0.30 μm). At 0.30–0.45 μm, the area fraction was approximately 65%. The grain-size distribution of WZ shows a dominant grain size of approximately 0.42 μm, which is also larger than the diameter, and its proportion is greater than 60%. Overall, both the ZB and WZ structures have grain-size distributions larger than their wire diameter (0.30–0.45 μm). This result suggests that most of the polytype grains are dispersed over the whole wire diameter in the horizontal direction.

In sample GaAs2 (Fig. 7.14b), the ZB distribution significantly changes, reflecting the variation in the NW diameter. The WZ polytypism shows a large fraction of the grain-size distribution at 0.4–0.6 μm, which is larger than the NW diameter. This result suggests that the grain sizes of both the ZB and WZ polytypes are larger than the NW diameter, which is similar to the aforementioned results of the GaAs1 sample.

Figure 7.14c shows the grain-size distribution of the GaAs/GaNAs/GaAs NW containing 2% N. The ZB distribution shows a Gaussian-like distribution with a peak at approximately 0.30–0.40 μm and an area fraction greater than 60%. This peak position is equivalent to the NW diameter. In contrast, the grain size of WZ exhibits a homogenous variation within 0.1–0.3 μm. This grain size of the WZ phase is significantly different from that of the other GaAs NW samples shown in Fig. 7.14a and b.

The phase map of the GaAs/GaNAs/GaAs core–shell NW also shows features that clearly differ from those of the GaAs NWs. The grains are horizontally distributed over the wire diameter in the pure GaAs NWs. In contrast, the horizontal dispersion of the WZ grain is mostly smaller than the wire diameter, and it exhibits a disordered dispersion in the GaNAs sample. The diameter of the GaNAs NW corresponds to >60% of the ZB grain size. However, the WZ grain size is smaller than the diameter. This result suggests that introducing the GaNAs shell reduces the WZ polytypism, which is in good agreement with previous studies, reporting a WZ-to-ZB transition observed in GaAs NWs buried by AlGaAs [80].

### 7.8.3 Polytypism Transfer Across the Nanowire Axis

Figure 7.15 shows the STEM results of the GaAs/GaNAs/GaAs core–multishell NWs [75]. As shown in Fig. 7.15a, the area near the top of the NW was observed from the [110] direction to determine the ZB and WZ polytypes based on the atomic arrangements in the NW. Crystalline defects are shown as dark lines in Fig. 7.15b. The borders of the dark lines are enlarged in Fig. 7.6c and d, which shows the filtered HAADF images at the area marked by rectangles in Fig. 7.15b. These images show that the dark lines separate the different crystalline polytypes in ZB and WZ. Most of Fig. 7.15c represents the GaNAs shell. The left side of the Fig. 7.15c represents the inner-core side of the NW, which exhibits a WZ structure. In the right side of the image, which corresponds to the outer sidewall side of the NW, the wire exhibits ZB segments. Therefore, the dark area observed in Fig. 7.15b is the area where a transition from a WZ polytypism to ZB occurred, whereas the polytype was preserved in the area toward the outer edge of the NW. Figure 7.15d shows the outer GaAs shell. NW can be distorted by strain when the shell layer contains lattice mismatch with the GaAs core [52]. The formation of WZ and ZB polytypism can be mainly attributed to the VLS growth of the GaAs core, and it is determined by the formation

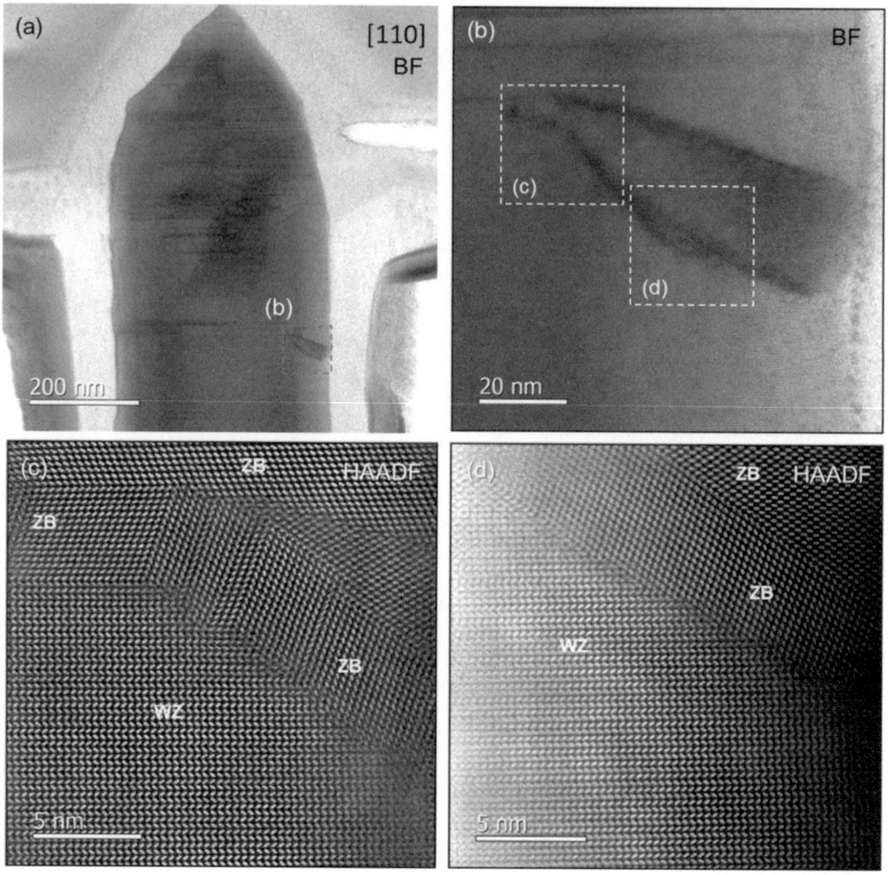


**Fig. 7.15** Scanning transmission electron microscopy (STEM) investigation of a GaAs/GaNAs/GaAs core–multishell NW: **a** Overview and observation areas, an enlarged image at the tip of the NW, and typical axial cross-sectional image of the NW. **b** Enlarged bright-field (BF)-STEM image for the area indicated by the rectangle in (**a**). The vertical dashed line indicates the interface between the GaNAs shell and the outermost GaAs shell, which is deduced from the contrast difference between them. **c**, **d** Filtered high-angle annular dark field (HAADF)-STEM images of the area indicated in (**b**) at the interface, showing the wurtzite and zinc blende areas from the inner core to the outer shell. Reproduced from Ref. [25] with a permission from the Institute of Physics

angle between the growth front and the liquid droplet at the tip [81, 82]. Based on these results, the decrease in the WZ polytypism can be attributed to the growth of the GaNAs shell, which has a small lattice constant, on the GaAs core. The resulting strain might decrease the WZ polytypism, which was observed by XRD, EBSD, and STEM (Fig. 7.16).

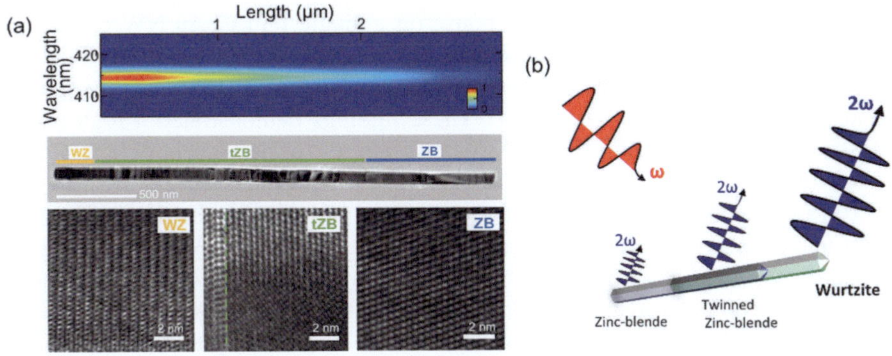


**Fig. 7.16** **a** Spatial and spectral distribution of the second-harmonic generation intensity measured along the single polytypic GaAs nanowire (NW) and transmission electron microscopy (TEM) micrograph of the studied GaAs NW. Yellow, green, and blue bars mark the NW regions with predominantly wurtzite (WZ), twinned zinc blende (tZB), and ZB crystal structure, respectively. The figure shows high-resolution TEM images obtained from WZ, tZB, and ZB regions. The dashed green line in the tZB image indicates the location of the twinned plane. Reproduced from Ref. [83] with a permission from Wiley

## 7.9 Functional Benefits of Polymorphism: Strong Second-Harmonic Generation in GaAs Nanowires Using Crystal-Structure Engineering

Using nanostructures for applications such as nanoscale light sources is crucial for the efficient miniaturization and allows tailoring and enhancing nonlinear responses [83–85]. Considering that a dielectric NW represents a naturally formed cavity, strong localization of the fundamental light and highly directional second harmonic generation (SHG) can be achieved using Mie resonances due to the interference of the cavity modes [84] and integration of the dielectric NWs into plasmonic structures [86, 87]. An additional degree of freedom in the engineering of the SHG response is its ability to grow NWs with different lattice structures, allowing the introduction of nonlinearity in crystallographic polytypes that cannot be fabricated in bulk under conventional growth conditions. This allows the tailoring of a polarimetric SHG response determined by crystal symmetry [88].

Determining the origin of SHG in thin NWs, including GaAs-based NWs, is important. The difficulty in this process is because of the high sensitivity of nonlinear properties of surfaces, interfaces, and defects in NWs. Random twins and stacking faults, which are often present in NWs, also enhance the spontaneous polarization in WZ NWs [89]. These built-in electric fields may significantly affect the SHG efficiency [90, 91]. However, this effect on the nonlinear properties of GaAs NWs was still largely unexplored and was first investigated in Ref. 83. SHG conversion efficiency in thin WZ GaAs NWs was found to be considerably enhanced (by approximately 7 times and reaching $2.5 \times 10^{-5}$ $W^{-1}$) compared with that in ZB GaAs NWs with the same diameter. The higher nonlinear response in WZ was independently confirmed by mapping the SHG response on a single thin GaAs NW containing WZ and ZB segments. The enhancement of the SHG efficiency was attributed to a reduced surface symmetry and a strong built-in electric field caused by spontaneous polarization in the WZ lattice. These findings demonstrated a high potential of subwavelength-scale WZ GaAs nanostructures for next-generation nonlinear nanophotonics.

To further confirm the effect of the crystalline structure on the SHG efficiency, SHG mapping was conducted on individual polytypic GaAs NWs and the crystal structure was identified using high-resolution TEM (HRTEM). Figure 7.15b shows a representative TEM image of a thin polytypic NW. The NW has an almost straight shape with a diameter of approximately 80 nm. The NW crystallized in the WZ crystal phase at the bottom and then transformed to twinned ZB (tZB) in the middle part and pure ZB phase at the top part, which are marked by yellow, green, and blue bars, respectively. Figure 7.2c–e depict representative HRTEM images of these segments. The transformation of the crystal structure can be clearly observed in their FFT patterns shown in the insets. Based on the knowledge of the crystalline structures of this NW, its nonlinear response was examined by scanning the tightly focused excitation/detection spot along the NW axis. Figure 7.15a shows the obtained SHG map, which reflects spatial and spectral distribution of the SHG intensity. It was observed that the SHG intensity depends on the crystalline structure and significantly increases (by approximately 6.7 times) in the WZ region. These results provided a clear experimental proof indicating that the nonlinear response of thin WZ GaAs NWs is superior to that of their ZB counterparts.

The findings introduced here provide an effective strategy for using lattice engineering to enhance and manipulate the nonlinear optical response in subwavelength NWs.

## 7.10 Conclusions

In this chapter, the microtexture and polymorphism observed during the molecular-beam epitaxial growth of group III–V semiconductor nanostructures were discussed. First, the technical details and development of molecular-beam epitaxy, which can be used to develop atomically precise and highly pure semiconductor materials for advanced devices, were introduced. The technique and understanding of the crystal growth mechanism by VLS, especially for thin films and NWs, allows the synthesis of various nanostructures. Some examples of the studies conducted by our group on this technique were discussed. Fundamental aspects of the growth of GaAs/GaAsN core–shell heterostructured NWs were studied based on molecular-beam epitaxy. The Ga catalyst crystallizes during the interruption of the growth process under a high As pressure, and the growth then mainly progresses to increase the NW diameter, forming a wire shell. The shell diameter linearly increases according to the growth time and group III flux, following the growth mechanism of planar layers, which results in a precisely controlled core–shell structure.

The formation of Bi-induced nanostructures was controlled during the growth of GaAs/GaAsBi core–shell NWs. In addition to being a component of the NWs, Bi also acts as a surfactant and a NW growth catalyst. This study offers a novel pathway to obtain unexplored group III–V nanostructures by employing characteristic supersaturation of the catalyst droplets as well as structural modifications induced by strain and incorporating Be and Si into the host GaAs matrix, which is also dependent on crystalline defects and orientations. Under Ga-deficient conditions during the growth process, Bi accumulates on the vertex of the GaAs NW core and serves as a NW growth catalyst for the branched structures to grow in the azimuthal $< 112 >$ direction. A strong correlation is observed between Bi accumulation and stacking faults. Bi is preferentially incorporated on the (112) B surface of GaAs, leading to a spatially selective Bi incorporation into a confined area with a Bi concentration higher than 7%. The obtained GaAs/GaAsBi/GaAs heterostructure, which has an interface defined by the crystalline twin defects in a ZB structure, can be potentially applied to a quantum confined structure. The GaAs/GaAsBi coaxial multishell NWs exhibited characteristic surface morphology with strong roughening. The NW exhibited a predominantly ZB structure from the bottom to the middle part. The polytypic WZ structure creates denser twin defects in the upper part than in the bottom and middle parts of the NW. When 2% Bi is introduced in the system, the GaAsBi alloy, which has a small bandgap energy, was obtained and cathodoluminescence from the NWs was observed at room temperature. The presence of localized states energetically and spatially dispersed throughout the NW is confirmed by these low-temperature cathodoluminescence spectra and images. These localized light emissions, which can be attributed to a quantum carrier confinement, may reflect the formation of a quantum structure.

The crystal structures of GaAs and GaAs/GaNAs/GaAs core–multishell NWs were also investigated. The EBSD patterns statistically and microscopically revealed the ZB and WZ polytypism within the NWs and their polytype switches along the length. Twin defects are observed in the ZB segments in all NWs. The unique grain map and grain-size distribution show a decrease in the WZ segments in the GaAs/GaNAs/GaAs NW. The local area where the polytype switches from WZ in the inner-core side to ZB toward the outer-shell surface is microscopically observed. Thus, it is suggested that the WZ polytype in the GaAs NWs decreases because of the strain induced by the growth of the GaNAs shell, which has a small lattice constant.

Highly efficient SHG (reaching $2.5 \times 10^{-5}\ W^{-1}$) is observed within the subwavelength WZ GaAs NWs, which is approximately 7 times higher than that of their ZB counterparts. This enhancement can be mainly attributed to an axial built-in electric field induced by spontaneous polarization in the WZ lattice due to electric field-induced second-order nonlinear susceptibility. This phenomenon can be optically and may be electrically controlled. Thus, these findings provide an effective strategy for enhancing and manipulating the nonlinear optical response in subwavelength NWs using lattice engineering.

In summary, the understanding and subsequent use of crystalline structures and growth mechanisms of the complex structures of crystals enable further advancements in several applications. This study suggests that new functional crystals can be fabricated by understanding the reasons for these phenomena and reliably controlling and recreating them. These new crystals can be used in different fields. This will also lead to several novel discoveries and achievements.

## References


1. A.Y. Cho, J.R. Arthur, Molecular beam epitaxy. Prog. Solid State Chem. **10**, 157–191 (1975). https://doi.org/10.1016/0079-6786(75)90005-9
2. A. Marian, M.A. Herman, H. Sitter, *Molecular beam epitaxy—Fundamentals and current status*. Springer Series in Materials Science. Springer
3. D.D. Davis, Space environment and its effects on materials. Materials for space operations. NASA SP, **27**, 1–11 (1962)
4. C.H.L. Goodman, C. H. L., & Pessa, M. V., Atomic layer epitaxy. J. Appl. Phys. **60**(3), R65–R82 (1986). https://doi.org/10.1063/1.337344
5. J.S. Blakemore, Semiconducting and other major properties of gallium arsenide. J. Appl. Phys. **53**(10), R123–R181 (1982). https://doi.org/10.1063/1.331665
6. Adachi, S, *'GaAs and Related Materia"s.' GaAs and related materials*. in A. Sadao (Ed.). World Scientific Publishing, Pt. Ltd.
7. I. Vurgaftman, J.R. Meyer, L.R. Ram-Mohan, Band parameters for III–V compound semiconductors and their alloys. J. Appl. Phys. **89**(11), 5815–5875 (2001). https://doi.org/10.1063/1.1368156
8. I. Vurgaftman, J.R.N. Meyer, Band parameters for nitrogen-containing semiconductors. J. Appl. Phys. **94**(6), 3675–3696 (2003). https://doi.org/10.1063/1.1600519
9. Harrison P, Valavanis A, Quantum wells, wires and dots: Theoretical and computational physics of semiconductor nanostructures. John Wiley & Sons
10. F. Ishikawa, M. Höricke, U. Jahn, A. Trampert, K.H. Ploog, Molecular beam epitaxial growth window for high-quality (Ga, In)(N, As) quantum wells for long wavelength emission. Appl. Phys. Lett. **88**(19), 191115 (2006). https://doi.org/10.1063/1.2202113
11. R.S. Wagner, W.C. Ellis, Vapor–liquid-solid mechanism of single crystal growth. Appl. Phys. Lett. **4**(5), 89–90 (1964). https://doi.org/10.1063/1.1753975
12. M. Yazawa, M. Koguchi, K. Hiruma, Heteroepitaxial ultrafine wire-like growth of InAs on GaAs substrates. Appl. Phys. Lett. **58**(10), 1080–1082 (1991). https://doi.org/10.1063/1.104377
13. M. Yazawa, M. Koguchi, A. Muto, M. Ozawa, K. Hiruma, Effect of one monolayer of surface gold atoms on the epitaxial growth of InAs nanowhiskers. Appl. Phys. Lett. **61**(17), 2051–2053 (1992). https://doi.org/10.1063/1.108329
14. M. Yazawa, K. Haraguchi, M. Koguchi, T. Katsuyama, K. Hiruma, H. Ohta, Nanocolumns composed of GaAs-InAs jointed whiskers and SiO 2 covers. Appl. Phys. Lett. **65**(9), 1157–1158 (1994). https://doi.org/10.1063/1.113010
15. K. Hiruma, M. Yazawa, T. Katsuyama, K. Ogawa, K. Haraguchi, M. Koguchi, H. Kakibayashi, Growth and optical properties of nanometer-scale GaAs and InAs whiskers. J. Appl. Phys. **77**(2), 447–462 (1995). https://doi.org/10.1063/1.359026
16. T. Sato, K. Hiruma, M. Shirai, K. Tominaga, K. Haraguchi, T. Katsuyama, T. Shimada, Site-controlled growth of nanowhiskers. Appl. Phys. Lett. **66**(2), 159–161 (1995). https://doi.org/10.1063/1.113549
17. M. Koguchi, H. Kakibayashi, M. Yazawa, K.H. Kenji Hiruma, T.K. Toshio Katsuyama, Crystal structure change of GaAs and InAs whiskers from zinc-blende to wurtzite type. Jpn. J. Appl. Phys. **31**(7R), 2061 (1992). https://doi.org/10.1143/JJAP.31.2061
18. K.A. Dick, K. Deppert, M.W. Larsson, T. Mårtensson, W. Seifert, L.R. Wallenberg, L. Samuelson, Synthesis of branched 'nanotrees' by controlled seeding of multiple branching events. Nat. Mater. **3**(6), 380–384 (2004). https://doi.org/10.1038/nmat1133

19. M.J. Bierman, Y.K. Lau, A.V. Kvit, A.L. Schmitt, S. Jin, Dislocation-driven nanowire growth and Eshelby twist. Science **320**(5879), 1060–1063 (2008). https://doi.org/10.1126/science.1157131
20. B. Tian, P. Xie, T.J. Kempa, D.C. Bell, C.M. Lieber, Single-crystalline kinked semiconductor nanowire superstructures. Nat. Nanotechnol. **4**(12), 824–829 (2009). https://doi.org/10.1038/nnano.2009.304
21. L.J. Lauhon, M.S. Gudiksen, D. Wang, C.M. Lieber, Epitaxial core-shell and core-multishell nanowire heterostructures. Nature **420**(6911), 57–61 (2002). https://doi.org/10.1038/nature01141
22. M.S. Gudiksen, L.J. Lauhon, J. Wang, D.C. Smith, C.M. Lieber, Growth of nanowire superlattice structures for nanoscale photonics and electronics. Nature **415**(6872), 617–620 (2002). https://doi.org/10.1038/415617a
23. E. Garnett, L. Mai, P. Yang, Introduction: 1D nanomaterials/nanowires. Chem. Rev. **119**(15), 8955–8957 (2019). https://doi.org/10.1021/acs.chemrev.9b00423
24. M. Law, J. Goldberger, P. Yang, Semiconductor nanowires and nanotubes. Annu. Rev. Mater. Res. **34**(1), 83–122 (2004). https://doi.org/10.1146/annurev.matsci.34.040203.112300
25. P. Yang, R. Yan, M. Fardy, Semiconductor nanowire: What's next? Nano Lett. **10**(5), 1529–1536 (2010). https://doi.org/10.1021/nl100665r
26. R. Yan, D. Gargas, P. Yang, Nanowire photonics. Nat. Photonics **3**(10), 569–576 (2009). https://doi.org/10.1038/nphoton.2009.184
27. J.C. Johnson, H.J. Choi, K.P. Knutsen, R.D. Schaller, P. Yang, R.J. Saykally, Single gallium nitride nanowire lasers. Nat. Mater. **1**(2), 106–110 (2002). https://doi.org/10.1038/nmat728
28. E. Barrigón, M. Heurlin, Z. Bi, B. Monemar, L. Samuelson, Synthesis and applications of III–V nanowires. Chem. Rev. **119**(15), 9170–9220 (2019). https://doi.org/10.1021/acs.chemrev.9b00075
29. C. Jia, Z. Lin, Y. Huang, X. Duan, Nanowire electronics: From nanoscale to macroscale. Chem. Rev. **119**(15), 9074–9135 (2019). https://doi.org/10.1021/acs.chemrev.9b00164
30. K. Tomioka, M. Yoshimura, T. Fukui, A III–V nanowire channel on silicon for high-performance vertical transistors. Nature **488**(7410), 189–192 (2012). https://doi.org/10.1038/nature11293
31. C. Thelander, T. Mårtensson, M.T. Björk, B.J. Ohlsson, M.W. Larsson, L.R. Wallenberg, L. Samuelson, Single-electron transistors in heterostructure nanowires. Appl. Phys. Lett. **83**(10), 2052–2054 (2003). https://doi.org/10.1063/1.1606889
32. P. Krogstrup, H.I. Jørgensen, M. Heiss, O. Demichel, J.V. Holm, M. Aagesen, J. Nygard, Fontcuberta i Morral, A, Single-nanowire solar cells beyond the Shockley-Queisser limit. Nat. Photonics **7**(4), 306–310 (2013). https://doi.org/10.1038/nphoton.2013.32
33. P. Yu, J. Wu, S. Liu, J. Xiong, C. Jagadish, Z.M. Wang, Design and fabrication of silicon nanowires towards efficient solar cells. Nano Today **11**(6), 704–737 (2016). https://doi.org/10.1016/j.nantod.2016.10.001
34. Y. Araki, M. Yamaguchi, F. Ishikawa, Growth of dilute nitride GaAsN/GaAs heterostructure nanowires on Si substrates. Nanotechnology **24**(6), 065601 (2013). https://doi.org/10.1088/0957-4484/24/6/065601
35. N. Ahn, Y. Araki, M. Kondow, M. Yamaguchi, F. Ishikawa, Effects of growth interruption, As and Ga fluxes, and nitrogen plasma irradiation on the molecular beam epitaxial growth of GaAs/GaAsN core–shell nanowires on Si(111). Jpn. J. Appl. Phys. **53**(6), 065001 (2014). https://doi.org/10.7567/JJAP.53.065001
36. H. Hibi, M. Yamaguchi, N. Yamamoto, F. Ishikawa, Selective synthesis of compound semiconductor/oxide composite nanowires. Nano Lett. **14**(12), 7024–7030 (2014). https://doi.org/10.1021/nl503385g
37. J.H. Paek, T. Nishiwaki, M. Yamaguchi, N. Sawaki, Catalyst free MBE-VLS growth of GaAs nanowires on (111)Si substrate. Phys. Status Solidi C **6**(6), 1436–1440 (2009). https://doi.org/10.1002/pssc.200881520
38. O. Demichel, M. Heiss, J. Bleuse, H. Mariette, Fontcuberta i Morral, A., Impact of surfaces on the optical properties of GaAs nanowires. Appl. Phys. Lett. **97**(20), 201907 (2010). https://doi.org/10.1063/1.3519980

39. E. Dimakis, U. Jahn, M. Ramsteiner, A. Tahraoui, J. Grandal, X. Kong, O. Marquardt, A. Trampert, H. Riechert, L. Geelhaar, Coaxial multishell (In, Ga)As/GaAs nanowires for near-infrared emission on Si substrates. Nano Lett. **14**(5), 2604–2609 (2014). https://doi.org/10.1021/nl500428v
40. T. Tauchnitz, T. Nurmamytov, R. Hübner, M. Engler, S. Facsko, H. Schneider, M. Helm, E. Dimakis, Decoupling the two roles of Ga droplets in the self-catalyzed growth of GaAs nanowires on SiO x/Si (111) substrates. Cryst. Growth Des. **17**(10), 5276–5282 (2017). https://doi.org/10.1021/acs.cgd.7b00797
41. T. Tauchnitz, Y. Berdnikov, V.G. Dubrovskii, H. Schneider, M. Helm, E. Dimakis, A simple route to synchronized nucleation of self-catalyzed GaAs nanowires on silicon for sub-Poissonian length distributions. Nanotechnology **29**(50), 504004 (2018). https://doi.org/10.1088/1361-6528/aae361
42. M. Yoshikawa, K. Miura, Y. Iguchi, Y. Kawamura, Optical characterization of InGaAsN layers grown on InP substrates. J. Cryst. Growth **311**(7), 1745–1747 (2009). https://doi.org/10.1016/j.jcrysgro.2008.10.117
43. M. Uchiyama, F. Ishikawa, M. Kondow, Effect of plasma conditions on the growth of GaNAs by plasma-assisted molecular-beam epitaxy. Jpn. J. Appl. Phys. **48**(8), 081102 (2009). https://doi.org/10.1143/JJAP.48.081102
44. J. Paek, M. Yamaguchi, H. Amano, MBE–VLS growth of catalyst-free III–V axial heterostructure nanowires on (1 1 1)Si substrates. J. Cryst. Growth **323**(1), 315–318 (2011). https://doi.org/10.1016/j.jcrysgro.2010.11.124
45. A. Fontcuberta i Morral, C. Colombo, G. Abstreiter, J. Arbiol, J.R. Morante, Nucleation mechanism of gallium-assisted molecular beam epitaxy growth of gallium arsenide nanowires. Appl. Phys. Lett. **92**(6), 063112 (2008). https://doi.org/10.1063/1.2837191
46. M. Yukimune, R. Fujiwara, H. Ikeda, K. Yano, K. Takada, M. Jansson, W.M. Chen, I.A. Buyanova, F. Ishikawa, GaAs/GaNAs core-multishell nanowires with nitrogen composition exceeding 2%. Appl. Phys. Lett. **113**(1), 011901 (2018). https://doi.org/10.1063/1.5029388
47. M.C. Plante, R.R. LaPierre, Control of GaAs nanowire morphology and crystal structure. Nanotechnology **19**(49), 495603 (2008). https://doi.org/10.1088/0957-4484/19/49/495603
48. S. Chen, M. Jansson, J.E. Stehr, Y. Huang, F. Ishikawa, W.M. Chen, I.A. Buyanova, Dilute nitride nanowire lasers based on a GaAs/GaNAs core/shell structure. Nano Lett. **17**(3), 1775–1781 (2017). https://doi.org/10.1021/acs.nanolett.6b05097
49. S. Chen, M. Yukimune, R. Fujiwara, F. Ishikawa, W.M. Chen, I.A. Buyanova, Near-infrared lasing at 1 μm from a dilute-nitride-based multishell nanowire. Nano Lett. **19**(2), 885–890 (2019). https://doi.org/10.1021/acs.nanolett.8b04103
50. T. Matsuda, K. Takada, K. Yano, R. Tsutsumi, K. Yoshikawa, S. Shimomura, Y. Shimizu, K. Nagashima, T. Yanagida, F. Ishikawa, Controlling Bi-provoked nanostructure formation in GaAs/GaAsBi core-shell nanowires. Nano Lett. **19**(12), 8510–8518 (2019). https://doi.org/10.1021/acs.nanolett.9b02932
51. F. Ishikawa, Y. Akamatsu, K. Watanabe, F. Uesugi, S. Asahina, U. Jahn, S. Shimomura, Metamorphic GaAs/GaAsBi heterostructured nanowires. Nano Lett. **15**(11), 7265–7272 (2015). https://doi.org/10.1021/acs.nanolett.5b02316
52. T. Matsuda, K. Takada, K. Yano, S. Shimomura, F. Ishikawa, Strain deformation in GaAs/GaAsBi core-shell nanowire heterostructures. J. Appl. Phys. **125**(19), 194301 (2019). https://doi.org/10.1063/1.5092524
53. G. Zha, M. Li, Y. Yu, L. Wang, J. Xu, X. Shang, H. Ni, Z. Niu, Strain-driven synthesis of self-catalyzed branched GaAs nanowires. Appl. Phys. Lett. **102**(16), 163115 (2013). https://doi.org/10.1063/1.4803028
54. H.J. Joyce, J. Wong-Leung, Q. Gao, H.H. Tan, C. Jagadish, Phase perfection in zinc blende and wurtzite III−V nanowires using basic growth parameters. Nano Lett. **10**(3), 908–915 (2010). https://doi.org/10.1021/nl903688v
55. J.A. Steele, R.A. Lewis, J. Horvat, M.J.B. Nancarrow, M. Henini, D. Fan, Y.I. Mazur, M. Schmidbauer, M.E. Ware, S.Q. Yu, G.J. Salamo, Surface effects of vapour–liquid-solid driven Bi surface droplets formed during molecular-beam-epitaxy of GaAsBi. Sci. Rep. **6**, 28860 (2016). https://doi.org/10.1038/srep28860

56. R.B. Lewis, M. Masnadi-Shirazi, T. Tiedje, Growth of high Bi concentration GaAs 1−x Bi x by molecular beam epitaxy. Applied Phy. Lett, **101**(8). https://doi.org/10.1063/1.4748172.
57. W. Shu, X. Zhang, X. Liu, H. Huang, Y. Huang, X. Ren, First-principles in-vestigations of GaAs (112)-(2×2) surface reconstruction. IEEE Adv. Optoelec. Micro/Nano-optics. **1**
58. K. Jacobi, J. Platen, C. Setzer, J. Maárquez, L. Geelhaar, C. Meyne, W. Richter, A. Kley, P. Ruggerone, M. Scheffler, Morphology, surface core-level shifts and surface energy of the faceted GaAs(112)A and ($\bar{1}\bar{1}\bar{2}$)B surfaces. Surf. Sci. **439**(1–3), 59–72 (1999). https://doi.org/10.1016/S0039-6028(99)00713-X
59. A. Jenichen, C. Engler, Reconstructions and surface facets of the GaAs(112)A and (112)B surfaces: First-principles DFT supercell calculations. Surf. Sci. **608**, 204–211 (2013). https://doi.org/10.1016/j.susc.2012.10.011
60. F. Bastiman, A.R.B. Mohmad, J.S. Ng, J.P.R. David, S.J. Sweeney, Non-stoichiometric GaAsBi/GaAs (100) molecular beam epitaxy growth. J. Cryst. Growth **338**(1), 57–61 (2012). https://doi.org/10.1016/j.jcrysgro.2011.07.036
61. M. Yoshimoto, S. Murata, A. Chayahara, Y. Horino, J. Saraie, K. Oe, Metasta-ble GaAsBi alloy grown by molecular beam epitaxy. Jpn. J. Appl. Phys., **42**(2, No. 10B), L1235–L1237 (2003). https://doi.org/10.1143/JJAP.42.L1235
62. L. Wang, L. Zhang, L. Yue, D. Liang, X. Chen, Y. Li, P. Lu, J. Shao, S. Wang, Novel dilute bismide, epitaxy, physical properties and device application. Crystals **7**(3), 63 (2017). https://doi.org/10.3390/cryst7030063
63. U. Jahn, J. Lähnemann, C. Pfüller, O. Brandt, S. Breuer, B. Jenichen, M. Ramsteiner, L. Geelhaar, H. Riechert, Luminescence of GaAs nanowires consisting of wurtzite and zinc-blende segments. Phys. Review. Part B **85**(4), 045323 (2012). https://doi.org/10.1103/PhysRevB.85.045323
64. Y.I. Mazur, V.G. Dorogan, M. Benamara, M.E. Ware, M. Schmidbauer, G.G. Tarasov, S.R. Johnson, X. Lu, S.-Q. Yu, T. Tiedje, G.J. Salamo, Effects of spatial confinement and layer disorder in photoluminescence of GaAs 1–x Bi x /GaAs heterostructures. J. Physics. Part D **46**(6), 065306 (2013). https://doi.org/10.1088/0022-3727/46/6/065306
65. V. Bahrami-Yekta, T. Tiedje, M. Masnadi-Shirazi, MBE growth optimization for GaAs 1–x Bi x and dependence of photoluminescence on growth temperature. Semicond. Sci. Technol. **30**(9), 094007 (2015). https://doi.org/10.1088/0268-1242/30/9/094007
66. S. Mazzucato, H. Lehec, H. Carrère, H. Makhloufi, A. Arnoult, C. Fontaine, T. Amand, X. Marie, Low-temperature photoluminescence study of exciton recombination in bulk GaAsBi. Nanoscale Res. Lett. **9**(1), 19 (2014). https://doi.org/10.1186/1556-276X-9-19
67. H.J. Joyce, J. Wong-Leung, Q. Gao, H.H. Tan, C. Jagadish, Phase perfection in zinc Blende and Wurtzite III-V nanowires using basic growth parameters. Nano Lett. **10**(3), 908–915 (2010). https://doi.org/10.1021/nl903688v
68. M. Hjort, S. Lehmann, J. Knutsson, R. Timm, D. Jacobsson, E. Lundgren, K.A. Dick, A. Mikkelsen, Direct imaging of atomic scale structure and electronic properties of GaAs wurtzite and zinc blende nanowire surfaces. Nano Lett. **13**(9), 4492–4498 (2013). https://doi.org/10.1021/nl402424x
69. T. Rieger, M.I. Lepsa, T. Schäpers, D. Grützmacher, Controlled wurtzite inclusions in self-catalyzed zinc blende III–V semiconductor nanowires. J. Cryst. Growth **378**, 506–510 (2013). https://doi.org/10.1016/j.jcrysgro.2012.12.035
70. S. Lehmann, D. Jacobsson, K.A. Dick, Crystal phase control in GaAs nanowires: Opposing trends in the Ga- and As-limited growth regimes. Nanotechnology **26**(30), 301001 (2015). https://doi.org/10.1088/0957-4484/26/30/301001
71. S. Assali, I. Zardo, S. Plissard, D. Kriegner, M.A. Verheijen, G. Bauer, A. Meijerink, A. Belabbes, F. Bechstedt, J.E. Haverkort, E.P.A.M. Bakkers, Direct band gap wurtzite gallium phosphide nanowires. Nano Lett. **13**(4), 1559–1563 (2013). https://doi.org/10.1021/nl304723c
72. E.M.T. Fadaly, A. Dijkstra, J.R. Suckert, D. Ziss, M.A.J. Van Tilburg, C. Mao, Y. Ren, V.T. van Lange, K. Korzun, S. Kölling, M.A. Verheijen, D. Busse, C. Rödl, J. Furthmüller, F. Bechstedt, J. Stangl, J.J. Finley, S. Botti, J.E.M. Haverkort, E.P.A.M. Bakkers, Direct-bandgap emission from hexagonal Ge and SiGe alloys. Nature **580**(7802), 205–209 (2020). https://doi.org/10.1038/s41586-020-2150-y

73. X. Cartoixà, M. Palummo, H.I.T. Hauge, E.P.A.M. Bakkers, R. Rurali, Optical emission in hexagonal SiGe nanowires. Nano Lett. **17**(8), 4753–4758 (2017). https://doi.org/10.1021/acs.nanolett.7b01441
74. C. Rödl, J. Furthmüller, J.R. Suckert, V. Armuzza, F. Bechstedt, S. Botti, Accurate electronic and optical properties of hexagonal germanium for optoelectronic applications. Phys. Rev. Mater. **3**(3), 034602 (2019). https://doi.org/10.1103/PhysRevMaterials.3.034602
75. M. Yukimune, R. Fujiwara, T. Mita, F. Ishikawa, Polytypism in GaAs/GaNAs core–shell nanowires. Nanotechnology **31**(50), 505608 (2020). https://doi.org/10.1088/1361-6528/abb904
76. D. Spirkoska, J. Arbiol, A. Gustafsson, S. Conesa-Boj, F. Glas, I. Zardo, M. Heigoldt, M.H. Gass, A.L. Bleloch, S. Estrade, M. Kaniber, J. Rossler, F. Peiro, J.R. Morante, G. Abstreiter, L. Samuelson, Fontcuberta i Morral, A., Structural and optical properties of high quality zinc-blende/wurtzite GaAs nanowire heterostructures. Physical Review. Part B **80**(24), 245325 (2009). https://doi.org/10.1103/PhysRevB.80.245325
77. S.I. Wright, M.M. Nowell, EBSD image quality mapping. Microsc. Microanal. **12**(1), 72–84 (2006). https://doi.org/10.1017/S1431927606060090
78. M. Hjort, P. Kratzer, S. Lehmann, S.J. Patel, K.A. Dick, C.J. Palmstrøm, R. Timm, A. Mikkelsen, Crystal structure induced preferential surface alloying of Sb on wurtzite/zinc blende GaAs nanowires. Nano Lett. **17**(6), 3634–3640 (2017). https://doi.org/10.1021/acs.nanolett.7b00806
79. J.A. Czaban, D.A. Thompson, R.R. LaPierre, GaAs core–shell nanowires for photovoltaic applications. Nano Lett. **9**(1), 148–154 (2009). https://doi.org/10.1021/nl802700u
80. G. Patriarche, F. Glas, M. Tchernycheva, C. Sartel, L. Largeau, J.C. Harmand, G.E. Cirlin, Wurtzite to zinc blende phase transition in GaAs nanowires induced by epitaxial burying. Nano Lett. **8**(6), 1638–1643 (2008). https://doi.org/10.1021/nl080319y
81. F. Glas, J.C. Harmand, G. Patriarche, Why does wurtzite form in nanowires of III-V zinc blende semiconductors? Phys. Rev. Lett. **99**(14), 146101 (2007). https://doi.org/10.1103/PhysRevLett.99.146101
82. L. GüniatGüniat, P. Caroff, I. Fontcuberta, A. Morral, Vapor Phase growth of semiconductor nanowires: Key developments and open questions. Chem. Rev. **119**(15), 8958–8971 (2019). https://doi.org/10.1021/acs.chemrev.8b00649
83. B. Zhang, J.E. Stehr, P.P. Chen, X. Wang, F. Ishikawa, W.M. Chen, I.A. Buyanova, Anomalously strong second-harmonic generation in GaAs nanowires via crystal-structure engineering. Adv. Func. Mater. **31**(36), 2104671 (2021). https://doi.org/10.1002/adfm.202104671
84. L. Xu, G. Saerens, M. Timofeeva, D.A. Smirnova, I. Volkovskaya, M. Lysevych, R. Camacho-Morales, M. Cai, K. Zangeneh Kamali, L. Huang, F. Karouta, H.H. Tan, C. Jagadish, A.E. Miroshnichenko, R. Grange, D.N. Neshev, M. Rahmani, Forward and backward switching of nonlinear unidirectional emission from GaAs nanoantennas. ACS Nano **14**(2), 1379–1389 (2020). https://doi.org/10.1021/acsnano.9b07117
85. G. Saerens, I. Tang, M.I. Petrov, K. Frizyuk, C. Renaut, F. Timpu, M. Reig Escalé, I. Shtrom, A. Bouravleuv, G. Cirlin, R. Grange, M. Timofeeva, Engineering of the second-harmonic emission directionality with III–V semiconductor rod nanoantennas. Laser Photonics Rev. **14**(9), 2000028 (2020). https://doi.org/10.1002/lpor.202000028
86. X. Liu, Q. Zhang, W.K. Chong, J.N. Yip, X. Wen, Z. Li, F. Wei, G. Yu, Q. Xiong, T.C. Sum, Cooperative enhancement of second-harmonic generation from a single CdS nanobelt-hybrid plasmonic structure. ACS Nano **9**(5), 5018–5026 (2015). https://doi.org/10.1021/nn5072045
87. J. Shi, Y. Li, M. Kang, X. He, N.J. Halas, P. Nordlander, S. Zhang, H. Xu, Efficient second harmonic generation in a hybrid plasmonic waveguide by mode interactions. Nano Lett. **19**(6), 3838–3845 (2019). https://doi.org/10.1021/acs.nanolett.9b01004
88. M. Timofeeva, A. Bouravleuv, G. Cirlin, I. Shtrom, I. Soshnikov, M. Reig Escalé, A. Sergeyev, R. Grange, Polar second-harmonic imaging to resolve pure and mixed crystal phases along GaAs nanowires. Nano Letters, **16**(10), 6290–6297 (2016). https://doi.org/10.1021/acs.nanolett.6b02592
89. B. Bauer, J. Hubmann, M. Lohr, E. Reiger, D. Bougeard, J. Zweck, Direct detection of spontaneous polarization in wurtzite GaAs nanowires. Appl. Phys. Lett. **104**(21), 211902 (2014). https://doi.org/10.1063/1.4880209

90. T.A. Germer, K.W. Kołasin-acuteski, J.C. Stephenson, L.J. Richter, Depletion-electric-field-induced second-harmonic generation near oxidized GaAs(001) surfaces. Physical Review. Part B **55**(16), 10694–10706 (1997). https://doi.org/10.1103/PhysRevB.55.10694
91. Y.-M. Chang, Y.-L. Hong, S. Gwo, Direct probe of the built-in electric field of Mg-doped a-plane wurtzite InN surfaces with time-resolved electric-field-induced second harmonic generation. Appl. Phys. Lett. **93**(13), 131106 (2008). https://doi.org/10.1063/1.2979238

# Chapter 8
# Crystal Phase Engineering in Self-Assisted GaAs Nanowires

## Self-Assisted GaAs Nanowires

**T. Dursap, C. Botella, P. Regreny, N. Chauvin, M. Bugnet, A. Danescu, and J. Penuelas**

**Abstract** The study focuses on the significance of cubic Zinc-Blende (ZB) and hexagonal wurtzite (WZ) structures in GaAs nanowires. GaAs typically crystallizes in the ZB structure in its bulk phase, while the WZ phase is observed only in NWs. This structural variation is crucial for applications like integrated light sources on silicon. However, the gold-catalyzed growth method for WZ GaAs NWs is incompatible with silicon electronics due to gold-induced trap states, necessitating a self-assisted growth method to avoid contamination. Recent advancements in molecular beam epitaxy (MBE) have enabled the controlled formation of WZ phases in self-assisted GaAs NWs. The chapter highlights the use of reflection high-energy electron diffraction (RHEED) for real-time monitoring of ZB and WZ phase growth. The nucleation mechanisms, influenced by the catalyst droplet's size and contact angle, are explored, establishing critical angles for phase determination. The chapter is divided into experimental results utilizing RHEED for WZ structure formation and a growth model to estimate the contact angle of the liquid phase. The findings underscore the potential for controlled WZ segment growth, supported by numerical simulations.

The occurrence of the cubic Zinc-Blende (ZB) and hexagonal wurtzite (WZ) structures appears as an important feature of III-V semiconductors, which determines their properties. It is now well known that GaAs, a technologically relevant and massively employed material, crystallizes in the ZB structure in its bulk phase, whereas WZ nucleation has only been demonstrated to occur in nanowires (NWs) [1]. Despite this

T. Dursap · C. Botella · P. Regreny · N. Chauvin · A. Danescu · J. Penuelas (✉)
UMR 5270, Univ Lyon, Ecole Centrale de Lyon, CNRS, INSA Lyon, Université Claude Bernard Lyon 1, CPE Lyon, INL, 69130 Ecully, France
e-mail: jose.penuelas@ec-lyon.fr

M. Bugnet
CNRS, INSA Lyon, Université Claude Bernard Lyon 1, MATEIS, UMR 5510, 69621 Villeurbanne, France



F. Ishikawa et al. (eds.), *Engineering Crystal Habit*, Topics in Applied Physics 152,
https://doi.org/10.1007/978-981-96-0266-7_8

limitation, the growth of GaAs NWs represents an advantageous outcome for applications such as integrated light sources on silicon. Moreover, hexagonal GaAs NWs have recently been shown to be crucial building blocks for the growth of hexagonal Ge and light emitting SiGe alloys [2].

As promising as these materials might be, the current reports show that GaAs NWs face a bottleneck limiting their upgrade from prototypes to applied devices. Indeed, WZ GaAs NWs are obtained using the Au-catalyzed growth [3–7], which is incompatible with the silicon electronic industry, due to trap states induced by gold in silicon [8, 9], thus preventing the integration of these NWs on Si substrates. Self-assisted growth thus appears as the optimal solution to overcome the gold contamination of the NWs. Recent prototypical growth reactors have been used to highlight the nucleation mechanism of the different crystal structures, and to control the phase on a single self-assisted GaAs NW [10]. The extended growth of the WZ phase in self-assisted GaAs NWs by molecular beam epitaxy (MBE) has only been investigated in recent studies, leading to the achievement of a 1.8 μm-long pure WZ segment by Jansen's group [11] on pre-patterned substrates by controlling the Ga flux during the growth. The ability to control and form on-demand the WZ phase in self-assisted GaAs NWs has become a major challenge in the NW community.

Most of the in situ and real time tools used to characterize and tune the crystal phases of the self-assisted GaAs NWs during the growth are often unique and only available in few selected laboratories, such as in situ X-ray diffraction (XRD) [12] or in situ transmission electron microscope (TEM) [10, 13, 14]. Such prototypical instruments are undeniably primordial to foster the understanding of the growing mechanisms of individual nano-objects, but they are not adaptable to the case of MBE growth on large substrates. Nevertheless, the reflection high-energy electron diffraction (RHEED) technique is commonly coupled with MBE reactors, and can be used as a powerful tool to characterize and follow the structural evolution during the growth of NWs using the distinct ZB and WZ diffraction diagrams. Despite this capability, only a few recent studies reported substantial RHEED observations applied to self-assisted GaAs NWs growth [15–18]. In this chapter, the RHEED technique is employed as a tool to monitor in real time the growth of the cubic ZB and hexagonal WZ phases in self-assisted GaAs NWs.

The formation of the ZB or WZ phase in GaAs NWs is dictated by the nucleation of the new monolayer (ML), and more particularly by its localization, either inside the liquid droplet or at the triple phase line (TPL), respectively [19–23]. The localization of nucleation depends on the size of the catalyst droplet, and thus on its contact angle $\beta$ formed regarding the top facet of the NW [10, 13, 14, 24]. The existence of two critical angles $\beta_{c1}$ and $\beta_{c2}$, between which the formation of WZ is observed and equal to about 125° and 90°, respectively, was then reported. The ZB nucleation occurs when the droplet size leads to angles larger than $\beta_{c1}$ or smaller than $\beta_{c2}$.

This chapter is dedicated to results obtained in the framework of already-published studies [25, 26] and are divided into two parts. Firstly, experimental results showing the RHEED analysis and control operated to form the WZ structure will be presented, while the second part will be focused on the growth model developed to approximate the value of the contact angle $\beta_{c1}$. The controlled growth of an extended, pure WZ segment, supported by numerical simulations is presented and discussed.

## 8.1 RHEED Applied to the Nanowire Growth

All the NWs presented in this chapter were synthetized on $1 \times 1$ cm$^2$ Si(111) substrates by MBE.

Each substrate was cleaned during 5 min in both acetone and ethanol, and degassed at 200°C in ultra-high vacuum before introduction inside the MBE reactor. The native oxide was preserved to enable the self-assisted growth. To form the Ga droplets, the substrate was heated up to 450 °C and 1 monolayer (ML) of Ga was pre-deposited [27, 28] at a rate of 0.5 ML.s$^{-1}$, quoted in units of equivalent growth rates of GaAs 2D layers measured by RHEED oscillations on a GaAs substrate. Then, the substrate temperature was increased up to the growth temperature. Finally, the opening of the Ga and As fluxes initiated the growth of the NWs.

The whole growth process was systematically monitored using in situ RHEED. Indeed, a RHEED diagram can provide a pattern showing the distinctive contribution of both ZB and WZ phases, along the appropriate azimuth. Figure 8.1 shows the typical RHEED pattern measured along the $[1\bar{1}0]$azimuth of self-assisted GaAs NWs. The indexed WZ and ZB (including the twinned ZB) spots are superimposed on the image, highlighted with the red squares and green circles, respectively. The position of each spot is in agreement with the epitaxial growth of NWs on Si(111). Indeed, the GaAs [111] and $[1\bar{1}0]$ directions for the ZB and the GaAs [0001] and $[10\bar{1}0]$ directions for the WZ are parallel to the Si [111] and $[1\bar{1}0]$ directions, respectively.

Despite the fact that self-assisted GaAs NWs grow in the ZB phase, it should be noticed that a contribution of the WZ structure is visible on the RHEED diagram shown in Fig. 8.1. Indeed, a short WZ segment can be synthetized at the end of the growth by consuming the liquid catalyst and is responsible of the RHEED WZ spots. This Ga droplet consumption occurs when the Ga shutter is closed, supplying only the As flux, thus leading to a progressive diminution of the droplet size and of the contact angle, as illustrated in the simplified schematic in Fig. 8.2a. The formation of this small WZ segment is visible in the RHEED diagram. This phenomenon is illustrated in Fig. 8.2c and d, showing the patterns obtained during the growth where only the ZB spots are visible, and after the droplet consumption where the WZ and ZB spots are observed, respectively.

By monitoring the evolution of this RHEED diagram during the entire growth of self-assisted GaAs NWs using a framerate of 30 images per seconds, the real-time detection of the WZ formation becomes accessible. Stopping the rotation of the sample during the growth allows the measurement of the evolution of specific diffraction spot intensities, namely the $(10\bar{1}2)$ WZ and $(002)_t$ ZB spots, which provides information about the structural evolution of self-assisted GaAs NWs.

Figure 8.2b shows the evolution of the $\frac{I_{ZB}}{I_{ZB}+I_{WZ}}$ (black curve) and $\frac{I_{WZ}}{I_{ZB}+I_{WZ}}$ (red curve) intensity ratios (IRs) as a function of the growth time. The parameters $I_{ZB}$ and $I_{WZ}$

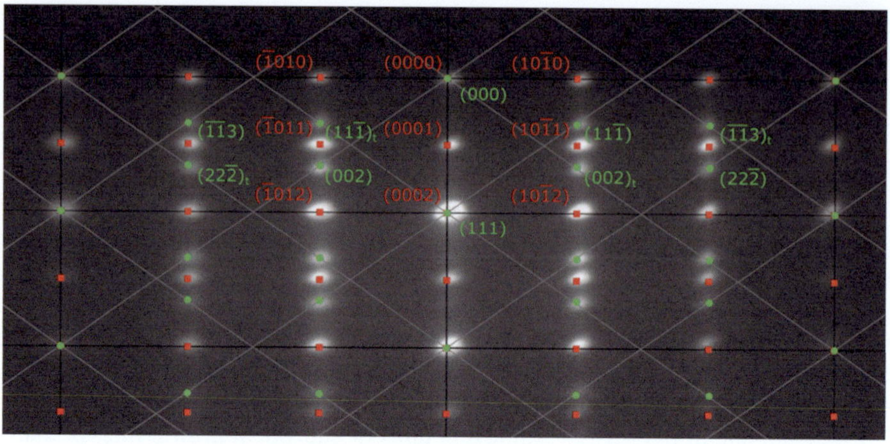


**Fig. 8.1** RHEED pattern of GaAs NWs recorded along the $[1\bar{1}0]$ azimuth, where the ZB and WZ spots are visible. An indexed diagram is surimposed on the RHEED pattern, showing the two ZB variant plans (green circles) and the WZ plans (red squares). Reproduced from Ref. [25] with the permission from the Royal Society of Chemistry under CC BY 3.0 license

represent the ZB and WZ measured spot intensities, respectively. These IRs will be respectively referred to as $IR_{ZB}$ and $IR_{WZ}$ in the rest of this chapter. The blue area corresponds to the standard growth, under Ga and As fluxes, associated to the stages (i) of Fig. 8.2a.

Due to the absence of NWs at the time $t = 0$ s, $IR_{ZB}$ and $IR_{WZ}$ are equal to 0.5. At this moment, the intensities measured correspond to the diffraction lines of the Si(111) substrate surface. As soon as the Ga and As fluxes are supplied to the surface, an increase of $IR_{ZB}$, and subsequently a decrease of $IR_{WZ}$ is observed, mainly due to the growth of parasitic nanocrystals, in the ZB phase. After about 10–20 s [16], the nucleation and growth of the NWs starts (zone ($i_1$)). After about 400 s of growth, the IRs tend to stabilize to values close to 1 and 0 for $IR_{ZB}$ and $IR_{WZ}$, respectively (zone ($i_3$)). This stabilization indicates that the probed parts of the NWs are entirely or quasi-entirely in the ZB phase (stage ($i_3$) in Fig. 8.2a).

The growth was interrupted after 600 s by closing the Ga shutter. The droplet is consumed under As atmosphere, leading to the sequence schematized in stages (ii) to (v) of Fig. 8.2a. A time delay, highlighted by zone (ii), is observed between the closure of the Ga shutter and the beginning of the $IR_{WZ}$ increase. This delay of about 20 s can be attributed to the time needed for the droplet contact angle to reach its critical value $\beta_1 = 125°–127°$, responsible of the WZ phase nucleation (stage (ii) in Fig. 8.2a). Then, an increase of $IR_{WZ}$ for about 70 s (zone (iii), corresponding to stage (iii) in Fig. 8.2a) is associated to the growth of a mixed-ZB/WZ segment followed by a pure-WZ segment. Finally, a slight increase of the $IR_{ZB}$ (zone (iv), corresponding to stage (iv) in Fig. 8.2a) is observed, and is the consequence of a final short ZB segment growth. This portion is induced by a contact angle lower than the critical value $\beta_2$ of about 90°, as illustrated in stage (v) of Fig. 8.2a. It should be noted that this final

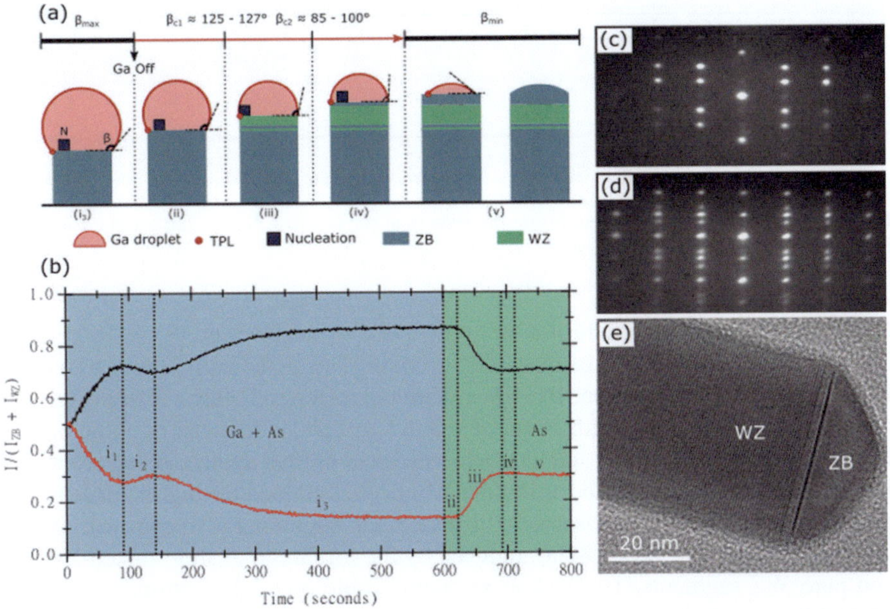


**Fig. 8.2** **a** Schematic of the Ga catalyst droplet evolution: ($i_3$) during the growth of the GaAs NWs, with a constant amount of Ga and As supplied to the droplet, leading to a wetting angle $\beta = \beta_{max}$ larger than $\beta_{c1}$. The ZB phase is nucleated during this period. After interrupting the Ga flux, a decrease of the droplet volume is observed, leading first to (ii) a nucleation inside the droplet (ZB phase) until (iii) the wetting angle $\beta$ becomes lower than $\beta_{c1}$, then leading to a nucleation at the TPL and thus to a transition zone *t* with a ZB/WZ phase mixing, followed by a pure-WZ section. Then, the wetting angle continues to decrease, becoming lower than $\beta_{c2}$, inducing (iv) a nucleation inside the droplet (short ZB segment) until (v) it reaches $\beta_{min}$, thus ending the growth with the final consumption of the droplet. **b** $IR_{ZB}$ (black curve) and $IR_{WZ}$ (red curve) as a function of the growth time. The growth under Ga and As flux is shown in blue while the consumption of the Ga droplet is shown in green. RHEED pattern recorded along the [1$\bar{1}$0] azimuth **c** during the NW growth and **d** after ending the growth by interrupting the Ga flux. The RHEED diagrams in **c** and **d** show the ZB pattern and the mixed ZB/WZ pattern, respectively. **e** Bright field TEM image of the top of a nanowire along the [1$\bar{1}$0] zone axis, showing the WZ and ZB structure induced by the consumption of the Ga catalyst droplet

transition from the WZ to the ZB phase is abrupt without any structural defects, as shown on the HRTEM image provided in Fig. 8.2e. The zone (v) plotted at the very end of the graphic in Fig. 8.2b is associated with the end of the NW growth, when $\beta = \beta_{min}$, with the final consumption of the Ga droplet. These measurements obtained via RHEED analysis are perfectly in line with the structural evolution commonly observed on NWs by TEM characterizations. Therefore, these IR variations can be associated with the size and contact angle modifications occurring on the Ga droplets, as illustrated in Fig. 8.2a.

It should be noted that the density and the length of the NWs, as well as the incidence angle of the electron beam regarding the substrate surface, can affect the

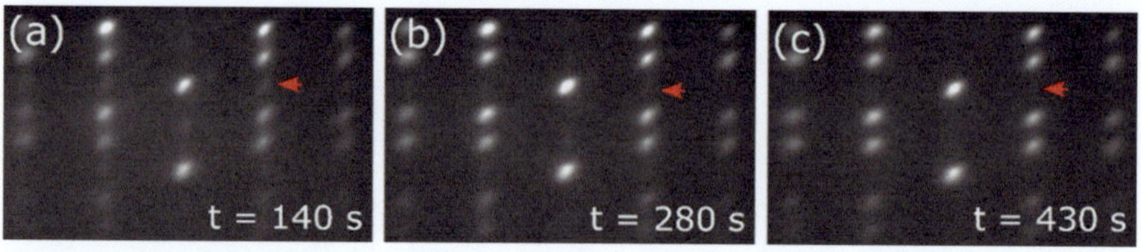


**Fig. 8.3** RHEED snapshots taken during the growth at **a** $t = 140$ s, **b** $t = 280$ s and **c** $t = 430$ s. Red arrows point the $(10\overline{1}2)$ WZ spot

RHEED diagram obtained during the growth due to shadowing effects. Indeed, the depth probed by the electron beam could be modified by all these different parameters. In this specific configuration, the NWs were characterized after a 10-min growth, which corresponds to a length of about 1.0–1.1 μm and a diameter of about 50 nm, for a density of 1 NW/$\mu m^2$, and led to an estimated probed depth around 400 to 500 nm. This depth was estimated using the IR curves obtained during the growth. An increase of the $IR_{WZ}$ was observed 100 s after the beginning of the growth, during about 40 s (zone ($i_2$) in Fig. 8.2a), induced by the formation of short mixed-ZB/WZ and pure-WZ segments. These short segments could be used to estimate the depth probed by the electron beam. The $IR_{WZ}$ peak reached a maximum once the WZ segment ended, and when the ZB to WZ transition occurred (Fig. 8.3a). Then these WZ spots slowly vanished (Fig. 8.3b) until their complete disappearance once the segment was no longer probed by the electron beam (Fig. 8.3c). Approximatively 290 s were needed for the RHEED WZ spots induced by this segment to disappear completely, which corresponds to about 400–500 nm given the NW growth rate.

The purpose of this study was to monitor the evolution of the RHEED diagram during the MBE growth of self-assisted GaAs NWs, and specifically to correlate the time dependence of the RHEED spots with the nucleation of the different crystal phases. The shadowing effects have only a slight influence on this analysis.

## 8.2 The Zinc Blende/Wurtzite Alternation

The previous section showed that RHEED patterns provide real-time information on the crystal structure evolution of the self-assisted GaAs NWs during growth. Therefore, RHEED could be used to control the formation of a WZ segment, inside ZB NWs. To do so, the droplet consumption phenomenon is mimicked by interrupting the incoming Ga flux for a short period. The purpose here was to estimate an appropriate time to reduce the size of the Ga droplets, inducing the nucleation of a WZ phase, without totally consuming the catalysts. Based on the previous results, the Ga flux was interrupted for 60 s after about 240 s of growth, to avoid the complete consumption of the droplets. The expected modifications of the Ga droplet are depicted in Fig. 8.4a, and the IR curves obtained during this growth are reported Fig. 8.4b, showing $IR_{ZB}$ and $IR_{WZ}$ in black and red curves, respectively.

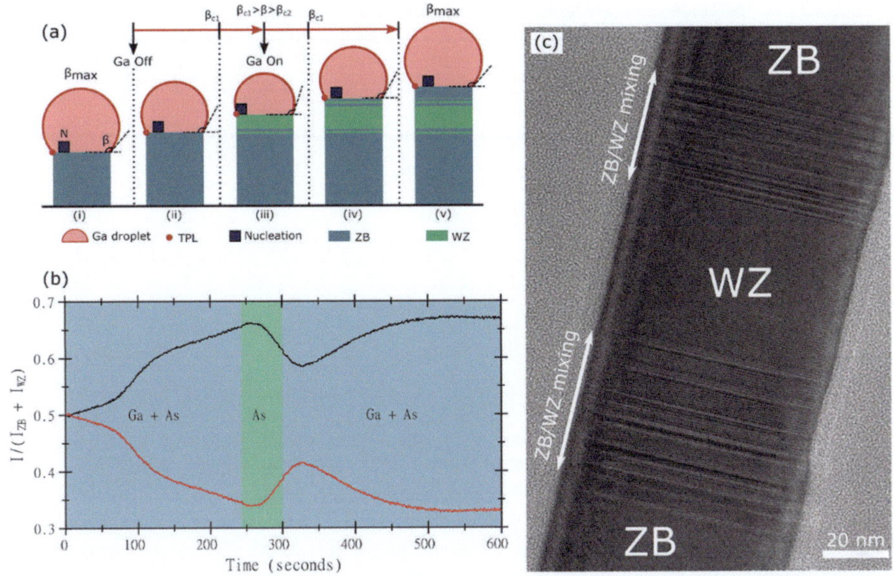


**Fig. 8.4** **a** Schematic of the Ga droplet evolution and corresponding crystal structure during the growth, showing: (i) the growth of GaAs NWs under Ga and As flux, with a wetting angle $\beta = \beta_{\mathrm{max}}$, leading to a nucleation inside the droplet and (ZB phase). The Ga flux was then interrupted, leading to a (ii) decrease of the wetting angle with a nucleation inside the droplet (ZB phase), followed by (iii) a nucleation at the TPL (ZB/WZ mixing followed by pure-WZ) once $\beta < \beta_{\mathrm{c1}}$. The droplet was then re-exposed to a Ga flux, leading to the progressive increase of the wetting angle up to (iv, v) $\beta_{\mathrm{max}}$. **b** $\mathrm{IR_{ZB}}$ (black curve) and $\mathrm{IR_{WZ}}$ (red curve) as a function of the growth time. The growth under Ga and As flux is shown in blue while the green part represents the growth under As only. **c** HRTEM image along the $[1\bar{1}0]$ zone axis of the WZ insertion grown in the ZB GaAs NW [25]

The ZB structure was first formed during the growth of the NWs (blue area), followed by the interruption of the Ga flux for 1 min after 4 min of growth (green area). The Ga flux was then re-supplied to the droplet by re-opening the Ga shutter. A delay was observed between the interruption of the Ga flux and the minimum of the $\mathrm{IR_{WZ}}$. In order to precisely estimate the duration of the WZ nucleation, the first derivative of the $\mathrm{IR_{WZ}}$ between 200 and 400 s was plotted (Fig. 8.5). The transitions from the ZB to the WZ phases and from the WZ to the ZB phases are indicated by the two vertical black lines, corresponding to the first derivative equals to zero. Thus, the first latency was estimated to about 17 s and corresponds to the time needed for the contact angle of the droplet to change from its maximal value $\beta_{\mathrm{max}} \approx 135°$ to the critical angle $\beta_{\mathrm{c1}}$. Similarly, the second delay of 26 s was observed for the contact angle between a local minimum $\beta_{\mathrm{Lmin}}$, reached the instant before the re-opening of the Ga flux, and the critical angle $\beta_{\mathrm{c1}}$ necessary to allow the transition to the ZB phase.

These phase transitions were used to estimate the lengths of the bottom ZB and WZ segments (including the ZB/WZ phase mixing segments visible in Fig. 8.4c). Taking into account the length and the growth time used, an average axial growth

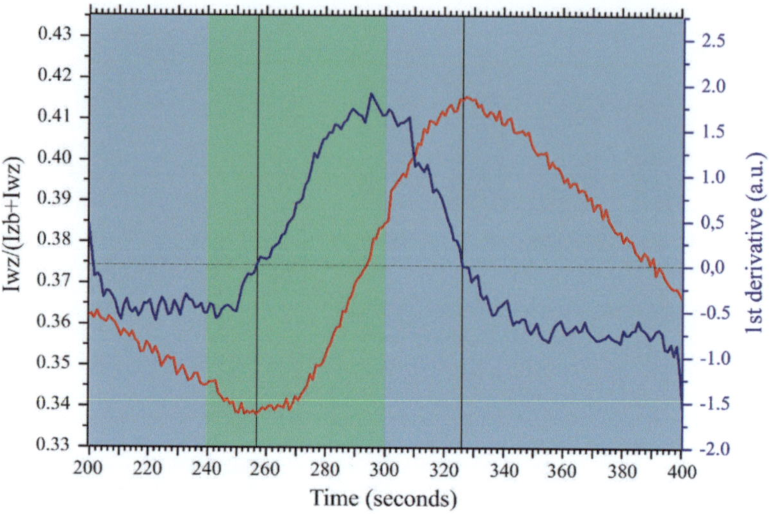


**Fig. 8.5** Enlarged image of the $IR_{WZ}$ (red curve) corresponding to the WZ segment growth. Blue areas correspond to the growth under Ga and As fluxes while green area is the growth under As flux only. The blue curve, linked to the right axis, corresponds to the first derivative of the $IR_{WZ}$. The black dashed line corresponds to the zero of the derivative [25]

rate of $v = 1.7$ nm/s can be extracted from the average length of the bottom ZB + WZ segments measured in the TEM on around twenty NWs. From the first derivative shown in Fig. 8.5, the growth duration of the WZ segment was estimated to 69 s, corresponding to an expected length of about 117 nm. Similarly, the length for the bottom ZB segment was calculated to be about 437 nm. The comparison between the bottom ZB and WZ lengths calculated from RHEED patterns and measured in the TEM is reported in Table 8.1, and validates the viability of the RHEED measurements to characterize the NWs.

**Table 8.1** Reported values of the growth time, the calculated lengths obtained from RHEED measurements (with an average growth rate of 1.7 nm/s) and the measured lengths extracted from TEM analysis. The growth time were obtained from Fig. 8.5 [25]

| | Growth time (s) | Calculated length (nm) (from RHEED) | Measured length (nm) (from TEM) |
|---|---|---|---|
| Bottom ZB segment | 257 | $\sim 437$ | $440 \pm 70$ |
| WZ segment (including the ZB/WZ phase mixing segments) | 69 | $\sim 117$ | $120 \pm 20$ |

## 8.3 Simulation of the Wetting Angle and Length Evolution

By extending the semi-empirical growth model for Au-catalyzed GaAs [29–31] and InAs [32] NWs, a model for the self-assisted GaAs NWs [33] is proposed to account for the Ga droplet evolution as a function of two parameters: (1) the Ga and As atoms entering the droplets by direct impingement of the corresponding fluxes, and (2) the Ga atoms entering the droplets by diffusion on the $SiO_2$-terminated silicon substrate (for short NWs) and on the NW facets. This model is described in the following paragraphs.

Assuming a growth time $t$, a NW of length $L(t)$, and a radius under the droplet $r(t)$, the Ga droplet at the top of the NW can be described with the following parameters: the critical concentration $c^*$, the values for the Ga and As nominal fluxes (*i.e.,* the number of atoms/(unit time) * (unit surface normal to the flux direction)), the radius and As concentration denoted $R(t)$ and $c(t)$, respectively, such as $c(t) \leq c^*$ (subcritical concentration). The amount of Ga and As atoms in the droplet can be written as $Q_{Ga}(t)$ and $Q_{As}(t)$, respectively.

Therefore, the structural evolution of the NW is driven by the amount of Ga and As atoms feeding the Ga droplet, and follows two requirements: (1) the solidification occurs only when the droplet concentration reaches its critical value $c^*$, and (2) the droplet contact angle is such that $\beta_{min} \leq \beta(t) \leq \beta_{max}$, where the two boundary contact angles $\beta_{min}$ and $\beta_{max}$ depend on the crystallographic properties of the NW.

The NW growth process can be described as follows: in a small interval of time $(t, t + \delta t)$, the droplet incorporates an amount of Ga atoms $\Delta Q_{Ga}(t)$, coming either from direct impingement, diffusion along the NW facets with a diffusion length $\lambda_{facet}$, or from diffusion on the substrate characterized by its diffusion length $\lambda_{substrate}$. The last contribution is lost when the NW length overcomes the diffusion length on the NW facets ($L(t) > \lambda_{facet}$). Meanwhile, the amount of As atoms incorporated in the droplet, defined as $\Delta Q_{As}(t)$, is assumed to be only due to direct impingement. Consequently, two situations may occur. Firstly, the As "potential concentration" in the droplet at $t + \delta t$, is defined as:

$$\frac{Q_{As}(t) + \Delta Q_{As}(t)}{Q_{As}(t) + \Delta Q_{As}(t) + Q_{Ga}(t) + \Delta Q_{Ga}(t)} \leq c^*. \tag{8.1}$$

In this situation, the solidification at the droplet / NW interface does not occur. Two distinct sub-cases can be distinguished: (1) the droplet is pinned on the top of the NW with a contact angle $\beta \leq \beta_{max}$, leading to an increase of the droplet volume with a constant NW length (no axial growth). (2) If the droplet volume induces a contact angle $\beta > \beta_{max}$, the computation is stopped by considering the droplet stability.

The second situation presents an As "potential concentration" in the droplet defined as:

$$\frac{Q_{As}(t) + \Delta Q_{As}(t)}{Q_{As}(t) + \Delta Q_{As}(t) + Q_{Ga}(t) + \Delta Q_{Ga}(t)} > c^*. \tag{8.2}$$

Otherwise stated, only the excess of Ga and As atoms, due to supersaturation, will contribute to the solidification process. However, the volume of solid material can contribute to both axial and radial growth, depending on the remaining volume of the Ga droplet, which includes the total amount of Ga and As atoms defined as $Q_{As}(t) + \Delta Q_{As}(t) + Q_{Ga}(t) + \Delta Q_{Ga}(t) - 2Q$. If this volume can be pinned atop the NW with a contact angle $\beta(t + \delta t)$ such as $\beta_{min} \leq \beta(t + \delta t) \leq \beta_{max}$, only the axial growth is observed. In the other case, two distinct sub-cases may occur.

If the contact angle $\beta(t+\delta t) > \beta_{max}$, an inverse tapered axial growth is observed. The inverse tapering leads to a modification of the NW radius to $r(t + \delta t)$, corresponding to the unique value able to support the fixed droplet volume at contact angle $\beta_{max}$. Subsequently, the axial growth will place the solid material in a truncated cone geometry with lower radius $r(t)$, upper radius $r(t + \delta t)$, and height determined by the amount of equal solid quantities $Q$ of Ga and As atoms.

If the contact angle $\beta(t + \delta t) < \beta_{min}$ the situation is similar to the previous one, with an axial growth and a direct tapering. A unique value of $r(t+\delta t)$ able to support the droplet at a contact angle $\beta_{min}$ exists. The solid material will then be placed in a truncated cone geometry with lower radius $r(t)$, upper radius $r(t + \delta t) < r(t)$, and height determined by the amount of equal solid quantities $Q$ of Ga and As atoms.

Finally, the algorithm can stop in the unrealistic situation when the NW radius becomes too large or too small. Assuming a critical concentration $c^* = 0.01$, and contact angles $\beta_{min} = 50°$ and $\beta_{max} = 140°$ under Ga and As fluxes of 0.5 and 1.15 ML/s, respectively, the numerical results are plotted in Fig. 8.6.

Besides the classical axial NW growth, the described model is able to predict the conditions triggering both axial and lateral growths, depending on the droplet evolution. The model is expected to provide and confirm the existence of a unique critical value of the contact angle, where the transition from the cubic ZB to the hexagonal WZ phases occurs. The results of the numerical simulations show the time evolution of the Ga and As amounts of atoms supplying the droplets, the droplet and NW radii, the contact angle of the droplets, and the NW length in Fig. 8.6a–d, respectively. The interruption of the Ga flux between 240 and 300 s induced a decrease of the droplet radius from about 25 nm down to 18 nm (Fig. 8.6b), and therefore of the contact angle from 134° down to 98° (Fig. 8.6c). Moreover, a reduction of the droplet capture surface for As atoms was induced by the decrease of the contact angle, leading to a slightly lower axial growth rate, as shown in Fig. 8.6d. An increase of the droplet radius and contact angle were observed immediately after the re-opening of the Ga shutter, after 300 s. The maximum contact angle $\beta_{max}$ was reached about 70 s after the re-feeding of the droplet in Ga atoms, leading to an increase of the NW radius (Fig. 8.6b), needed to accommodate the higher droplet volume.

The vertical dashed lines in Fig. 8.6c and d correspond to the time $t_1 = 257$ s and $t_2 = 326$ s, highlighting the transition times identified in Fig. 8.5. Therefore, the corresponding bottom ZB and WZ segment lengths predicted by the numerical simulations are 450 nm and 106 nm, respectively (Fig. 8.6d). The lengths obtained using this model are in good agreement with the TEM measurements of 440 ± 70 nm and 120 ± 20 nm, respectively, as reported in Table 8.1. Similarly, contact angles were identified and predicted for the ZB to WZ transition (126°) and for the WZ

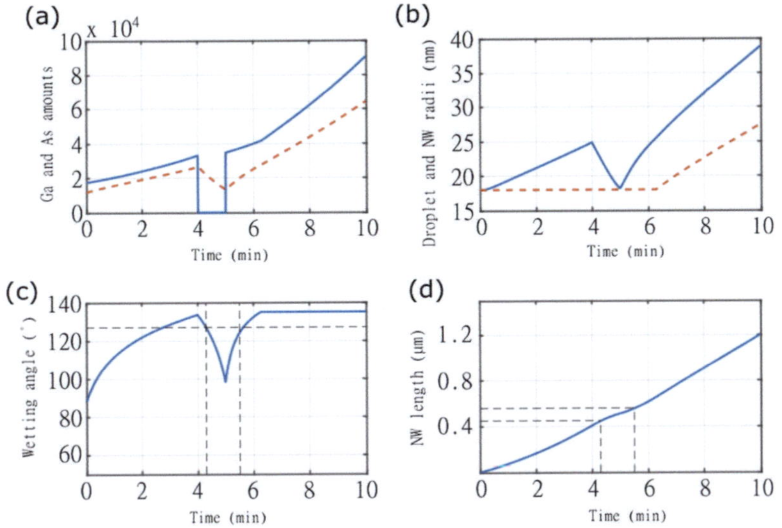


**Fig. 8.6** Curves representing the time evolution of **a** the number of Ga atoms (blue line) and As atoms (dashed orange line) feeding the droplet, **b** the droplet (blue line) and NW (dashed orange line) radii, **c** the wetting angle and **d** the NW length. The numerical results were obtained using: Ga and As sources located at incidence angles of 27.9° and 41°, respectively, with respect to the normal of the substrate. The nominal Ga and As fluxes used: $F_{Ga} = 3.53$ atoms/sec.nm$^2$ and $F_{As} = 27.3$ atoms/sec.nm$^2$. The best fit was obtained using diffusion lengths on $SiO_2$ and on NW facets of 38 nm and 1.4 μm, respectively. The As concentration threshold in the droplet was fixed at 1%. The initial radius and wetting angle of the droplet were equal to 18 nm and 90°, respectively. Reproduced from Ref. [25] with the permission from the Royal Society of Chemistry under CC BY 3.0 license

to ZB transition (124°), as shown in Fig. 8.6c, in agreement with the experimental results recently reported by Kim et al. [24] by ex situ TEM and measured by Panciera et al. by in situ TEM [10] on self-assisted GaAs NWs. The lengths predicted by the numerical models are compared with the previously measured lengths in Table 8.2.

## 8.4 Simulation of the Wetting Angle as a Function of the III and V Fluxes

As far as the self-assisted GaAs growth is concerned, the cubic ZB and hexagonal WZ phases are determined by the contact angle $\beta$ between the droplet and the top facet of the NW [1]. For instance, when the Ga flux is interrupted and the As atoms feed the droplet by direct impingement, the contact angle of the droplet decreases and the crystal structure evolves from the ZB to the WZ phase, and then back to the ZB phase, as highlighted in the previous chapter. As discussed, the nucleation of the WZ phase occurs for a contact angle in the 90–125° range [10, 25]. Therefore, the main question to be addressed is the following: is it possible to provide values for

**Table 8.2** Reported values of the growth time, the calculated lengths obtained from RHEED measurements (with an average growth rate of 1.7 nm/s), measured lengths extracted from TEM analysis and simulated lengths predicted by the numerical simulations. The growth time was obtained from Fig. 8.5. The measured lengths are average values obtained by measurements on TEM bright field images of about twenty NWs. The WZ segment measured by TEM includes the ZB/WZ mixing phases [25]

| | Growth time (s) | Calculated length (nm) (from RHEED) | Measured length (nm) (from TEM) | Simulated length (nm) |
|---|---|---|---|---|
| Bottom ZB segment | 257 | ~ 437 | 440 ± 70 | 450 |
| WZ segment (including the ZB/WZ phase mixing segments) | 69 | ~ 177 | 120 ± 20 | 106 |

the V/III flux ratio that ensure a fixed contact angle in the 90–125° range to promote the growth of the WZ phase?

The positive answer to this question is a consequence of the general statement that the self-assisted NW growth is a self-regulated system [33]. Considering fixed values of the Ga and As fluxes, the radius of the NW, the size of the droplet and its contact angle evolve toward a stationary regime, in which the amounts of Ga and As atoms supplied to the droplet are equal. However, this so-called constant growth regime could only be reached once the length of the NWs overcome a first-stage of "transient' regime.

Previous studies on self-assisted GaAs NW growth [29, 30, 32–34], show that the Ga atoms are assumed to supply the liquid droplet through three distinct sources, as illustrated in Fig. 8.7.

Firstly, the amount of Ga atoms able to attain the droplet by surface diffusion on the $SiO_2$-terminated substrate, denoted $q_{Ga}^{sub}$ and defined by:

$$q_{Ga}^{sub}(t) = F_{Ga}S^{sub}(t) \tag{8.3}$$

where $F_{Ga}$ corresponds to the Ga flux on the $SiO_2$-terminated substrate (here 0.5 ML/s) and $S^{sub}(t) = \pi\left[\left(\lambda_{SiO_2} + r(t)\right) - r(t)\right]$ is the substrate capture area, with $\lambda_{SiO_2}$ the diffusion length on the $SiO_2$-terminated substrate and $r(t)$ the NW radius.

Secondly, the amount of Ga atoms able to reach the droplet by diffusion along the NW facets, denoted $q_{Ga}^{facet}$ and defined by:

$$q_{Ga}^{facet}(t) = F_{Ga}\tan(\alpha_{Ga})S^{facet}(t) \tag{8.4}$$

where $\alpha_{Ga}$ is the angle of the Ga source with respect to the normal of the substrate and $S^{facet}(t) = 2r(t)min\left(\lambda_{facet}, L(t)\right)$ represents the NW facet capture area projected on the plane normal to the Ga flux orientation. The *min* function accounts for the NW length only for NWs with length $L(t)$ smaller than $\lambda_{facet}$.

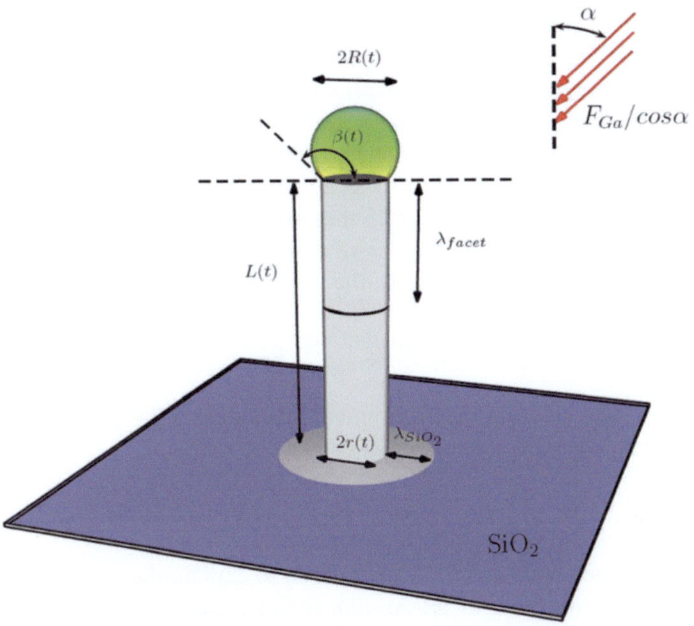


**Fig. 8.7** Geometry of the NW and droplet, and generic flux position at an angle $\alpha$ with respect to the substrate normal and diffusion lengths. Reproduced from Ref. [33] with the permission from the Royal Society of Chemistry under CC BY 3.0 license

Thirdly, the amount of Ga atoms feeding the droplet by direct impingement is written as:

$$q_{Ga}^{droplet}(t) = \frac{F_{Ga}}{\cos\alpha_{Ga}} S(\alpha_{Ga}, \beta(t), r(t)) \tag{8.5}$$

where $S(\alpha_{Ga}, \beta(t), r(t))$ is the value of the droplet area projected in the direction normal to the flux, when the droplet has a contact angle $\beta(t)$ with the top of a NW with a radius $r(t)$.

Meanwhile, assuming that the As atoms only feed the catalyst through direct impingement on the droplet surface, and thus give a quantity $q_{As}^{droplet}$ of As atoms supplying the droplet defined by:

$$q_{As}^{droplet}(t) = q_{As} S(\alpha_{As}, \beta(t), r(t)) \tag{8.6}$$

where $q_{As}$ corresponds to the nominal flux of As and $\alpha_{As}$ represents to the angle of the As source with respect to the normal of the substrate.

By considering the stationary regime introduced at the beginning of this section, in which the amounts of Ga and As atoms supplied to the droplet are equal:

$$q_{Ga}^{sub} + q_{Ga}^{facet} + q_{Ga}^{droplet} = q_{As}^{droplet} \tag{8.7}$$

Therefore, recent experimental results obtained by using two Ga sources with different positions with respect to the substrate [33] show a disappearance of the substrate diffusion contribution $q_{Ga}^{sub}$ when the NW length $L(t)$ overcomes the diffusion length $\lambda_{facet}$ on the NW facets, which is typically in the 1–2 μm range. By growing self-assisted GaAs NWs for 25 min using the standard growth parameters, the NWs are longer than 2 μm. This 25-min growth will hereafter be referred to as "standard growth". In these conditions, the previous equation can be reduced to:

$$q_{Ga}^{facet} + q_{Ga}^{droplet} = q_{As}^{droplet} \tag{8.8}$$

Subsequently considering the Ga nominal flux $q_{Ga} = \frac{F_{Ga}}{\cos\alpha_{Ga}}$, the equal amount of Ga and As atoms supplied to the droplet can be expressed as:

$$q_{Ga}\sin(\alpha_{Ga})2r\lambda_{facet} + q_{Ga}S(\alpha_{Ga}, \beta, r) = q_{As}S(\alpha_{As}, \beta, r) \tag{8.9}$$

Furthermore, the above equality can be used to explain the self-regulation of the system. Indeed, in the case of an As-rich regime, the equation's right-hand side (RHS) becomes larger than the left-hand side (LHS), inducing a decrease of the droplet volume. The contact angle of the droplet decreases to a value for which the surface diffusion term balances the difference between the direct impingement terms. On the contrary, in the Ga-rich regime, the volume of the droplet will tend to increase leading to a rise of its contact angle. As both direct impingement terms increase and the V/III flux ratio is larger than 1, the RHS will increase faster than the LHS. In both cases the evolution of the system tends to a stationary growth regime, presenting a constant contact angle.

With known orientations of the Ga and As sources ($\alpha_{Ga}$ and $\alpha_{As}$, respectively), a contact angle $\beta$ in the 90–125° range can be determined using the V/III flux ratio given by:

$$\frac{q_{As}}{q_{Ga}} = \frac{\sin(\alpha_{Ga})2r\lambda_{facet} + S(\alpha_{Ga}, \beta, r)}{S(\alpha_{As}, \beta, r)} = \frac{\sin(\alpha_{Ga})2\gamma + S(\alpha_{Ga}, \beta, 1)}{S(\alpha_{As}, \beta, 1)}, \tag{8.10}$$

where the last equality holds since $S(\alpha, \beta, r)$ is quadratic with respect to$r$, and $\gamma$ represents the number$\frac{\lambda_{facet}}{r}$. Moreover, this equation shows that a fixed V/III flux ratio can lead to different contact angles for different NW radii, so that the ideal situation is the monodisperse case. Using the model presented in the work of M. Vettori et al. [33], which accounts for the droplet evolution and a variable NW radius, a numerical value for the diffusion length on the facets $\lambda_{facet} = 1.8$ μm, and the reactor configuration ($\alpha_{Ga} = 28°$ and $\alpha_{As} = 41°$), the values of the contact angle $\beta$ *are* as a function of the radius $r$ of the NW and the V/III flux ratio$\frac{q_{As}}{q_{Ga}}$. These contact

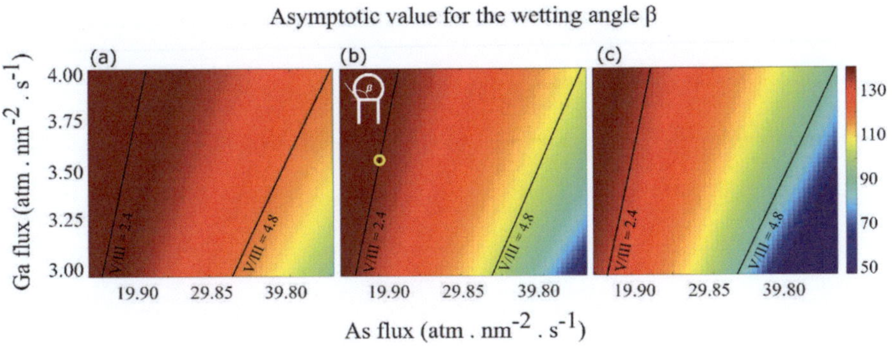


**Fig. 8.8** Values of the wetting angle $\beta$ obtained for NWs with a radius $r$ equal to **a** 40 nm, **b** 50 nm and **c** 60 nm for various V/III flux ratio. The area highlighted with the yellow circle represents the conditions used for the NW growth (V/III flux ratio = 2.4 at $q_{\mathrm{Ga}}$ = 3.23 atm.nm$^{-2}$.s$^{-1}$) to overcome the transient regime where the diffusion on the $SiO_2$-terminated substrate contributes to the Ga supply. Values lower than 50° are not represented as the droplet disappears and the VLS growth stops. The black lines represent V/III flux ratios equal to (left) 2.4 and (right) 4.8 [26]

angles are called asymptotic, as they equilibrate the amount of Ga and As atoms for various radii and V/III flux ratios.

Figure 8.8a–c illustrate the asymptotic values of the contact angle $\beta$ for a large range of ($q_{\mathrm{As}}$, $q_{\mathrm{Ga}}$) couples and for different radius. The yellow circle on the left side of Fig. 8.8b (V/III flux ratio = 2.4 and $q_{\mathrm{Ga}} = 0.5$ ML/s = 3.53 atm/(nm.s)) represents the ($x$, $y$) position of the standard growth conditions. As illustrated in Fig. 8.8, V/III flux ratio larger than 3.5–4.0 can lead to asymptotic values of the contact angle lower than 125°. On the contrary, too large values of the V/III flux ratio will stop the VLS growth process due to the extinction of the droplet. The oblique black lines visible on the color maps correspond to V/III flux ratios of 2.4 (left) and 4.8 (right).

In order to experimentally verify these simulations, NWs obtained after the standard growth were characterized by SEM, as shown in Fig. 8.9a. The NWs were about 2.7 ± 0.2 μm long, with a diameter in the 90–100 nm range. The contact angle was measured about 137°, in remarkable agreement with Panciera et al. [10] and with the expected asymptotic value highlighted in Fig. 8.8b by the yellow circle. Furthermore, the RHEED diagram recorded along the $[1\bar{1}0]$ azimuth at the end of the growth highlights the presence of the ZB structure only, as expected by the measured contact angle (Fig. 8.9b).

The numerical implementation of the model suggests the following growth recipe: a first VLS stage using the standard conditions, leading to a ZB phase, used to overcome the transient regime due to the on—substrate diffusion. This stage will be followed by an increase of the V/III flux ratio from 2.4 to a value between 4.0 and 7.0. It is important to notice that for large diameter NWs (typically larger than about 120 nm) the VLS growth process stops, while for NWs with diameters smaller than a critical threshold (typically 80 nm) the VLS growth continues with a contact angle larger than 125°. However, considering a typical diameter distribution, about 90%

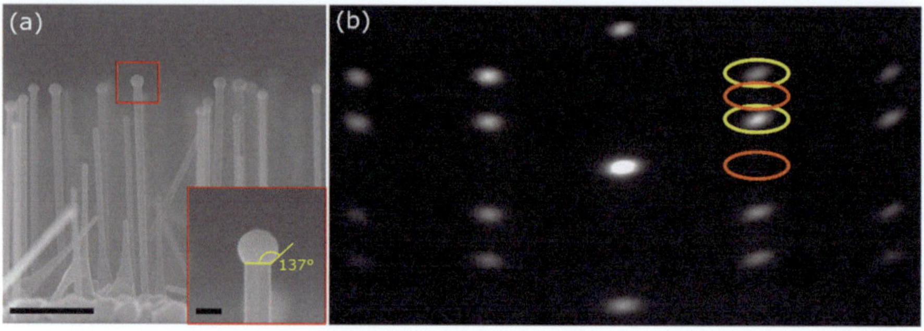


**Fig. 8.9** **a** Side view SEM image showing the NW morphology after the standard growth of 25 min. Scale bar is 1 µm. The inset corresponds to a higher magnification image of the droplet at the end of a NW. Scale bar is 100 nm. **b** RHEED pattern recorded at the end of the growth along the $[1\bar{1}0]$ azimuth, where only the ZB spots (highlighted by the yellow circles) are visible. The orange circles show the position of the WZ spots

of the sample is estimated to switch from ZB phase to the stationary growth in the WZ phase.

Therefore, a first series of samples was grown starting with a standard growth, followed by the growth using higher V/III flux ratio. The contact angle induced by these increases of V/III flux ratio was systematically measured using SEM. Each value was compared with the expected angles given by the simulation and reported in Table 8.3. The correlation between the measurements and the simulations is strong, showing an optimal V/III flux ratio of 4.0, which led to a contact angle of 120°.

In order to investigate more in depth the impact of the V/III flux ratio on the contact angle, a series of four samples were grown using four different As flux to

**Table 8.3** Measured and simulated values of the contact angle $\beta$ for different growth conditions. The first row corresponds to the standard growth of the self-assisted GaAs NWs, with a V/III flux ratio of 2.4. The other rows correspond to a standard growth, followed by an extended growth with higher As flux. The hypothetical optimal growth is written in blue

| | Growth time under high V/III (in mins) | Measured contact angle (°) | Simulated contact angle (°) |
|---|---|---|---|
| Standard growth reference | | 137 | 140 |
| Standard growth followed by V/III = 3.0 | 20 | 135 | 135 |
| Standard growth followed by V/III = 3.7 | 20 | 127 | 125 |
| **Standard growth followed by V/III = 4.0** | **10** | **120** | **120** |
| Standard growth followed by V/III = 5.1 | 5 | 107 | 110 |

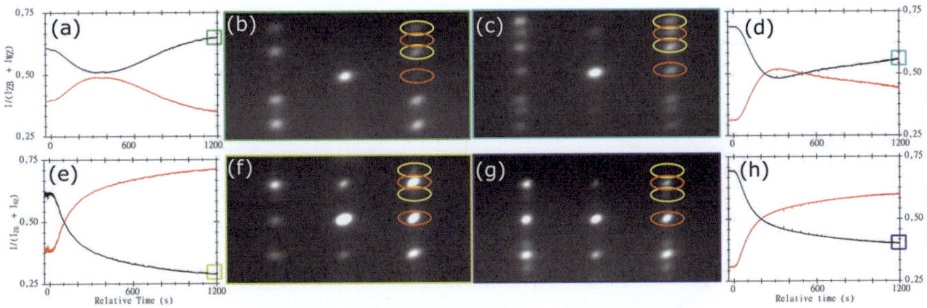


**Fig. 8.10** $IR_{ZB}$ (black line) and $IR_{WZ}$ (red line) obtained during 20-min growth and RHEED patterns taken at the end of the growth for a V/III flux ratio of **a** and **b** 3.0, **c** and **d** 3.7, **e** and **f** 4.0, **g** and **h** 5.1. On the IR *x*-axis, $t = 0$ corresponds to the increase of the As flux. The time indicates the growth time under high As flux, which follows a first stage of standard growth. The yellow and orange circles indicate the ZB and WZ spots, respectively

obtain V/III flux ratios of 3.0, 3.7, 4.0 and 5.1. The IRs and RHEED diffraction patterns obtained are reported in Fig. 8.10.

With a low V/III flux ratio (3.0 and 3.7), an increase of the $IR_{WZ}$ (and decrease of the $IR_{ZB}$) was first observed, indicating the nucleation of a WZ segment, and was followed by an increase of the $IR_{ZB}$. The increase of the As flux led to a decrease of the contact angle, and consequently to the formation of a faulty section. The droplet then tends to stabilize with a contact angle larger than 125°, as shown in Table 8.3. On the contrary, higher V/III flux ratio (4.0 and 5.1) led to a constant increase of the $IR_{WZ}$. The change of the slope observed after about 200 s is due to the loss of the ZB contribution, due to a probed depth of a few hundreds of nanometers. Thus, it appears that a V/III flux ratio of 4.0 (or 5.1) starts as the optimal parameter to grow an extended WZ segment in self-assisted GaAs NWs.

## 8.5 Evidences of the Pure-WZ Phase

The stationary growth of the WZ phase was followed in real-time using in situ RHEED analysis method discussed in the previous section. The evolution of the RHEED pattern was continuously recorded during the entire growth under high As pressure (V/III = 4.0).

Figure 8.11a illustrates the $IR_{ZB}$ and $IR_{WZ}$ obtained during a 20 min growth with a V/III flux ratio of 4.0 that follows a standard growth. The *x*-axis represents the growth time under high As flux, such that $t = 0$ s corresponds to the increase of the As pressure. An increase of the $IR_{WZ}$ was observed around 20 s after the increase of the V/III flux ratio, as shown in the previous section [25]. A progressive disappearance of the ZB signal during the growth is then clearly visible in the IRs lines (Fig. 8.11a) as well as on the RHEED pattern shown in Fig. 8.11b–e, thus proving the growth of a WZ phase. The growth ended by stopping the Ga and As fluxes simultaneously,

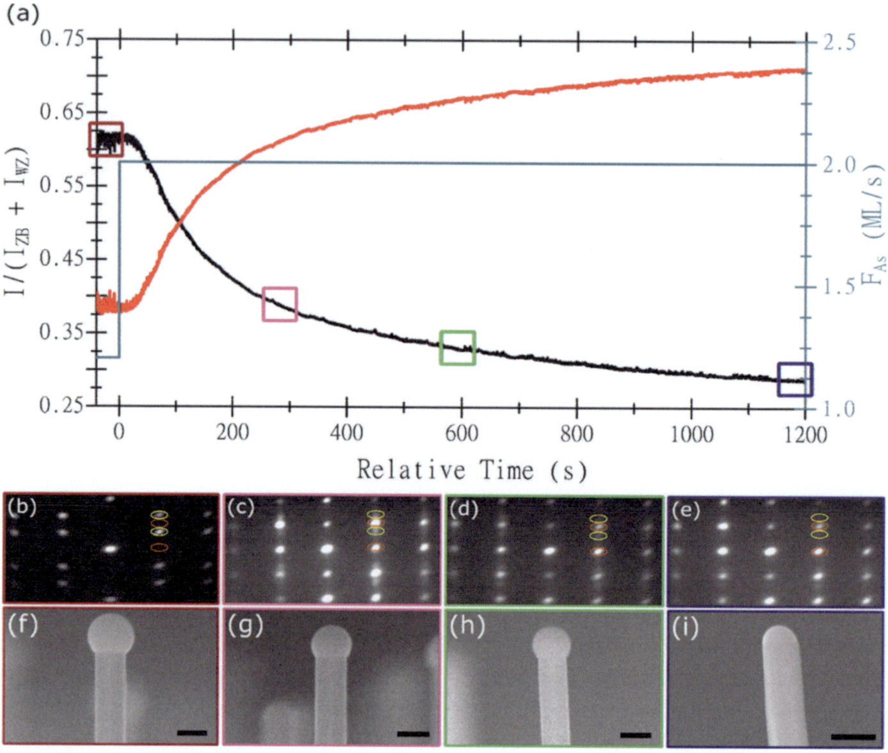


**Fig. 8.11** **a** $IR_{ZB}$ (black line) and $IR_{WZ}$ (red line), associated to the left y-axis, as a function of the relative growth time ($t = 0$ s corresponds to the increase of the As flux, after the 25 min growth under the standard conditions). The light blue line, associated to the right y-axis, corresponds to the As flux in ML/s. RHEED patterns recorded along the [1$\bar{1}$0] azimuth and SEM images of the catalyst droplet **b**, **f** before the increase of the As flux, and after **c**, **g** 5 min, **d**, **h** 10 min and **e**, **i** 20 min of growth under high As flux. The ZB and WZ spots on the RHEED patterns are highlighted with the yellow and orange circles, respectively. Scale bars in **f**–**i** are 100 nm. Reproduced from Ref. [26] with the permission from Institute of Physics

to maintain the droplet at the top of the NWs. The contact angle of the Ga droplet was measured on tens of NWs before the increase of As pressure and after 5 min, 10 min and 20 min (Fig. 8.11f–i, respectively) of growth using SEM. After 5 and 10 min, the contact angle was measured close to 120°, as predicted by the simulation in Fig. 8.8b. However, a slight decrease of the contact angle is visible for the 20-min growth with a value measured around 100°. It is important to mention that these SEM characterizations were performed on four distinct samples, grown successively on the same day, using exactly the same growth recipe. The growth was stopped by interrupting the III and V fluxes, and lowering the substrate temperature to 330 °C. It is noteworthy that this procedure could modify the catalyst droplet due to oxidation in contact with air. However, the strong correlation between the experiments and simulations validates the relevance of the procedure.

Coupling these measurements to the RHEED IRs analysis provides indirect but strong evidences of the growth of an extended WZ segment. The presence of the droplet revealed by SEM imaging after a 20-min growth, using a V/III flux ratio of 4.0, suggests that the growth of the WZ segment could be extended even further.

In order to investigate further the presence of the WZ structure and the phase transitions between ZB and WZ in a single NW, high-resolution (aberration-corrected) high angle annular dark field (HAADF) imaging in scanning transmission electron microscopy (STEM) mode, as well as dark-field (DF)-TEM imaging were employed. Figure 8.12.a shows a HAADF-STEM overview of a single NW after the 45-min growth (first stage standard growth followed by 20 min with a V/III flux ratio of 4.0). The NW initially crystallizes in the ZB phase as evidenced by the atomic structure revealed in the HAADF-STEM image in Fig. 8.12c, in agreement with the simulations in Fig. 8.8b and the RHEED observation in Fig. 8.9b. The transition region between the ZB and the WZ phases consists of alternate WZ and ZB domains, as highlighted by the sharp contrast variations in the DF-TEM image shown in Fig. 8.12b and by the HAADF—STEM image in Fig. 8.12d, where the domains are delimited by the yellow lines. This specific transition was the last one observed before the 1.3 $\mu$m long WZ segment evidenced in the DF-TEM image obtained using the $[1\bar{1}00]$ g vector, specific to the WZ phase, as shown in Fig. 8.12b. Only two stacking faults are visible in the whole segment and are pointed with red arrows, dividing the long WZ segment into 3 sub-segments of 620 nm, 180 nm and 480 nm. Nevertheless, considering the total length of the extended WZ and the presence of only 2 stacking faults inside it, this segment is considered as pure-WZ. Furthermore, the WZ structure is confirmed by high-resolution HAADF-STEM imaging, as shown in Fig. 8.12e and f. However, it is important to mention the presence of few atomic planes corresponding to the ZB structure under the droplet as visible on the right side of Fig. 8.12f. These ZB MLs were expected due to the lower contact angle, close to the critical value $\beta_2$, in the 85—100° range [10].

To investigate the NW atomic structure further, the position of the Ga and As atomic columns was identified from the relative HAADF intensity across the Ga—As pairs dumbbell (illustrated in Fig. 8.12g and h). Indeed, the HAADF imaging conditions result in contrasted images in which the intensity scales proportionally to ~ $Z^2$ where Z is the atomic number of the elements [36]. A higher intensity is expected for As atoms compared to Ga ones, as shown in Fig. 8.12g and h, providing evidence that As atoms are positioned toward the top of the NW. Therefore these atomic-resolution HAADF images directly show that this NW is As-polarized, in agreement with the expected polarity of the ZB and WZ crystal phases of the GaAs NWs [37].

The last part of this work consists in the photoluminescence (PL) study of the sample. It should be noted that the specific high refractive index of the NWs, leading to waveguiding properties, induces a strong dependence of the absorption efficiency on the incident light wavelength $\lambda$ [38].

In order to understand the absorption phenomenon occurring in a single NW as a function of $\lambda$, finite difference time domain (FDTD) simulations were performed using Rsoft FullWAVE software. Considering a single GaAs NW, indicated by the

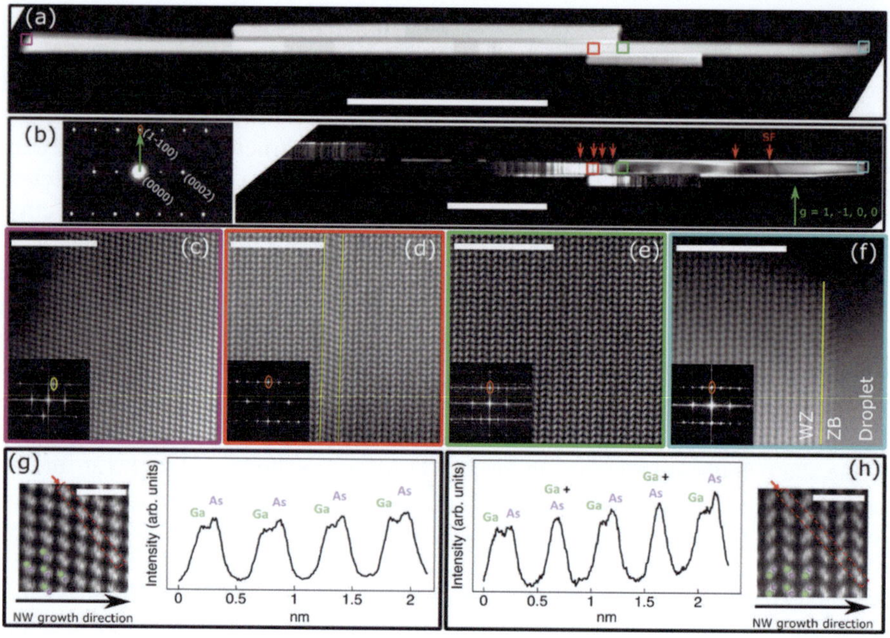


**Fig. 8.12** **a** HAADF-STEM overview of a single NW. Broken pieces of other NWs are visible at the top and bottom. Scale bar is 1 μm. **b** Diffraction pattern and DF-TEM image of the extended WZ segment with red arrows indicating stacking faults in the WZ segment. Scale bar is 500 nm. High-resolution HAADF-STEM images showing **c** the foot of the NW with a ZB crystalline structure, **d** the beginning of the extended WZ segment with a 3MLs long ZB inclusion, **e** a portion of the extended WZ segment, and **f** the end of the extended WZ segment with the Ga droplet/NW interface. **c**–**f** correspond to the regions highlighted with colored squares, magenta, red, green, and cyan, respectively, in (**a**) and (**b**). The ZB and WZ spots of the FFTs are highlighted with the yellow and orange circles, respectively. Scale bars in **c**–**f** are 5 nm. The ZB structure in **c**, **d** and **g** is observed in the $[1\bar{1}0]$ zone axis, and the WZ structure in **e**, **f** and **h** is seen in the $[11\bar{2}0]$ zone axis. **g** and **h** High-resolution HAADF-STEM images and intensity profiles of the ZB segment near the NW foot and of the WZ segment near the NW head, respectively. Scale bars in **g** and **h** are 1 nm. The overlayed atomic models of WZ and ZB highlight the Ga (green) and As (purple) atomic column positions, and were made using the VESTA software [35]. Reproduced from Ref. [26] with the permission from Institute of Physics

white dashed lines in Fig. 8.13, illuminated with a plane wave propagating along a direction parallel to the NW axis ($z$-axis) with a transverse polarization (along the $x$-axis), an absorption profile was determined. As shown in Fig. 8.13, the absorption is strongly localized in the upper part of the NW (WZ section) when $\lambda = 532$ nm, whereas the absorption is mainly observed in the bottom part (ZB section) when $\lambda = 671$ nm.

Low temperature (12 K) PL measurements were performed on the extended WZ sample, using a 532 nm and 671 nm continuous wave optical excitation at low excitation power (21 W/cm$^2$), revealing a broad emission in the 1.46–1.49 eV range, as shown in Fig. 8.14a. This energy is already known to be related to a type II ZB/WZ

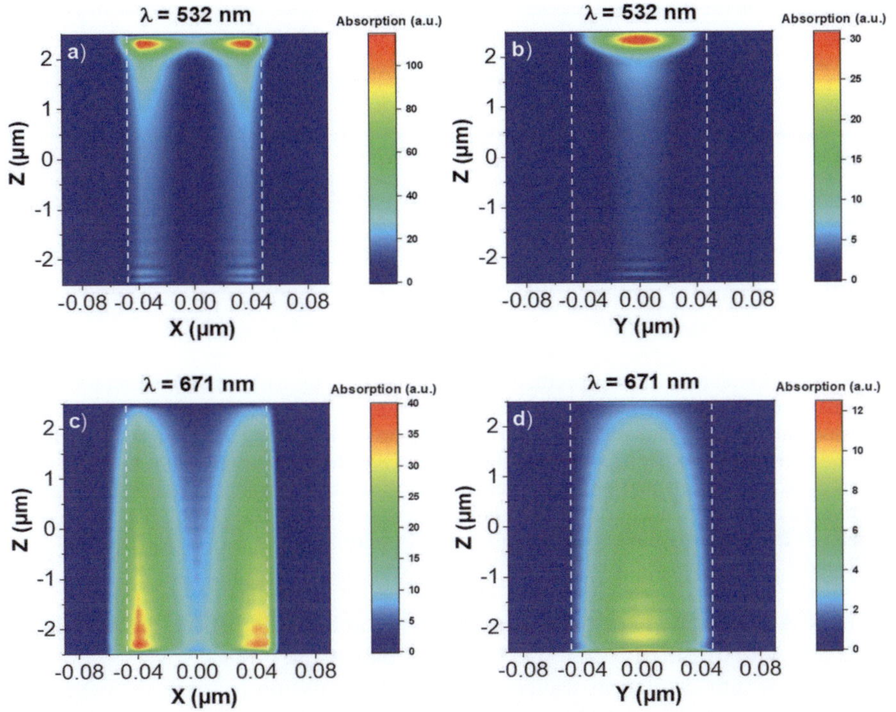


**Fig. 8.13** Absorption profile for a 2D—cut along the NW growth axis, illuminated with plane waves of **a, c** 532 nm and **b, d** 671 nm. The laser excitation is polarized along the x axis

emission [39, 40]. By exciting the sample with a $\lambda = 671$ nm laser, a peak at 1.516 eV is observed (peak A on the red line), which is 3 meV below the low temperature band gap of the ZB structure of GaAs (1.519 eV). If peak A is related to the recombination of free excitons, then their recombination energy should be equal to:

$$E = E_{gap} + E_{strain} + E_{QC} - E_{binding} \tag{8.11}$$

where $E_{strain}$, $E_{QC}$ and $E_{binding}$ are the energy shifts induced by the strain, the quantum confinement and the exciton binding energy, respectively. The strain in the NWs can be neglected due to the absence of shell used to passivate the GaAs. Furthermore, the NWs are standing straight, and not lying on the substrate, thus avoiding any substrate-induced strain during the sample cooling [41, 42]. Meanwhile, the quantum confinement is weak, or inexistent: a 1–2 meV confinement energy is expected for a cylindrical shaped ZB GaAs NW, with a 90–100 nm diameter, according to the formula developed by Yamamoto et al. [43]. Furthermore, an exciton binding energy in the order of 4.2 meV has been reported for the bulk ZB GaAs [44, 45]. Therefore,

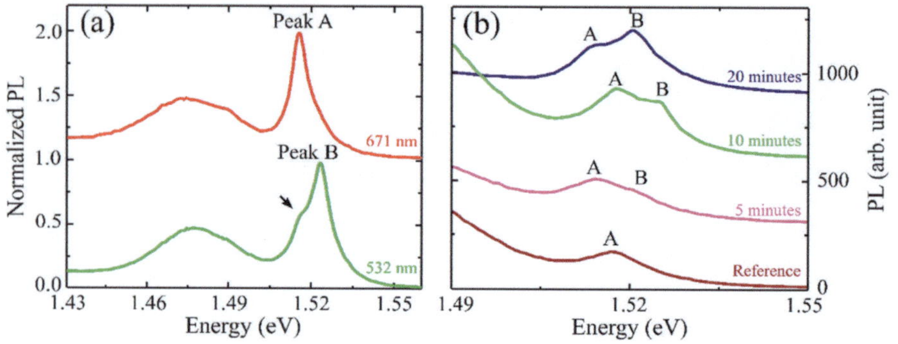


**Fig. 8.14** **a** Photoluminescence spectra of the sample at 12 K for a 671 nm excitation wavelength (red line) and a 532 nm excitation wavelength (green line). The spectra were normalized to the strongest emission peak and offset for clarity. **b** Close-up view on the peaks related to free excitons measured for different samples. Dark red line corresponds the standard growth, exhibiting only a ZB structure. Pink, green and purple lines represents the standard growth followed by a WZ growth of 5 min, 10 min and 20 min, respectively. The spectra were offset vertically for clarity. Reproduced from Ref. [26] with the permission from Institute of Physics

the peak A emission energy agrees with the recombination of free excitons in ZB GaAs and with a laser absorption in the lower part of the NWs.

For a 532 nm excitation wavelength, the PL emission is quite different, it is dominated by a peak at 1.5235 eV (peak B), whereas the peak A appears as a low-energy shoulder, highlighted by the arrow in Fig. 5.8a. Indeed, the peak B is only observed when a WZ structure is grown on the upper part of the NWs as shown in Fig. 5.8b, where a single peak (A) is observed on the reference ZB sample at 1.517 eV, which is closed to the ZB GaAs reference band gap of 1.519 eV [46]. When a WZ segment is grown on the upper part of the NWs, a second peak (B) appears, 7–8 meV above peak A. The intensity of the peak B, compared to the intensity of peak A increases along with the length of the grown WZ section. Consequently, peak B is assumed to be related to the free exciton recombination in the WZ segment of the NWs, which is in agreement with the strong absorption of the 532 nm laser in the upper part of the NWs, as shown in Fig. 5.7a. Furthermore, these results are in agreement with low temperature PL studies performed by other groups on WZ GaAs NWs grown by metalorganic vapor-phase epitaxy (catalyst-free) or by MBE (gold or manganese catalysts) [47–49], where WZ emissions were observed at 1.522 eV [47], 1.519 eV [48] and 1.518 eV. However, to determine the low temperature band gap of WZ GaAs, the quantum confinement and the exciton binding energy contributions must be considered. Performing the calculations using the dielectric constant$\varepsilon_0 = \sqrt{\varepsilon_0^{\perp}\varepsilon_0^{\parallel}} = 12.77$, determined by De's group [50] and an exciton reduced mass $\mu = 0.05$–$0.06$ from De Luca's group [51] led to a quantum confinement (1–2 meV) and a binding energy (4–5 meV) for the WZ GaAs very similar to the ZB GaAs ones. From these values, a band gap was estimated for the WZ GaAs located about 6–9 meV above the gap of its cubic counterpart. It is worth noting the narrowness of the measured peaks. Indeed, linewidths of around 7 meV for the peak A and 8.5 meV for the peak B, are comparable or better than the 7 meV [47] and 18 meV [48] values reported for non–passivated WZ GaAs NWs. Furthermore, these results are close to the 4 meV linewidth obtained on single passivated NWs [49, 52].

## 8.6 Conclusion

In summary, this work shows that it is possible to track the formation of ZB and WZ structures through the indexation of the ZB and WZ diffraction spots observed in a RHEED pattern during the growth of self-assisted GaAs NWs, in real time. Using this approach, the complete consumption of the Ga catalyst droplets was followed at the end of the growth, which was induced by the interruption of the Ga flux. A procedure was established to produce extended WZ segments inside ZB GaAs NWs using only an in situ characterization technique such as RHEED. Finally, this work demonstrates the strong capabilities of the RHEED to characterize the NW crystal structure during the growth, by confirming the observations using numerical simulations and *post mortem* TEM measurements. A value of the critical contact angle $\beta_1 = 125°$ was obtained by combining the RHEED analysis with a numerical model. Furthermore, numerical simulations provided a set of parameters intended to stabilize the contact angle of the Ga droplet in self-assisted GaAs nanowires grown by MBE for an extended time. An extended pure WZ segment longer than one micrometer was obtained, experimentally demonstrating the applicability of the theoretical model. Evidences of a 1.3–1.5 μm-long extended WZ segment, grown using a V/III flux ratio of 4.0, were provided using high-resolution HAADF-STEM and DF-TEM, and supported by RHEED analysis. Furthermore, the luminescence properties of the NWs were investigated by PL measurements, showing the spectral dependence on the crystalline phase.

This chapter demonstrates that the precise tuning of the As flux opens the way to control the length of pure WZ crystalline segments in self-assisted GaAs nanowires. The growth of pure WZ self-assisted GaAs nanowires could potentially be achieved with the constant tuning of the As flux over time, modified through a RHEED feedback loop to compensate the variation of the amount of Ga supplied to the droplet during the transient regime. This combined experimental and numerical study is of major interest to tune the properties of III-V nanostructures on demand, and to fabricate semiconducting heterostructures with novel functionalities, such as the growth of hexagonal Ge by MBE for photonic applications [53], or GaAs-$TiO_2$ nanostructures for photo-driven hydrogen production [54].

## References

1. F. Glas, J.C. Harmand, G. Patriarche, Why does wurtzite form in nanowires of III-V zinc blende semiconductors? Phys. Rev. Lett. **99**(14), 146101 (2007). https://doi.org/10.1103/PhysRevLett.99.146101
2. E.M.T. Fadaly, A. Dijkstra, J.R. Suckert, D. Ziss, M.A.J. van Tilburg, C. Mao, Y. Ren, V.T. van Lange, K. Korzun, S. Kölling, M.A. Verheijen, D. Busse, C. Rödl, J. Furthmüller, F. Bechstedt, J. Stangl, J.J. Finley, S. Botti, J.E.M. Haverkort, E.P.A.M. Bakkers, Direct bandgap emission from hexagonal Ge and SiGe alloys. Nature **580**(7802), 205–209 (2020). https://doi.org/10.1038/s41586-020-2150-y
3. M.C. Plante, R.R. LaPierre, Control of GaAs nanowire morphology and crystal structure. Nanotechnology **19**(49), 495603 (2008). https://doi.org/10.1088/0957-4484/19/49/495603
4. P. Caroff, J. Bolinsson, J. Johansson, Crystal phases in III—V nanowires: From random toward engineered polytypism. IEEE J. Sel. Top. Quantum Electron. **17**(4), 829–846 (2011). https://doi.org/10.1109/JSTQE.2010.2070790
5. D.L. Dheeraj, D.L. Dheeraj, A.M. Munshi, M. Scheffler, A.T. van Helvoort, H. Weman, B.O. Fimland et al., Controlling crystal phases in GaAs nanowires grown by Au-assisted molecular beam epitaxy. Nanotechnology **24**(1), 015601 (2013). https://doi.org/10.1088/0957-4484/24/1/015601
6. M. Soda, M. Soda, A. Rudolph, D. Schuh, J. Zweck, D. Bougeard, E. Reiger et al., Transition from Au to pseudo-Ga catalyzed growth mode observed in GaAs nanowires grown by molecular beam epitaxy. Physical Review. Part B **85**(24), 245450 (2012). https://doi.org/10.1103/PhysRevB.85.245450
7. S. Lehmann, D. Jacobsson, K.A. Dick, Crystal phase control in GaAs nanowires: Opposing trends in the Ga- and As-limited growth regimes. Nanotechnology **26**(30), 301001 (2015). https://doi.org/10.1088/0957-4484/26/30/301001
8. J.B. Jackson, D. Kapoor, S.-G. Jun, M.S. Miller, Integrated silicon nanowire diodes and the effects of gold doping from the growth catalyst. J. Appl. Phys. **102**(5), 054310 (2007). https://doi.org/10.1063/1.2778290
9. V. Schmidt, J.V. Wittemann, S. Senz, U. Gösele, Silicon nanowires: A review on aspects of their growth and their electrical properties. Adv. Mater. **21**(25–26), 2681–2702 (2009). https://doi.org/10.1002/adma.200803754
10. F. Panciera, Z. Baraissov, G. Patriarche, V.G. Dubrovskii, F. Glas, L. Travers, U. Mirsaidov, J.C. Harmand, Phase selection in self-catalyzed GaAs nanowires. Nano Lett. **20**(3), 1669–1675 (2020). https://doi.org/10.1021/acs.nanolett.9b04808
11. M.M. Jansen, P. Perla, M. Kaladzhian, N. von den Driesch, J. Janßen, M. Luysberg, M.I. Lepsa, D. Grützmacher, A. Pawlis, Phase-Pure wurtzite GaAs nanowires grown by self-catalyzed selective area molecular beam epitaxy for advanced laser devices and quantum disks. ACS Applied Nano Materials **3**(11), 11037–11047 (2020). https://doi.org/10.1021/acsanm.0c02241
12. P. Krogstrup, M. Hannibal Madsen, W. Hu, M. Kozu, Y. Nakata, J. Nygård, M. Takahasi, R. Feidenhans'l, *In-situ* x-ray characterization of wurtzite formation in GaAs nanowires. Appl. Phys. Lett., **100**(9), 093103. https://doi.org/10.1063/1.3688489
13. D. Jacobsson, F. Panciera, J. Tersoff, M.C. Reuter, S. Lehmann, S. Hofmann, K.A. Dick, F.M. Ross, Interface dynamics and crystal phase switching in GaAs nanowires. Nature **531**(7594), 317–322 (2016). https://doi.org/10.1038/nature17148
14. J.C. Harmand, G. Patriarche, F. Glas, F. Panciera, I. Florea, J.L. Maurice, L. Travers, Y. Ollivier, Atomic step flow on a nanofacet. Phys. Rev. Lett. **121**(16), 166101 (2018). https://doi.org/10.1103/PhysRevLett.121.166101
15. D. Scarpellini, A. Fedorov, C. Somaschini, C. Frigeri, M. Bollani, S. Bietti, R. Nöetzel, S. Sanguinetti, Ga crystallization dynamics during annealing of self-assisted GaAs nanowires. Nanotechnology **28**(4), 045605 (2017). https://doi.org/10.1088/1361-6528/28/4/045605
16. F. Bastiman, H. Küpers, C. Somaschini, V.G. Dubrovskii, L. Geelhaar, Analysis of incubation time preceding the Ga-assisted nucleation and growth of GaAs nanowires on Si(111). Phys. Rev. Mater., **3**(7), 3, 073401 (2019). https://doi.org/10.1103/PhysRevMaterials.3.073401

17. J. Jo, Y. Tchoe, G.C. Yi, M. Kim, Real-time characterization using in situ RHEED transmission mode and TEM for investigation of the growth behaviour of nanomaterials. Sci. Rep. **8**(1), 1694 (2018). https://doi.org/10.1038/s41598-018-19857-2
18. J. Jakob, P. Schroth, L. Feigl, D. Hauck, U. Pietsch, T. Baumbach, Quantitative analysis of time-resolved RHEED during growth of vertical nanowires. Nanoscale **12**(9), 5471–5482 (2020). https://doi.org/10.1039/c9nr09621c
19. G.E. Cirlin, V.G. Dubrovskii, Y.B. Samsonenko, A.D. Bouravleuv, K. Durose, Y.Y. Proskuryakov, B. Mendes, L. Bowen, M.A. Kaliteevski, R.A. Abram, D. Zeze, Self-catalyzed, pure zincblende GaAs nanowires grown on Si(111) by molecular beam epitaxy. Physical Review. Part B **82**(3), 035302 (2010). https://doi.org/10.1103/PhysRevB.82.035302
20. X. Yu, H. Wang, J. Lu, J. Zhao, J. Misuraca, P. Xiong, S. von Molnár, Evidence for structural phase transitions induced by the triple phase line shift in self-catalyzed GaAs nanowires. Nano Lett. **12**(10), 5436–5442 (2012). https://doi.org/10.1021/nl303323t
21. P. Krogstrup, S. Curiotto, E. Johnson, M. Aagesen, J. Nygård, D. Chatain, Impact of the liquid phase shape on the structure of III-V nanowires. Phys. Rev. Lett. **106**(12), 125505 (2011). https://doi.org/10.1103/PhysRevLett.106.125505
22. V.G. Dubrovskii, G.E. Cirlin, N.V. Sibirev, F. Jabeen, J.C. Harmand, P. Werner, New mode of vapor−liquid−solid nanowire growth. Nano Lett. **11**(3), 1247–1253 (2011). https://doi.org/10.1021/nl104238d
23. P. Krogstrup, H.I. Jørgensen, E. Johnson, M.H. Madsen, C.B. Sørensen, A.F. Morral, M. Aagesen, J. Nygård, F. Glas, Advances in the theory of III–V nanowire growth dynamics. J. Physics. Part D **46**(31), 313001 (2013). https://doi.org/10.1088/0022-3727/46/31/313001
24. W. Kim, W. Kim, V.G. Dubrovskii, J. Vukajlovic-Plestina, G. Tütüncüoglu, L. Francaviglia, L. Güniat, H. Potts, M. Friedl, J.B. Leran, I. Fontcuberta, A. Morral et al., Bistability of contact angle and its role in achieving quantum-thin self-assisted GaAs nanowires. Nano Lett. **18**(1), 49–57 (2018). https://doi.org/10.1021/acs.nanolett.7b03126
25. T. Dursap, M. Vettori, A. Danescu, C. Botella, P. Regreny, G. Patriarche, M. Gendry, J. Penuelas, Crystal phase engineering of self-catalyzed GaAs nanowires using a RHEED diagram. Nanoscale Advances **2**(5), 2127–2134 (2020). https://doi.org/10.1039/d0na00273a
26. T. Dursap, M. Vettori, C. Botella, P. Regreny, N. Blanchard, M. Gendry, N. Chauvin, M. Bugnet, A. Danescu, J. Penuelas, Wurtzite phase control for self-assisted GaAs nanowires grown by molecular beam epitaxy. Nanotechnology **32**(15), 155602 (2021). https://doi.org/10.1088/1361-6528/abda75
27. L. Fouquat, M. Vettori, C. Botella, A. Benamrouche, J. Penuelas, G. Grenet, X-ray photoelectron spectroscopy study of Ga nanodroplet on silica-terminated silicon surface for nanowire growth. J. Cryst. Growth, **514**, 83–88. (2019). https://doi.org/10.1016/j.jcrysgro.2019.03.003
28. L. Fouquat, X. Guan, C. Botella, G. Grenet, P. Regreny, M. Gendry, H. Yi, J. Avila, M. Bugnet, J. Penuelas, Insights into the arsenic shell decapping mechanisms in As/GaAs nanowires by x-ray and electron microscopy. J. Phys. Chem. C **125**(51), 28136–28142 (2021). https://doi.org/10.1021/acs.jpcc.1c09101
29. V.G. Dubrovskii, N.V. Sibirev, J.C. Harmand, F. Glas, Growth kinetics and crystal structure of semiconductor nanowires. Physical Review. Part B **78**(23), 235301 (2008). https://doi.org/10.1103/PhysRevB.78.235301
30. P. Krogstrup, R. Popovitz-Biro, E. Johnson, M.H. Madsen, J. Nygård, H. Shtrikman, Structural phase control in self-catalyzed growth of GaAs nanowires on silicon (111). Nano Lett. **10**(11), 4475–4482 (2010). https://doi.org/10.1021/nl102308k
31. V.G. Dubrovskiĭ, N.V. Sibirev, R.A. Suris, G.É. Cirlin, V.M. Ustinov, M. Tchernysheva, J.C. Harmand, The role of surface diffusion of adatoms in the formation of nanowire crystals. Semiconductors **40**(9), 1075–1082 (2006). https://doi.org/10.1134/S1063782606090168
32. M. Tchernycheva, M. L. Travers, G. Patriarche, F. Glas, J. Harmand, G.E. Cirlin, V.G. Dubrovskii, Au-assisted molecular beam epitaxy of InAs nanowires: Growth and theoretical analysis. J. Appl. Phys., **102**(9), 094313. https://doi.org/10.1063/1.2809417
33. M. Vettori, A. Danescu, X. Guan, P. Regreny, J. Penuelas, M. Gendry, Impact of the Ga flux incidence angle on the growth kinetics of self-assisted GaAs nanowires on Si(111). Nanoscale Advances **1**(11), 4433–4441 (2019). https://doi.org/10.1039/c9na00443b

34. V.G. Dubrovskii, V.G. Dubrovskii, G.E. Cirlin, I.P. Soshnikov, A.A. Tonkikh, N.V. Sibirev, Y.B. Samsonenko, V.M. Ustinov et al., Diffusion-induced growth of GaAs nanowhiskers during molecular beam epitaxy: Theory and experiment. Physical Review. Part B **71**(20), 205325 (2005). https://doi.org/10.1103/PhysRevB.71.205325
35. K. Momma, F. Izumi, VESTA *3* for three-dimensional visualization of crystal, volumetric and morphology data. J. Appl. Crystallogr. **44**(6), 1272–1276 (2011). https://doi.org/10.1107/S0021889811038970
36. S.J. Pennycook, B. Rafferty, P.D. Nellist, Z-contrast imaging in an aberration-corrected scanning transmission electron microscope. Microsc. Microanal. **6**(4), 343–352 (2000). https://doi.org/10.1007/s100050010045
37. M. de la Mata, C. Magen, J. Gazquez, M.I. Utama, M. Heiss, S. Lopatin, F. Furtmayr, C.J. Fernández-Rojas, B. Peng, J.R. Morante, R. Rurali, M. Eickhoff, Fontcuberta i Morral, A., Xiong, Q., & Arbiol, J., Polarity assignment in ZnTe, GaAs, ZnO, and GaN-AlN nanowires from direct dumbbell analysis. Nano Lett. **12**(5), 2579–2586 (2012). https://doi.org/10.1021/nl300840q
38. S. Mokkapati, D. Saxena, H.H. Tan, C. Jagadish, Optical design of nanowire absorbers for wavelength selective photodetectors. Sci. Rep. **5**, 15339 (2015). https://doi.org/10.1038/srep15339
39. D. Spirkoska, D. Spirkoska, A.L. Efros, W.R.L. Lambrecht, T. Cheiwchanchamnangij, Fontcuberta i Morral, A., & Abstreiter, G. et al., Valence band structure of polytypic zinc-blende/wurtzite GaAs nanowires probed by polarization-dependent photoluminescence. Physical Review. Part B **85**(4), 045309 (2012). https://doi.org/10.1103/PhysRevB.85.045309
40. N. Vainorius, D. Jacobsson, S. Lehmann, A. Gustafsson, K.A. Dick, L. Samuelson, M. Pistol, Observation of type-II recombination in single wurtzite/zinc-blende GaAs heterojunction nanowires. Physical Review. Part B **89**(16), 165423 (2014). https://doi.org/10.1103/PhysRevB.89.165423
41. R. Anufriev, R. Anufriev, N. Chauvin, H. Khmissi, K. Naji, M. Gendry, C. Bru-Chevallier et al., Impact of substrate-induced strain and surface effects on the optical properties of InP nanowires. Appl. Phys. Lett. **101**(7), 072101 (2012). https://doi.org/10.1063/1.4745608
42. A. Senichev, A. Senichev, P. Corfdir, O. Brandt, M. Ramsteiner, S. Breuer, J. Schilling, L. Geelhaar, P. Werner et al., Electronic properties of wurtzite GaAs: A correlated structural, optical, and theoretical analysis of the same polytypic GaAs nanowire. Nano Res. **11**(9), 4708–4721 (2018). https://doi.org/10.1007/s12274-018-2053-5
43. N. Yamamoto, S. Bhunia, Y. Watanabe, Polarized cathodoluminescence study of InP nanowires by transmission electron microscopy. Appl. Phys. Lett. **88**(15), 153106 (2006). https://doi.org/10.1063/1.2168043
44. D.D. Sell, Resolved free-exciton transitions in the optical-absorption spectrum of GaAs. Phys. Review. Part B **6**(10), 3750–3753 (1972). https://doi.org/10.1103/PhysRevB.6.3750
45. S.B. Nam, D.C. Reynolds, C.W. Litton, R.J. Almassy, T.C. Collins, C.M. Wolfe, Free-exciton energy spectrum in GaAs. Physical Review. Part B **13**(2), 761–767 (1976). https://doi.org/10.1103/PhysRevB.13.761
46. H. Ehrenreich, F. Spaepen, *Solid state physics, 61* (Academic Press, 2009)
47. F. Martelli, M. Piccin, G. Bais, F. Jabeen, S. Ambrosini, S. Rubini, A. Franciosi, Photoluminescence of Mn-catalyzed GaAs nanowires grown by molecular beam epitaxy. Nanotechnology **18**(12), 125603 (2007). https://doi.org/10.1088/0957-4484/18/12/125603
48. L.C. Chuang, M. Moewe, K.W. Ng, T.D. Tran, S. Crankshaw, R. Chen, W.S. Ko, C. Chang-Hasnain, GaAs nanoneedles grown on sapphire. Appl. Phys. Lett. **98**(12), 123101 (2011). https://doi.org/10.1063/1.3567492
49. S. Furthmeier, F. Dirnberger, J. Hubmann, B. Bauer, T. Korn, C. Schüller, J. Zweck, E. Reiger, D. Bougeard, Long exciton lifetimes in stacking-fault-free wurtzite GaAs nanowires. Appl. Phys. Lett. **105**(22), 222109 (2014). https://doi.org/10.1063/1.4903482
50. A. De, C.E. Pryor, Optical dielectric functions of wurtzite III-V semiconductors. Physical Review. Part B **85**(12), 125201 (2012). https://doi.org/10.1103/PhysRevB.85.125201

51. M. De Luca, S. Rubini, M. Felici, A. Meaney, P.C.M. Christianen, F. Martelli, A. Polimeni, Addressing the fundamental electronic properties of wurtzite GaAs nanowires by high-field magneto-photoluminescence spectroscopy. Nano Lett. **17**(11), 6540–6547 (2017). https://doi.org/10.1021/acs.nanolett.7b02189
52. Z. Lu, S. Shi, J. Lu, P. Chen, Photoluminescence of the single wurtzite GaAs nanowire with different powers and temperatures. J. Lumin. **152**, 258–261 (2014). https://doi.org/10.1016/j.jlumin.2014.01.066
53. I. Dudko, T. Dursap, A.D. Lamirand, C. Botella, P. Regreny, A. Danescu, S. Brottet, M. Bugnet, S. Walia, N. Chauvin, J. Penuelas, Hexagonal Ge Grown by molecular beam epitaxy on self-assisted GaAs nanowires. Cryst. Growth Des. **22**(1), 32–36 (2022). https://doi.org/10.1021/acs.cgd.1c00945
54. T. Dursap, M. Fadel, P. Regreny, C. Tapia Garcia, C. Chevalier, H.S. Nguyen, E. Drouard, S. Brottet, M. Gendry, A. Danescu, M. Koepf, V. Artero, M. Bugnet, J. Penuelas, Enhanced light trapping in GaAs/TiO2-based photocathodes for hydrogen production. ACS Appl. Mater. Interfaces. **15**(46), 53446–53454 (2023). https://doi.org/10.1021/acsami.3c11481

# Chapter 9
# Exploration of Nanowire Photonics Towards Advent of Super-Smart Societies

**Jun Tatebayashi and Yasufumi Fujiwara**

**Abstract** Herein, the growth and optical characteristics of ZnO/Tm,Yb-codoped ZnO core–shell nanowires (ZnO:Tm,Yb/ZnO NWs). The growth conditions of the NWs are optimized by systematically changing the growth temperature, VI/II ratio, and growth duration. Photoluminescence (PL) characterization of the ZnO:Tm,Yb/ZnO NWs with different laser sources is performed to investigate the mechanism of energy transfer between the $Tm^{3+}$ and $Yb^{3+}$ ions in the NWs. Resonant excitation of $Tm^{3+}$ ions at 488 nm results in luminescence from the $Yb^{3+}$ ions, whereas no luminescence is observed from $Yb^{3+}$ ions for non-resonant excitation. Furthermore, time-resolved PL characterization of the NWs reveals a shortening of the lifetime of $Tm^{3+}$ luminescence in the ZnO:Tm,Yb/ZnO NWs, as compared with that in ZnO:Tm,Yb/ZnO NWs. These results demonstrate that energy transfer from $Tm^{3+}$ to $Yb^{3+}$ ions in the NWs occurs by down-conversion.

## 9.1 Introduction

In 2015, the United Nations Member States adopted 17 interlinked global goals, which is called "the Sustainable Development Goals (SDGs)". The SDGs were developed in the Post-2015 Development Agenda as the future global development framework to succeed the Millennium Development Goals which ended in 2015, towards a plan of action for people, planet and prosperity in the next 15 years towards 2030, which is called "the 2030 Agenda". Japanese government proposed a new society, "Society 5.0", in the 5th Science and Technology Basic Plan that Japan should aspire to. The past and ongoing human societies can be categorized into 4 societies, the hunting society (Society 1.0), agricultural society (Society 2.0), industrial society (Society 3.0), and information society (Society 4.0) and the Society 5.0 will be the forthcoming "human-centered" society that balances economic advancement with the resolution of social problems by a system that highly integrates both cyberspace and physical space.

J. Tatebayashi (✉) · Y. Fujiwara
Graduate School of Engineering, Osaka University, Osaka 565-0871, Japan
e-mail: tatebaya@mat.eng.osaka-u.ac.jp

 
F. Ishikawa et al. (eds.), *Engineering Crystal Habit*, Topics in Applied Physics 152,
https://doi.org/10.1007/978-981-96-0266-7_9

As a science and technology supporting the "Internet of Things" that Society 5.0 is pursuing, semiconductor devices have evolved as an indispensable basis of science and technology in our daily life. Furthermore, Si-based photovoltaics (PVs) are the most prevalent solar cell technologies yielding electricity from sunlight, but have a problem of limited energy conversion efficiency to ~30%, known as the Shockley-Queisser limit, which refers to the maximum theoretical energy conversion efficiency of PVs using a single p–n junction [1]. This is predominantly due to the mismatch between wide range of solar spectrum and the absorption properties of Si, where the excessive energy of photons with much higher energy than the Si bandgap is lost as phonons. Down-conversion (DC), in which one high-energy photon is absorbed and two lower-energy photons are emitted [2–5], is a promising strategy for improving the energy conversion efficiencies of PVs, as it could theoretically circumvent such Shockley-Quiesser limit from 30 to ~40% [6]. Specifically, Tm,Yb-codoped ZnO (ZnO:Tm,Yb) is a promising combination of rare-earth elements for DC processes as a means to improve the energy conversion efficiencies of Si PVs. In this material, the ZnO host can efficiently absorb high-energy photons from the solar spectrum efficiently and transfer their energy to the intra-4f shell of $Tm^{3+}$ ions (the ${}^1G_4$-${}^3H_6$ transition) emitting at 489 nm. Next, the energy is transferred to the ${}^2F_{5/2}$-${}^2F_{7/2}$ transition of $Yb^{3+}$ ions referred as the DC process, where radiative recombination emits photons in the near-infrared region at 980 nm [7, 8]. Finally, photons from $Yb^{3+}$ ions can be efficiently absorbed by the Si PVs without producing phonons, and hence the quantum efficiency should exceed 100%.

Tatebayashi et al. have previously demonstrated the growth of ZnO:Tm,Yb films on sapphire substrates grown by sputtering-assisted metalorganic chemical vapor deposition (SA-MOCVD) and observed their luminescence due to intra-4f-shell transitions between $Tm^{3+}$ and $Yb^{3+}$ ions [9]. However, the highest ${}^1G_4$-${}^3H_6$ transition in the $Tm^{3+}$ ions, corresponding to luminescence at 489 nm and is essential for the DC process [7], has been very difficult to observe. This is due to the comparatively low crystallinity of the ZnO films studied, which is in turn due to lattice mismatch between the ZnO films and sapphire substrates.

Semiconductor nanowires (NWs), which are crystalline, highly oriented, and vertically aligned nanostructures, have been continuously exploited as key nanophotonic components, such as single-photon emitters [10], nanolasers [11–15], flexible light-emitting diodes [16–18], and PVs [19–23]. NW-based PVs have recently garnered considerable research attention because the light-trapping ability of NWs imparts significantly improved carrier-generation efficiency. This kind of NW structure is an effective architecture for creating ZnO materials with high purities and improved crystallinities. This is because the strain energy caused by lattice mismatch in such architectures can be relieved in the radial direction due to the unique boundary conditions resulted from the finite lateral dimensionality of the NW [24, 25]. Accordingly, Tatebayashi et al. have previously exploited NW architectures to improve the crystallinity of ZnO and demonstrated their luminescence at 489 nm based on the ${}^1G_4$-${}^3H_6$ transition by growing ZnO/ZnO:Tm,Yb core–shell (ZnO:Tm,Yb/ZnO) NW structures [26].

In this chapter, the growth and optical characteristics of ZnO:Tm,Yb/ZnO NWs by SA-MOCVD is discussed to obtain high-quality ZnO:Tm,Yb layers. The growth conditions of the core ZnO NWs are optimized by systematically altering the growth temperature, VI/II ratio, or growth duration and characterizing the resulting materials by scanning electron microscopy (SEM) and optical measurements. Photoluminescence (PL) characterization of the ZnO:Tm,Yb/ZnO NWs with different pump laser sources is performed to investigate the mechanism of energy transfer between the RE ions therein. Time-resolved PL (TR-PL) characterization of the ZnO:Tm,Yb/ZnO NWs is performed, revealing that the PL lifetimes of the RE ions vary depending on the presence of the co-dopants.

### *9.1.1 Growth and Optical Characteristics of Rare-Earth-Doped ZnO NWs*

Samples were grown by SA-MOCVD facilities (Fig. 9.1a) on c-plane sapphire (0001) substrates [9, 26–28]. 2-inch sapphire wafers were loaded on a horizontal substrate holder with a resistive heater in a bell-jar chamber. Doping the RE ions onto the ZnO host was achieved by sputtering with a sintered mixed-metal-oxide target (5 cm diameter) that was fixed above the substrate at a 90° angle and at a distance of 10 cm. The RE ion concentration was, in the first instance, controlled by adjusting the radio frequency (RF), although several different targets with various metal-oxide ratios were used to achieve the range of RE ion concentration ratios investigated in the present study.

Growth was initiated by the formation of ZnO NWs via the thermal reaction of diethylzinc (DEZn) in $O_2$ gas at a substrate temperature of 650 °C, with a fixed DEZn flow rate of 0.74 μmol/s and various $O_2$ flow rates ranging from 0.74 to 14.8 μmol/s, which correspond to a VI/II ratio from 1 to 20. This growth was then followed by the growth of ZnO:Tm,Yb/ZnO NWs. The incorporation of $Tm^{3+}$ and $Yb^{3+}$ ions was achieved by RF magnetron sputtering using different mix-sintered $Tm_2O_3/Yb_2O_3$ (1:5 or 1:1) targets for co-doping and sintered $Tm_2O_3$ or $Yb_2O_3$ targets for doping. The concentrations of Tm and Yb in ZnO:Tm,Yb with mixing ratio of 1:5 and 1:1 were estimated to be 0.1/0.3 at% and 0.2/0.2 at%, respectively, by the energy

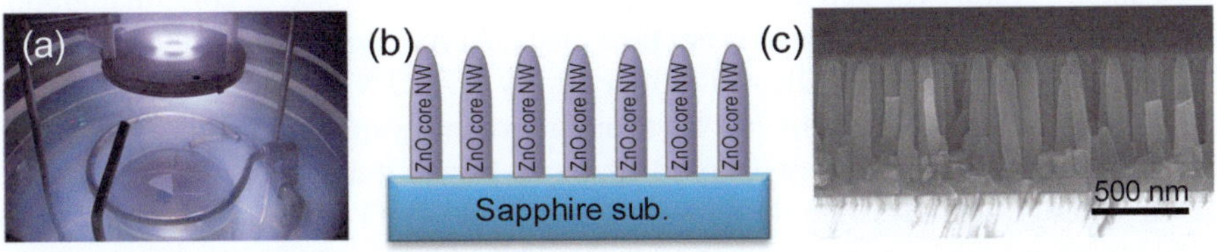


**Fig. 9.1** **a** Photo of the SA-MOCVD chamber. **b** Schematic and **c** cross sectional SEM images of typical ZnO NWs grown on sapphire substrates

dispersive X-ray (EDX) spectroscopy equipped with a scanning electron microscope (JEOL Ltd., EX-64175JMV) [9]. Tm- or Yb-doped ZnO films were also prepared for comparison by using a sintered $Tm_2O_3$ target instead of the mix-sintered target. The concentrations of Tm in ZnO:Tm was estimated to be 0.2 at% by the same EDX analysis. After the NW growth, samples were annealed at 600 °C for 30 min in ambient $O_2$ to activate the luminescence from the $Tm^{3+}$ ions in the ZnO. Structural characterization of the fabricated structures was performed by using scanning electron microscope (SEM) (Hitachi High-Tech SU-9000) and X-ray diffractometer (XRD) (Rigaku HyPix-3000, 45 kV acceleration voltage and 200 mA current) in the θ-2θ configuration at the (002) geometry. Optical characterization was performed for both indirect and direct excitation using a conventional PL setup equipped with an electrically-cooled Si photodiode array (Princeton Instruments, PIXIS 256) and 75-cm monochromator (Acton Research Corporation, SpectraPro-2300i). Indirect excitation of RE ions occurs via the energy transfer from the ZnO host excited by a He-Cd laser (325 nm) (Kimmon Koha, IK3452 R-F), and the resonant excitation of RE ions was performed by using a multi-line air-cooled Ar laser with different emission wavelengths (457, 488, and 514 nm) (Edmund, #58–453). The $^1G_4$-$^3H_6$ transition in the $Tm^{3+}$ ions can only be resonantly photoexcited using a laser wavelength of 488 nm. The PL was detected by the electrically-cooled Si photodiode array. TR-PL measurements were performed at 10 K with excitation by a pulsed He-Cd laser modulated with an acousto-optic modulator (Intra Action Corp. ME-802 J Modulator Driver), and detection with a silicon avalanche photodiode (Hamamatsu Photonics, R13456) equipped with 50-cm spectrometer (Princeton Instruments, Acton SpectraPro 2500). The spectrometer resolution for the TR-PL measurements was below 0.05 nm.

The aim of this study is to grow rare-earth-doped NWs, as illustrated in Fig. 9.1b. ZnO NWs can be obtained using suboptimal growth conditions as shown in Fig. 9.1c. However, the optimization of several growth parameters, including growth temperature, VI/II ratio, and growth duration is required for the growth of high-quality ZnO NWs with high aspect ratios. Therefore, we investigate the effects of temperature, VI/II ratio, and growth duration on the morphological properties of ZnO NWs by fixing the DEZn flow rate while varying the $O_2$ flow rate, growth temperature, and growth duration.

Figure 9.2a shows the SEM images of ZnO films grown for the same growth duration (60 min) at different growth temperatures ranging from 500 to 700 °C. For growth temperatures of 500 and 550 °C, the surface is comprised of columnar grains, but considerable film growth does not take place. At moderate growth temperatures, i.e., 600 and 650 °C, a vertically well-aligned array of ZnO NW structures is grown. At the high growth temperature of 700 °C, a rough surface morphology with large islands forms due to the desorption of Zn or O adatoms. Figure 9.2b shows the dependence of the average thickness/height and diameter of the ZnO structures (grains/NWs/islands) on growth temperature between 500 and 700 °C, as observed from cross-sectional SEM images. The thickness/height increases as the growth temperature increases up to 600 °C due to the enhanced chemical reaction of the growth species at higher temperatures. As the temperature increases beyond 600 °C the thickness

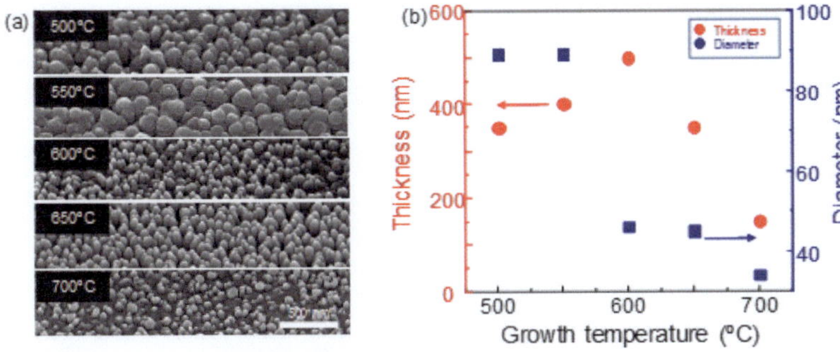


**Fig. 9.2** **a** SEM images of ZnO materials formed at different growth temperatures from 500 to 700 °C, showing the growth of ZnO structures with different configurations, i.e., grains at 500 and 550 °C, NWs at 600 and 650 °C, and islands at 700 °C. **b** Thickness (height) and average diameter of the ZnO structures as a function of growth temperature

decreases, which is likely due to desorption of the growth species. The diameter of the ZnO grains does not vary with the growth temperature below 600 °C. However, at 600 °C, the average diameter dramatically decreases owing to the structural transition from grains to NWs. The diameter is even smaller for growth at 700 °C owing to the transition from NWs to islands.

The optical characteristics of ZnO NWs/grains prepared at different growth temperatures are studied in order to optimize their PL properties. Figure 9.3a shows the low-temperature PL spectra of ZnO grains/NWs/islands prepared with a VI/II ratio of 15 at different temperatures from 500 to 700 °C. Each spectrum is normalized to its maximum PL intensity. The spectra obtained are dominated by the near-band-edge transition of the ZnO bulk at 3.364 eV, which is assigned to the recombination of donor-bound excitons ($D_0X$). However, the actual peak position and shape vary with growth temperature, which is likely due to the effect of strain between the ZnO structure and sapphire substrate. Accordingly, Fig. 9.3b shows the energy difference ($\Delta E$) between the theoretical $D_0X$ emission value and the actual peak as well as its full width at half maximum (FWHM) as a function of growth temperature. The relatively small $\Delta E$ and FWHM values for growth temperatures between 600 and 650 °C indicate that the compressive strain between the ZnO and sapphire substrate is relieved owing to the growth of ZnO NWs, leading to improved crystallinity.

Figure 9.4a–f show top-view and cross-sectional SEM images of ZnO NWs/grains grown at 650 °C for the same duration but with different VI/II ratios ranging from 25 to 1. Figure 9.5a shows a plot of the height and the diameter of the NWs/grains vs. the VI/II ratio used. At a ratio of 25, NWs are not formed, while an intermediate structure is formed at a ratio of 20 (Fig. 9.4b). However, the growth of NWs is observed at ratios below 15. The NW diameter decreases as the VI/II ratio falls from 15 to 5, but the height does not vary. The NW dimensions dramatically increase when the VI/II ratio is very low (~1) owing to a change in the growth mode from the growth of NWs with flat tops to the growth of NWs with hexagonal pyramids on their tips.

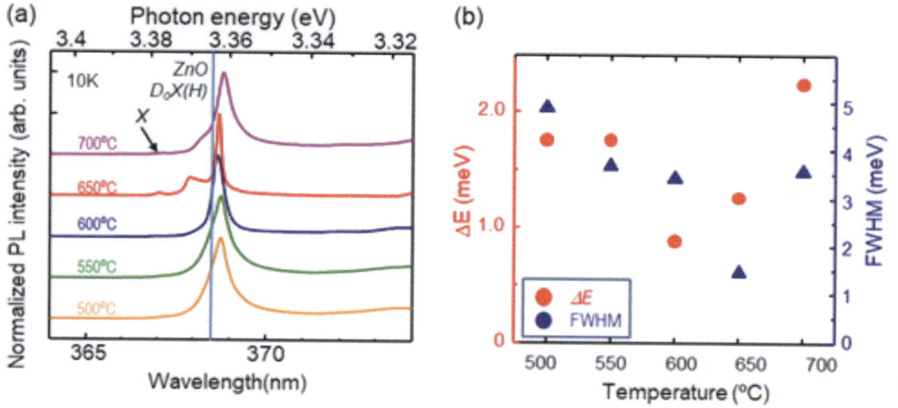


**Fig. 9.3** **a** PL spectra of ZnO structures grown on sapphire substrates at different growth temperatures from 500 to 700 °C. **b** Plots showing how the growth temperature affects energy difference (ΔE) between the theoretical $D_0X$ emission and that of the actual peak as well as its FWHM

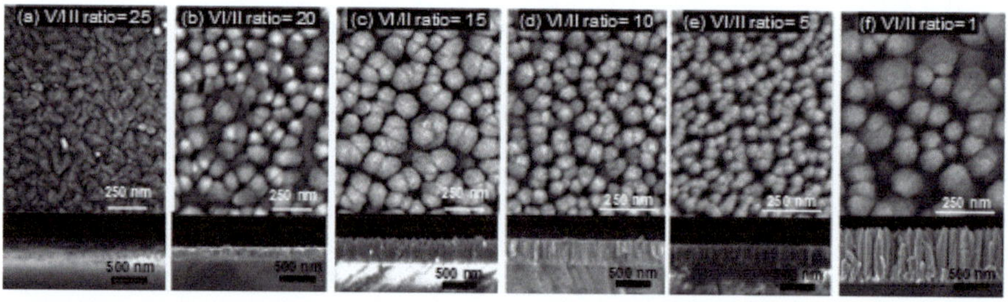


**Fig. 9.4** **a–f** SEM images of ZnO materials grown at various VI/II ratios ranging from 25 to 1

This behavior is consistent with a previous report of the growth of similar hexagonal pyramids on microcrystalline grains in ZnO layers [29].

Figure 9.5a Plot of the average diameter and height of ZnO NWs/grains as a function of VI/II ratio. (b) PL spectra of ZnO NWs/structures grown on sapphire substrates as a function of VI/II ratio ranging from 25 to 1. Figure 9.5b shows the PL spectra recorded at 10 K of ZnO grains/NWs formed at various VI/II ratios ranging from 25 to 1. All the samples exhibit light emission from the ZnO host at ~385 nm. However, the PL intensities are stronger when the ZnO forms as NWs at VI/II ratios between 1 and 15. The PL intensity of the ZnO NWs tends to decrease as the VI/II ratio decreases because the diameter of the NWs decreases accordingly, leading to an increase in the surface-to-volume ratio of the NWs and hence enhanced non-radiative recombination of carriers through surface states.

The results above reveal that a moderately high growth temperature (600–650 °C) and a low VI/II ratio (5–15) are required to form ZnO NWs on sapphire substrates. Thus, as well as the growth temperature and VI/II ratio, the growth duration needs to be optimized. Therefore, the structural and optical characteristics of ZnO NWs

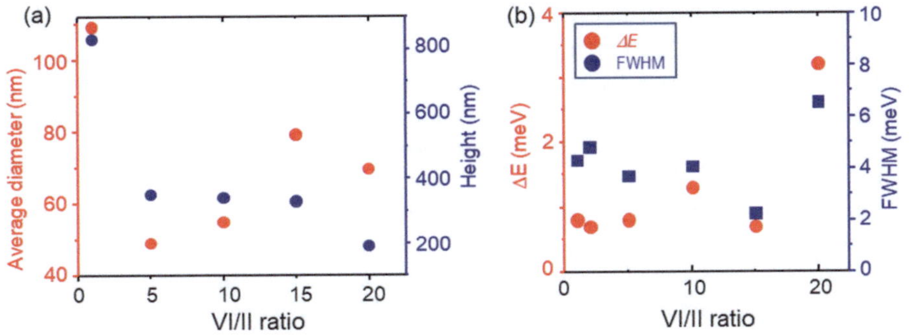


**Fig. 9.5** **a** Plot of the average diameter and height of ZnO NWs/grains as a function of VI/II ratio. **b** Plots showing how the VI/II ratio affects energy difference (ΔE) between the theoretical $D_0X$ emission and that of the actual peak as well as its FWHM

formed at the fixed growth temperature of 650 °C and VI/II ratio of 15 but different growth durations are closely studied.

Figure 9.6a and b show SEM images of typical ZnO NWs prepared with a growth duration of 2 h under the optimized growth conditions of 650 °C and a VI/II ratio of 15. The images reveal the growth of high-density ZnO NWs ($1.25 \times 10^8$ cm$^{-2}$) with an average diameter of ~100 nm. Figure 9.6c shows the average ZnO NW height versus growth duration. The average height of the ZnO NWs shows a monotonic but nonlinear increase with increasing growth duration above 2 h. This is a common behavior in the growth of NWs, where growth of the NW nucleation sites occurs in the initial stage. Subsequently, the nonlinear increase in the height of the NWs is observable as the second stage, which is likely due to the desorption of Zn/O adatoms form the sidewalls of the NWs, resulting in vertical NW growth [30].

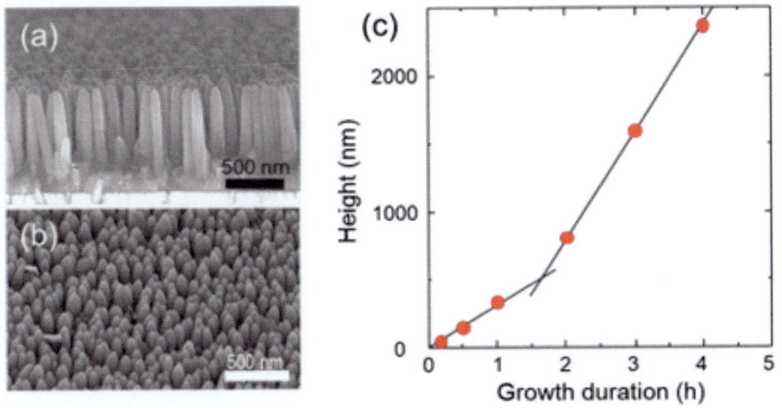


**Fig. 9.6** **a, b** SEM images of typical ZnO NWs on sapphire substrate taken from different angles. **c** Plot of average ZnO NW height as a function of growth duration ranging from 10 min to 4 h

The optical characteristics of ZnO NWs formed under different growth durations are investigated in order to determine the maximum PL intensity of the ZnO core NWs. Figure 9.7a shows PL spectra recorded at 10 K of the ZnO NWs formed at various growth durations ranging from 10 min to 4 h. Each spectrum is normalized by its maximum PL intensity. The obtained spectra are dominated by the near-band-edge transition of bulk ZnO at ~3.364 eV, which is ascribed to the recombination of donor-bound excitons ($D_0X$) [31]. However, the actual peak shape and position vary with the growth duration, which is likely due to variation of the strain field between the ZnO NWs and sapphire substrates. Figure 9.7b shows the energy difference (ΔE) between the actual peak and the theoretical D0X emission peak as well as its FWHM for different growth durations. The relatively large ΔE and FWHM at growth durations less than 1 h indicate lower crystal quality and a comparatively large strain field at the beginning of NW growth. Both the ΔE and FWHM values are minimal for a growth duration of 2 h and show only slight increases with increasing growth durations beyond this point, indicating that the ZnO NWs formed with a growth duration of 2 h have the highest crystallinity and the lowest strain relaxation.

The results above reveal that a moderately high growth temperature (600–650 °C), a low VI/II ratio (5–15) and a proper growth duration of 2 h are mandatory to form high-quality ZnO NWs on sapphire substrates. The structural properties of the fabricated ZnO NWs/grains are characterized by XRD to clarify the growth configuration of the ZnO structures. Figure 9.8a and b compare the pole figures of the ZnO NWs or grains grown at 650 °C and the grains grown at 550 °C under a VI/II ratio of 15 and growth duration of 2 h along the ZnO (101) Bragg reflections. The crystalline axes of ZnO NWs have regularly 60° spaced peaks, whereas that of ZnO columnar grains are randomly distributed in the in-plane direction. This result indicates that ZnO NWs are aligned toward the in-plane direction of the substrate, and

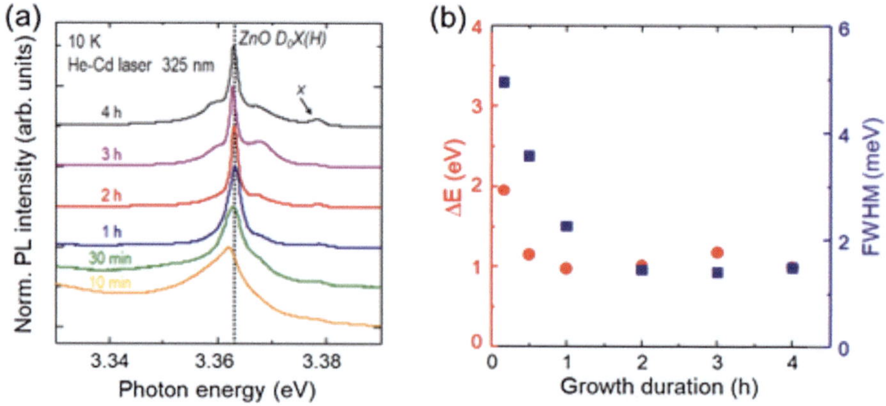


**Fig. 9.7** **a** PL spectra at 10 K of ZnO NWs grown on sapphire substrates for different durations ranging from 10 min to 4 h. **b** Plot showing how growth duration (from 10 min to 4 h) affects the energy difference (ΔE) between the actual peak and the theoretical $D_0X$ emission as well as the FWHM of the PL peak

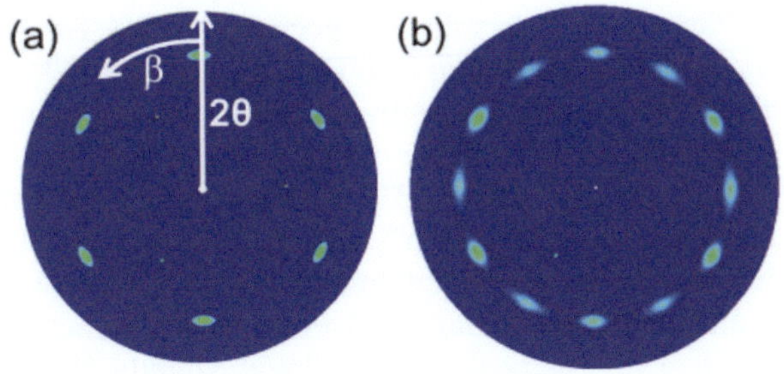


**Fig. 9.8** X-ray pole figures taken at the (101) Bragg reflection for **a** ZnO NWs grown at 650 °C and **b** ZnO grains grown at 550 °C with optimal VI/II ratio of 15 and growth duration of 2 h

agrees well with the previous report by Park et al. on a similar material by MOCVD [32]. The obtained results imply that highly vertically aligned ZnO-based hexagonal NW structures are formed. In terms of both the structural and optical properties, the optimized growth conditions to form ZnO NWs with a high quality and a high uniformity are a growth temperature of 650 °C, a VI/II ratio of 15 and a growth duration of 2 h.

The growth and optical characteristics of ZnO:Tm,Yb/ZnO NWs are studied by growing ZnO:Tm,Yb shell structures on optimized ZnO core NWs, as discussed above. Figure 9.9a shows a schematic and SEM image of ZnO:Tm,Yb/ZnO NWs grown on a sapphire (0001) substrate. Matchstick-shaped heterostructure NWs are formed based on needle-shaped ZnO NW cores with an average height and core diameter of > 1 μm and ~100 nm, respectively, that are grown at a high spatial density on the substrate. During the growth of the ZnO:Tm,Yb coating, the tips of the ZnO core NWs are exposed to the sputtering plasma. Since the obtained ZnO NWs are densely packed on the sapphire substrates, ZnO:Tm,Yb shells only form on the top of the ZnO NWs while the bottoms of the ZnO NWs remain unchanged.

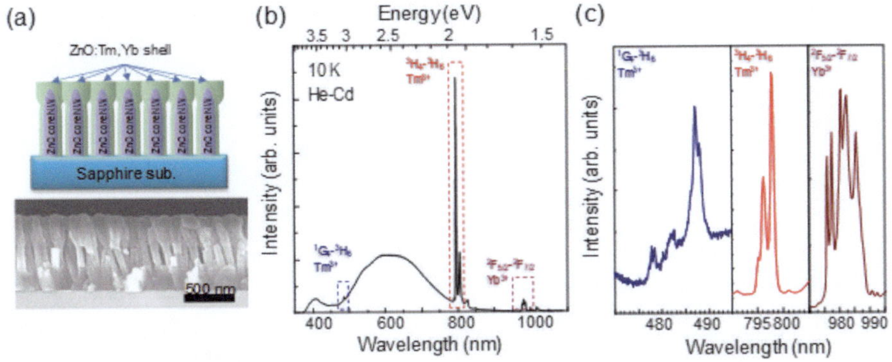


**Fig. 9.9** **a** (Top) Schematic and (bottom) cross-sectional SEM image of typical ZnO@ZnO:Tm,Yb NWs grown on a sapphire substrate. **b** PL spectra of ZnO@ZnO:Tm,Yb NWs recorded at 10 K under excitation by a He-Cd laser. **c** Magnifications of the indicated regions in **b** due to $Tm^{3+}$ and $Yb^{3+}$ luminescence from $^1G_4$–$^3H_6$ (~490 nm), $^3H_4$–$^3H_6$ (~800 nm), and $^2F_{5/2}$–$^2F_{7/2}$ (~980 nm) transitions

Figure 9.9b shows the PL spectrum of ZnO:Tm,Yb/ZnO NWs recorded at 10 K under indirect excitation by a He-Cd laser. A bunch of sharp peaks at ~980 nm that originate from $Yb^{3+}$ intra-4f-shell transitions ($^2F_{5/2}$–$^2F_{7/2}$) is observed, and there are several sharp peaks in the vicinity of 800 nm that originate from $Tm^{3+}$ intra-4f-shell transitions ($^3H_4$–$^3H_6$). Furthermore, a sharp peak originating from the highest $^1G_4$-$^3H_6$ transition in $Tm^{3+}$ ions is observed at ~490 nm, although a very broad emission originating predominantly from oxygen vacancies in the ZnO core NW is also observable. Magnified PL signals for each of these transitions are shown in Fig. 9.9c. The peaks observed with slightly different luminescence wavelengths originate from differences in the local environments of both $Tm^{3+}$ and $Yb^{3+}$ ions [26]. It should be noted here that the luminescence intensity of the highest $^1G_4$–$^3H_6$ transitions from the $Tm^{3+}$ ions is ~10-times stronger than that reported previously [26], which is due to the improved crystallinity of the ZnO core NWs.

In order to investigate the energy transfer between $Tm^{3+}$ and $Yb^{3+}$ ions in ZnO:Tm,Yb/ZnO NWs, PL characterization using direct excitation of $Tm^{3+}$ ions by a three-line Ar laser was performed. Figure 9.10a shows the relevant regions of the PL spectra for the $Yb^{3+}$ ions in ZnO:Tm,Yb/ZnO NWs under excitation by an Ar laser at different wavelengths and an He-Cd laser for comparison.

Under indirect excitation by the He-Cd laser, several sharp peaks from $Yb^{3+}$ ions in ZnO:Tm,Yb/ZnO NWs with different local environment structures are observed in the wavelength range 976–990 nm. In contrast, under excitation by the Ar laser, luminescence from the $Yb^{3+}$ ions is only observed upon excitation at 488 nm and no luminescence is observed when excited at 457 or 514 nm. This indicates the occurrence of energy transfer from $Tm^{3+}$ to $Yb^{3+}$ ions induced by resonant excitation of the $^3H_4$-$^3H_6$ transition in the $Tm^{3+}$ ions at 488 nm. It should be noted that only a

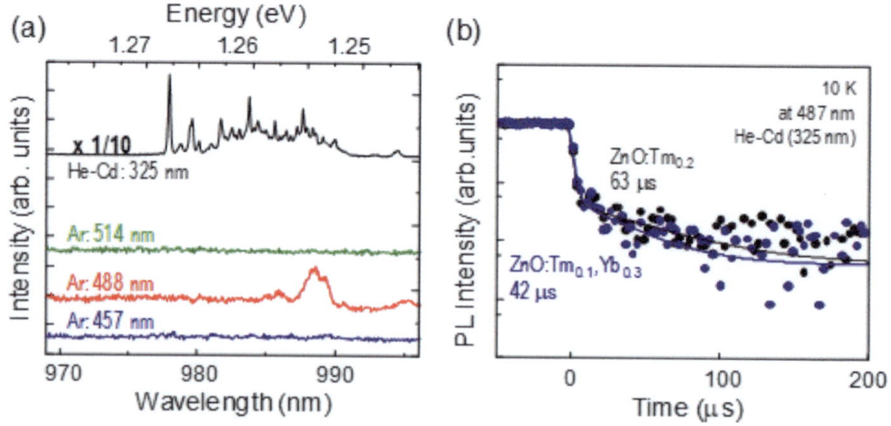


**Fig. 9.10** **a** Enlargement of the $Tm^{3+}$ luminescence region of the PL spectra of ZnO@ZnO:Tm,Yb NWs under excitation by He-Cd (325 nm) and three-line Ar (457, 488, and 514 nm) lasers. **b** Decay curve of $Tm^{3+}$ luminescence at 487 nm in ZnO:Tm (black) and ZnO:Tm,Yb (blue) grown on ZnO NWs. The black and blue solid lines are double exponential fits to the data

portion of the luminescence from $Yb^{3+}$ sites emerges under direct excitation of the $^1G_4$-$^3H_6$ transition in $Tm^{3+}$ ions, as shown in Fig. 9.10a. This phenomenon suggests that only $Yb^{3+}$ ions with specific local structures participate in the energy transfer from $Tm^{3+}$ ions. Detailed studies on this will be the subject of future publications.

TR-PL analysis is performed to investigate the energy transfer mechanism between the RE ions in ZnO:Tm,Yb/ZnO NWs, and the results are shown in Fig. 9.10b. Two different samples are prepared; ZnO:Tm,Yb/ZnO and ZnO:Tm/ZnO NWs. The PL lifetime of $Tm^{3+}$ was estimated by fitting the obtained decay curve with a double exponential curve. There are two decay components; one is a fast decay originating from luminescence related to defects and the other is a slow decay originating from luminescence from $Tm^{3+}$ ions. The lifetime of the $Tm^{3+}$ luminescence decreases from 63 to 42 μs when both Tm and Yb are co-doped into ZnO. This result demonstrates the existence of energy transfer from $Tm^{3+}$ to $Yb^{3+}$ ions in the ZnO:Tm,Yb shell.

## 9.2 Summary

In conclusion, the growth and optical properties of ZnO:Tm,Yb/ZnO grown on sapphire substrates by SA-MOCVD are reported for the demonstration of DC. The growth conditions of the ZnO core NWs are optimized by changing several growth parameters, including the growth temperature, VI/II ratio, and growth duration. PL characterization of ZnO:Tm,Yb/ZnO NWs with different pump laser sources is performed to investigate the energy transfer mechanism between the RE ions in the ZnO:Tm,Yb shell. Indirect excitation of the ZnO core produces several sharp peaks from both $Tm^{3+}$ and $Yb^{3+}$ ions as well as broad luminescence from the oxygen vacancies in the ZnO core. Most importantly, resonant excitation of $Tm^{3+}$ ions results in luminescence from $Yb^{3+}$ ions, while no luminescence is observed from $Yb^{3+}$ ions when pumped at 457 and 514 nm. TRPL characterization of ZnO:Tm,Yb/ZnO NWs reveals a shortening of the lifetime of $Tm^{3+}$ luminescence in the Tm,Yb-codoped ZnO NWs when compared with that of Tm-doped ZnO NWs.

Thus, both the PL characterization under direct excitation of $Tm^{3+}$ ions and TR-PL measurements confirm the occurrence of DC, but further study on the details, including pump-power dependence, are required to clarify the details of the DC process. Nevertheless, obtained demonstration of this energy transfer provides a potential strategy for the realization of PVs with improved energy conversion efficiencies utilizing wavelength conversion.

## References

1. W. Shockley, H.J. Queisser, Detailed balance limit of efficiency of p-n junction solar cells. J. Appl. Phys. **32**, 510–519 (1961)

2. R.T. Wegh, H. Donker, K.D. Oskam, A. Meijerink, Visible quantum cutting in LiGdF4:Eu3+ through downconversion. Science **283**, 663–666 (1999)
3. T. Trupke, M.A. Green, P. Würfel, Improving solar cell efficiencies by down-conversion of high-energy photons. J. Appl. Phys. **92**, 1668–1674 (2002)
4. P. Vergeer, T.J.H. Vlugt, M.H.F. Kox, M.I. den Hertog, J.P.J.M. van der Eerden, A. Meijerink, Quantum cutting by cooperative energy transfer in YbxY1−xPO4:Tb3+ Phys. Rev. B **71**, 014119 (2005)
5. D. Timmerman, I. Izeddin, P. Stallinga, I.N. Yassievich, T. Gregorkiewicz, Space-separated quantum cutting with silicon nanocrystals for photovoltaic applications. Nature Photon. **2**, 105–109 (2008)
6. S. Kurtz, D. Friedman, J. Geisz, W. McMahon, Using MOVPE growth to generate tomorrow's solar electricity. J. Cryst. Growth **298**, 748–753 (2007)
7. L. Xie, Y. Wang, H. Zhang, Near-infrared quantum cutting in YPO4:Yb3+, Tm3+ via cooperative energy transfer. Appl. Phys. Lett. **94**, 061905 (2009)
8. T. Tsuji, Y. Terai, M.H.B. Kamarudin, Y. Fujiwara, Formation of Eu3+ luminescent centers in Eu-doped ZnO grown by sputtering-assisted metalorganic chemical vapor deposition. Jpn. J. Appl. Phys. **52**, 111101 (2013)
9. J. Tatebayashi, G. Yoshii, T. Nakajima, H. Kamei, J. Takatsu, D.M. Lebrun, Y. Fujiwara, Control of the energy transfer between Tm3+ and Yb3+ ions in Tm, Yb-codoped ZnO grown by sputtering-assisted metalorganic chemical vapor deposition. J. Appl. Phys. **123**, 161409 (2018)
10. J. Tatebayashi, Y. Ota, S. Ishida, M. Nishioka, S. Iwamoto, Y. Arakawa, Site-controlled formation of InAs/GaAs quantum-dot-in-nanowires for single photon emitters. Appl. Phys. Lett. **100**, 263101 (2012)
11. M.H. Huang, S. Mao, H. Feick, H. Yan, Y. Wu, H. Kind, E. Weber, R. Russo, P. Yang, Room-temperature ultraviolet nanowire nanolasers. Science **292**, 1897–1899 (2001)
12. J.X. Ding, J.A. Zapien, W.W. Chen, Y. Lifshitz, S.T. Lee, Lasing in ZnS nanowires grown on anodic aluminum oxide templates. Appl. Phys. Lett. **85**, 2361–2363 (2004)
13. Z. Liu, L. Yin, H. Ning, Z. Yang, L. Tong, C.-Z. Ning, Dynamical color-controllable lasing with extremely wide tuning range from red to green in a single alloy nanowire using nanoscale manipulation. Nano Lett. **13**, 4945–4950 (2013)
14. K.H. Li, X. Liu, Q. Wang, S. Zhao, Z. Mi, Ultralow-threshold electrically injected AlGaN nanowire ultraviolet lasers on Si operating at low temperature. Nat. Nanotechnol. **10**, 140–144 (2015)
15. J. Tatebayashi, S. Kako, J. Ho, Y. Ota, S. Iwamoto, Y. Arakawa, Room-temperature lasing in a single nanowire with quantum dots. Nat. Photon. **9**, 501–505 (2015)
16. C. Lee, Y. Kim, Y.J. Hong, S. Jeon, S. Bae, B.H. Hong, G. Yi, Flexible inorganic nanostructure light-emitting diodes fabricated on graphene films. Adv. Mater. **23**, 4614–4619 (2011)
17. K. Chung, H. Beak, Y. Tchoe, H. Oh, H. Yoo, M. Kim, G.-C. Yi, Growth and characterizations of GaN micro-rods on graphene films for flexible light emitting diodes. APL Mater. **2**, 092512 (2014)
18. N. Guan, X. Dai, A. Messanvi, H. Zhang, J. Yan, E. Gautier, C. Bougerol, F.H. Julien, C. Durand, J. Eymery, M. Tchernycheva, Flexible white light emitting diodes based on nitride nanowires and nanophosphors. ACS Photon. **3**, 597–603 (2016)
19. M. Law, L.E. Greene, J.C. Johnson, R. Saykally, P. Yang, Nanowire dye-sensitized solar cells. Nat. Mater. **4**, 455–459 (2005)
20. T. Stelzner, M. Pietsch, G. Andrä, F. Falk, E. Ose, S. Christiansen, Silicon nanowire-based solar cells. Nanotechnology **19**, 295203 (2008)
21. P. Krogstrup, H.I. Jørgensen, M. Heiss, O. Demichel, J.V. Holm, M. Aagesen, J. Nygard, A.F.I. Morral, Single nanowire solar cells beyond the Shockley-Queisser limit. Nat. Photon. **7**, 306–310 (2013)
22. G. Mariani, A.C. Scofield, C.-H. Hung, D.L. Huffaker, GaAs nanopillar-array solar cells employing in situ surface passivation. Nat. Comm. **4**, 1497 (2013)

23. B.D. Yuhas, P. Yang, Nanowire-based all-oxide solar cells. J. Am. Chem. Soc. **131**, 3756–3761 (2009)
24. E. Ertekin, P.A. Greaney, D.C. Chrzan, T.D. Sands, Equilibrium limits of coherency in strained nanowire heterostructures. J. Appl. Phys. **97**, 114325 (2005)
25. F. Glas, Critical dimensions for the plastic relaxation of strained axial heterostructures in free-standing nanowires. Phys. Rev. B **74**, 121302 (2006)
26. J. Tatebayashi, G. Yoshii, T. Nakajima, M. Mishina, Y. Fujiwara, Formation and optical properties of Tm, Yb-codoped ZnO nanowires grown by sputtering-assisted metalorganic chemical vapor deposition. J. Cryst. Growth **503**, 13–19 (2018)
27. Y. Terai, K. Yoshida, M.H. Kamarudin, Y. Fujiwara, Photoluminescence properties of Eu3+ ions in Eu-doped ZnO grown by sputtering-assisted metalorganic chemical vapor deposition. Phys. Status Solidi C **8**, 519–521 (2011)
28. R. Okada, W. Miao, Y. Terai, T. Tsuji, Y. Fujiwara, Sputtering-assisted metal-organic chemical vapor deposition of Yb-doped ZnO for photonic conversion in Si solar cells. Phys. Status Solidi C **11**, 1292–1295 (2014)
29. W.I. Park, D.H. Kim, S.-W. Jung, G.-C. Yi, Metalorganic vapor-phase epitaxial growth of vertically well-aligned ZnO nanorods. Appl. Phys. Lett. **80**, 4232–4234 (2002)
30. K. Ikejiri, T. Sato, H. Yoshida, K. Hiruma, J. Motohisa, S. Hara, T. Fukui, Growth characteristics of GaAs nanowires obtained by selective area metal–organic vapour-phase epitaxy. Nanotechnology **19**, 265604 (2008)
31. S.F. Chichibu, T. Sota, G. Cantwell, D.B. Eason, C.W. Litton, Polarized photoreflectance spectra of excitonic polaritons in a ZnO single crystal. J. Appl. Phys. **93**, 756–758 (2003)
32. J.Y. Park, D.J. Lee, Y.S. Yun, J.H. Moon, B.-T. Lee, S.S. Kim, Temperature-induced morphological changes of ZnO grown by metalorganic chemical vapor deposition, J. Cryst. Growth., **276**, 158–164 (2005)

# Chapter 10
# Phase-Selective Synthesis of Oxide Materials via Mist Chemical Vapor Deposition

**Hitoshi Takane, Takeru Wakamatsu, Katsuhisa Tanaka, and Kentaro Kaneko**

**Abstract** A mist chemical vapor deposition (CVD) method, which is inexpensive and environmentally friendly, thus a sustainable method for thin film growth, has provided us with opportunities to fabricate various oxide thin films. This chapter reviews our previous work on the mist CVD growth of r-$GeO_2$ and r-$Ge_xSn_{1-x}O_2$ alloy films. We also discuss the synthesis of materials that are difficult to crystallize, like volatile or non-equilibrium-phased oxides by the mist CVD method.

## 10.1 Introduction

A mist chemical vapor deposition (mist CVD) method, which emerged in the mid-2000s [1], has been gaining great attention as a cost-effective and environmentally friendly, thus a sustainable method for thin film growth. As displayed in Fig. 10.1, the whole process in the mist CVD is performed under atmospheric pressure. The outline of the process is as follows;

1. Source solution where precursors are dissolved in a solvent is atomized into mist particles by using an ultrasonic vibrator.

H. Takane (✉) · T. Wakamatsu (✉) · K. Tanaka (✉)
Department of Material Chemistry, Kyoto University, Kyoto 615-8510, Japan
e-mail: takane.hitoshi.33v@st.kyoto-u.ac.jp

T. Wakamatsu
e-mail: wakamatsu.takeru.45e@st.kyoto-u.ac.jp

K. Tanaka
e-mail: tanaka.katsuhisa.4n@kyoto-u.ac.jp

K. Kaneko (✉)
Research Organization of Science and Technology, Ritsumeikan University, Shiga 525-8577, Japan
e-mail: ken0710@fc.ritsumei.ac.jp

 
F. Ishikawa et al. (eds.), *Engineering Crystal Habit*, Topics in Applied Physics 152,
https://doi.org/10.1007/978-981-96-0266-7_10

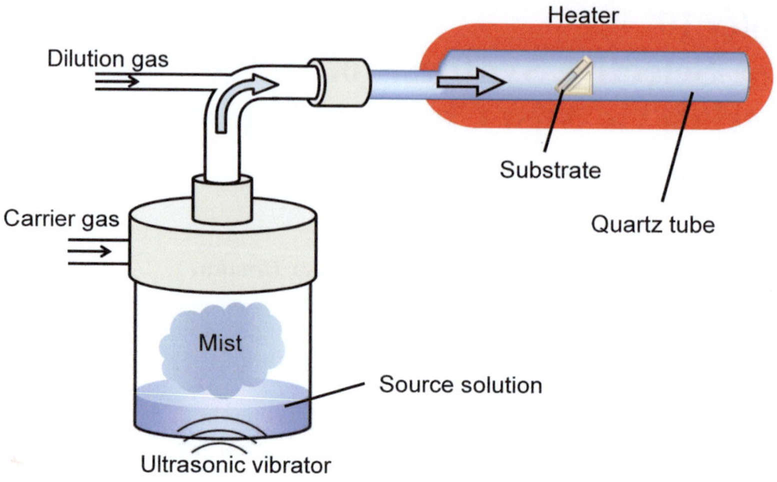


**Fig. 10.1** Schematic image of hot-wall-type mist chemical vapor deposition (CVD) equipment

2. The mist particles are transferred to a substrate in the heated quartz tube by carrier and dilution gas.
3. The chemical reactions occur and thin films are grown on the heated substrate.

So far, the mist CVD has been mainly used for the growth of oxide thin films, such as ZnO [1–4], $SnO_2$ [5], and β-$Ga_2O_3$ [6]. In addition to these oxides, the growth of various metastable oxides, including α-$Ga_2O_3$ [7], α-$In_2O_3$ [8, 9], α-$Ir_2O_3$ [10, 11], by the mist CVD method has been reported, which is one of largest features of the method. However, although there have been some reports on growth dynamics and mechanisms of the mist CVD [12–15], those remain controversial. In this chapter, we present our previous work about the mist CVD growth of r-$GeO_2$ and r-$Ge_xSn_{1-x}O_2$ alloy films, and discuss the synthesis of materials hard to crystallize, like volatile or non-equilibrium-phased oxides by the mist CVD method.

## 10.2 Rutile-Type Germanium Oxide

Germanium oxides ($GeO_2$) has been widely investigated for a wide variety of applications, such as optical glasses [16–19], gate dielectric layers [20, 21], and so on. Recently, however, a lot of attentions have been concentrated into rutile-type $GeO_2$ (r-$GeO_2$), which is one of two predominant polymorphous at ambient conditions (rutile [22] and α-quartz [23]), as a promising material for future power-electronics devices. This is because r-$GeO_2$, with an ultra-wide bandgap of 4.68 eV [24], has been theoretically and experimentally predicted to have excellent physical properties, including ambipolar dopability [25–27], high carrier mobilities [26, 28], water-insolubility

[29], and a higher thermal conductivity than β-$Ga_2O_3$ [30]. Single-crystalline r-$GeO_2$ bulk crystals can be also synthesized [27, 31–33], indicating possibility of homoepitaxy of r-$GeO_2$ thin films in the future. Moreover, like III-V and nitride semiconductors, a rutile-structured semiconductor alloy system ($GeO_2$-$SnO_2$-$SiO_2$ and $TiO_2$) has been proposed as shown in Fig. 10.2 [34], which is base for heterostructures contributing to various optical and electrical devices. In spite of its remarkable properties, r-$GeO_2$ has a large challenge, namely, thin film growth of r-$GeO_2$ has been difficult. Chae et al. pointed out that its difficulty is due to the high vapor pressure of volatile GeO as well as the existence of some deeply metastable polymorphs for $GeO_2$, that is, although r-$GeO_2$ is the lowest free energy phase (stable phase) at the ambient condition, both amorphous-type and quartz-type $GeO_2$ are deeply metastable polymorphs as shown in Fig. 10.3 [35, 36]. In addition, it is kinetically easiest for $GeO_2$ to form the glass phase along with $SiO_2$ and $BO_3$, owning to the high valence state (+4) of germanium that bonds to oxygens forming the interconnected backbone of the glass network that is close in energy to the rutile phase [37]. These days, some attempts have been made for fabrication of r-$GeO_2$ thin film using different growth techniques, including molecular beam epitaxy (MBE) [35], pulsed laser deposition (PLD) [38, 39], and mist CVD [34, 40].

In this section, we mainly describe the results obtained in the framework of already-published studies, where we tried epitaxy of r-$GeO_2$ thin films by using a mist CVD [34, 40]. We present experimental results of growth of r-$GeO_2$ and r-$Ge_xSn_{1-x}O_2$ thin films by the mist CVD. Afterward, we discuss the characteristics of crystallization of r-$GeO_2$ by the mist CVD.

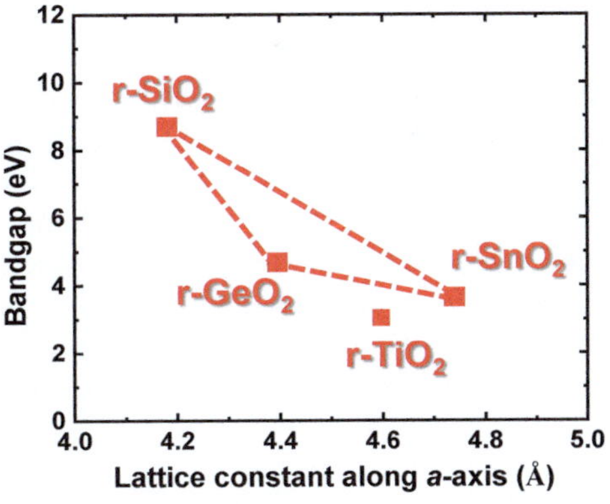


**Fig. 10.2** Relationship between band gap and lattice constant along the *a*-axis of an alloy system based on rutile-structured oxides (r-$GeO_2$, r-$SnO_2$, r-$SiO_2$, and r-$TiO_2$). Reproduced from Takane et al., Phys. Rev. Materials 6, 084,604 (2022). Copyright 2022 The American Physical Society

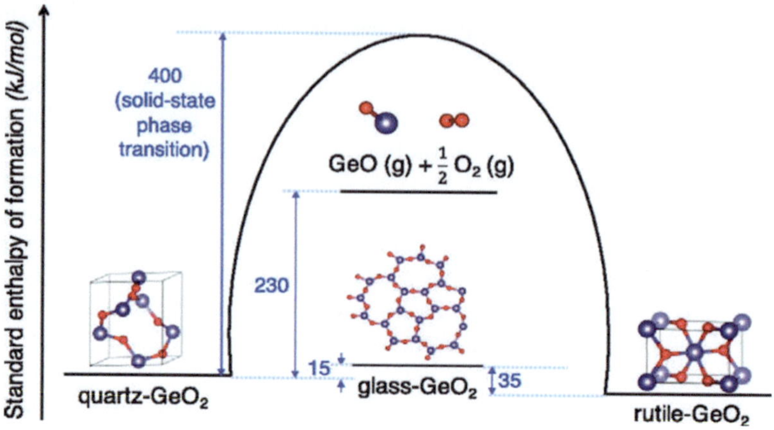


**Fig. 10.3** Reaction coordinate-energy diagram for different polymorphs of $GeO_2$. At the atmospheric condition, the thermodynamic stability of both metastable quartz and glass $GeO_2$ is competitive to the thermodynamically stable r-$GeO_2$ phase. Solid-state reaction of quartz to rutile requires traversing a large energy barrier (400 kJ/mol at 1 atm). Much less energy is required to sublimate $GeO_2$ as GeO (g) + 1/2 $O_2$ (g) and re-condensate it into the rutile phase. For the film growth of r-$GeO_2$, quartz-$GeO_2$ powders are evaporated using the effusion cell [quartz-$GeO_2$ → GeO (g) + 1/2 $O_2$ (g)] and GeO and $O_2$ molecules crystallize into the rutile phase on the substrates [GeO (g) + 1/2 $O_2$ (g) → rutile-$GeO_2$]. Reproduced from Chae et al., Appl. Phys. Lett. 117, 072,105 [35] with the permission of AIP Publishing

## 10.3 Growth of r-$GeO_2$ Thin Film by Mist CVD

Here, we presents the mist CVD growth of r-$GeO_2$ thin films. The isostructural (001)-plane $TiO_2$ (r-$TiO_2$) was used as a substrate. Besides, bis [2-carboxyethylgermanium (IV)] sesquioxide ($C_6H_{10}Ge_2O_7$) was adapted as a Ge precursor and was dissolved in deionized water ($H_2O$) with the addition of a small amount of hydrochloric acid (HCl), which helped to solve the precursors completely. The growth conditions except for the growth temperature ($T_g$) are listed in Table 10.1.

In order to determine the growth temperature, thermogravimetry–differential thermal analysis (TG–DTA) was conducted for the source solution in aqueous vapor, and the decomposition processes of the precursor was analyzed. Figure 10.4 shows

**Table 10.1** Growth conditions for r-$GeO_2$ thin films

| | |
|---|---|
| Ge precursor | Bis [2-carboxyethylgermanium (IV)] sesquioxide: $C_6H_{10}Ge_2O_7$ |
| Solvent | $H_2O$ + HCl |
| Substrate | r-$TiO_2$ (001) |
| Carrier gas (flow rate) | $O_2$ (3.0 L/min) |
| Dilution gas (flow rate) | $O_2$ (0.5 L/min) |

the TG–DTA data, which indicates that the weight decreased at 321–495, 553–783, and 1150–1200 °C after the water completely evaporated at the temperature of ~300 °C. The produced substances at each temperature range were estimated from the weight loss as summarized in Table 10.2. From Table 10.2, $GeO_2$ was estimated to be produced in the temperature range from 553 to 783 °C in aqueous vapor. Thus, we set the growth temperatures of the r-$GeO_2$ thin film to be 550–850 °C, which covers the entire temperature range where $GeO_2$ could be produced.

We tried fabricating r-$GeO_2$ thin films on r-$TiO_2$ substrates at $T_g = 550$–$850$ °C and $C_6H_{10}Ge_2O_7$ molar concentration of 0.25 mol/L. The x-ray diffraction (XRD) 2θ/ω scan profiles of the films grown at $T_g = 675$–$800$ °C is shown in Fig. 10.5. Note that we confirmed that r-$GeO_2$ was not grown at $T_g < 675$ °C and $T_g > 800$ °C by the XRD 2θ/ω scan profiles. At $T_g = 675$ and 800 °C, only peaks of r-$TiO_2$ 002 diffraction are visible. On the contrary, at $T_g = 700$–$775$ °C, peaks of r-$GeO_2$ 002 diffraction as well as those of r-$TiO_2$ 002 diffraction are visible, indicating that crystallized (001)-oriented r-$GeO_2$ thin films were grown on r-$TiO_2$ (001) substrates. Note that no distinct peaks are visible except for the diffraction peaks by the films and

**Table 10.2** Relationships between temperature and weight of the source solution and estimated reaction products corresponding to each temperature range. Reproduced from Takane et al., Appl. Phys. Lett. 119, 062,104 [40] with the permission of AIP Publishing

| Experiment | | | Calculation | | |
|---|---|---|---|---|---|
| Temperature (°C) | Weight (mg) | Weight loss rate (%) | Substances produced at each range | Molar weight (g/mol) | Molar weight loss rate (%) |
| 289 | 0.259 | 0.0 | $Ge_2(O)_3(CH_2CH_2COOH)_2$ | 339.4 | 0.00 |
| 321–495 | 0.213 | 17.8 | $Ge_2(O)_3(COOH)_2$ | 283.3 | 16.5 |
| 553–783 | 0.160 | 38.2 | 2 $GeO_2$ | 209.3 | 38.3 |
| 1150–1200 | 0.141 | 45.6 | 2 GeO | 177.3 | 47.7 |

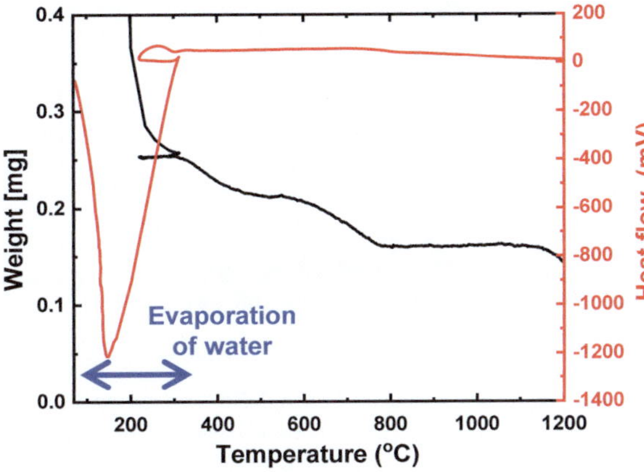


**Fig. 10.4** TG–DTA data for $C_6H_{10}Ge_2O_7$ dissolved in mixed solution ($H_2O$ and HCl) in aqueous vapor. Black and red lines indicate the weight of the sample and heat flow, respectively. Reproduced from Takane et al., Appl. Phys. Lett. 119, 062,104 (2021) [40] with the permission of AIP Publishing

substrates in the entire range of the spectra. The XRD rocking curve full width at half maximums (FWHMs) of 002 diffraction of the crystallized r-$GeO_2$ thin films ($T_g$ = 700–775 °C) were 900–1500 arc sec. The growth rates of all the crystallized r-$GeO_2$ thin films, measured by cross-sectional scanning electron microscopy (SEM), were ≥ 1 μm.

In order to obtain a r-$GeO_2$ thin film with better crystalline quality, we used source solution with $C_6H_{10}Ge_2O_7$ molar concentration of 0.001 mol/L for a lower growth rate. The growth temperature was set to $T_g$ = 725 °C. Figure 10.6a and b show the XRD 2θ/ω and rocking curve ω scan profiles of r-$GeO_2$ grown under the condition. The rocking curve FWHM of the r-$GeO_2$ 002 diffraction was 560 arc sec. For the film with relatively better crystallinity, we conducted a transmission electron microscopy (TEM) observation. The cross-sectional TEM image and selected area

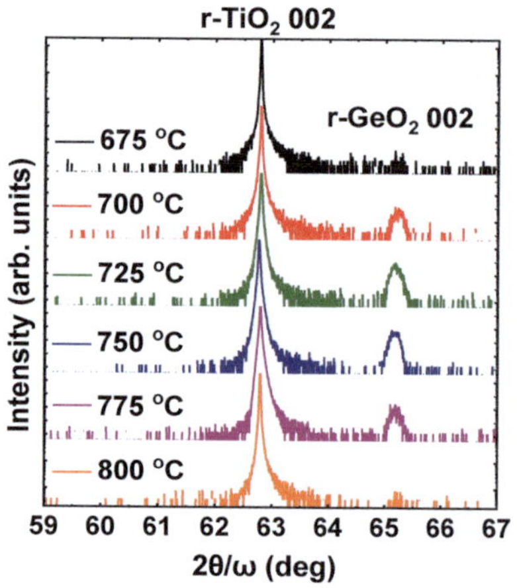


**Fig. 10.5** XRD 2θ/ω scan profiles of the r-$GeO_2$ films on (001) r-$TiO_2$ substrates grown at $T_g$ = 550–850 °C and $C_6H_{10}Ge_2O_7$ molar concentration of 0.25 mol/L. Reproduced from Takane et al., Appl. Phys. Lett. 119, 062,104 [40] with the permission of AIP Publishing

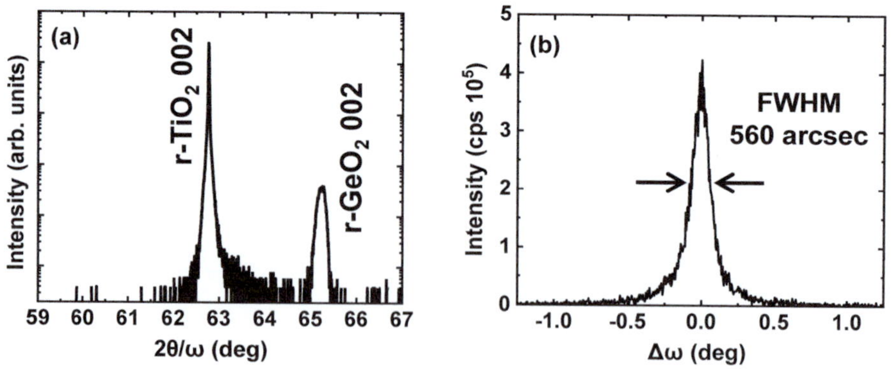


**Fig. 10.6** **a** XRD 2θ/ω scan and **b** XRD ω scan profiles of the r-$GeO_2$ films on (001) r-$TiO_2$ substrates grown at $T_g$ = 725 °C and $C_6H_{10}Ge_2O_7$ molar concentration of 0.001 mol/L. Reproduced from Takane et al., Appl. Phys. Lett. 119, 062,104 [40] with the permission of AIP Publishing

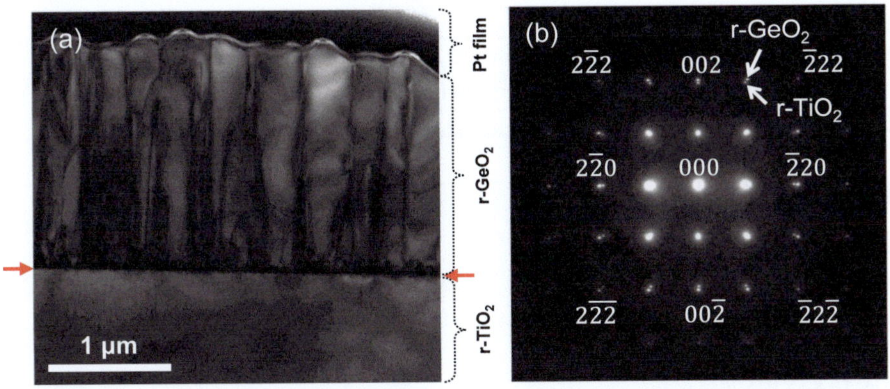


**Fig. 10.7** **a** Cross-sectional TEM image at the r-$GeO_2$/r-$TiO_2$ interface viewed along the < 110 > zone axis. **b** SAED patterns at the interface. The patterns are viewed along the < 110 > zone axis. Reproduced from Takane et al., Appl. Phys. Lett. 120, 099,903 [41] with the permission of AIP Publishing

electron diffraction (SAED) pattern at the r-$GeO_2$/r-$TiO_2$ interface are displayed in Fig. 10.7a and b, respectively. It was confirmed that obtained r-$GeO_2$ has almost relaxed introducing threading dislocations and has the same crystal structure as that of r-$TiO_2$ without any distinct rotational domain. However, it was found that the $GeO_2$ film, whose TEM and SAED patterns were shown in Fig. 10.7, fluctuated in its thickness and crystallinity in plane, as also reported in Ref. [38]. The detail of the fluctuation in the film is described later in this chapter.

## 10.4 Growth of r-$Ge_xSn_{1-x}O_2$ Alloy Thin Film by Mist CVD

Then, we fabricated r-$Ge_xSn_{1-x}O_2$ alloy thin films by mist CVD. Bis [2-carboxyethylgermanium (IV)] sesquioxide ($C_6H_{10}Ge_2O_7$) and tin (II) chloride dihydrate ($SnCl_2 \cdot 2H_2O$) were adapted as Ge and Sn precursor, respectively. In order to obtain r-$Ge_xSn_{1-x}O_2$ alloy thin films with different solidus compositions of Ge ($x$), we used a mixed solution of the two precursors with different Ge concentration ratios in the source solutions. The growth conditions are listed in Table 10.3. Figure 10.8 shows the XRD $2\theta/\omega$ scan profiles of the fabricated r-$Ge_xSn_{1-x}O_2$ alloy thin films, indicating that (001)-oriented r-$Ge_xSn_{1-x}O_2$ alloy thin films were grown on (001) r-$TiO_2$ substrates. No distinct peaks are visible except for the diffraction peaks by the films and substrates in the entire range of the spectra. Using energy disperse x-ray spectroscopy (EDS), Ge compositions (x) in each film were found to be 0.00, 0.41, 0.57, 0.66, 0.70, 0.96, and 1.00. The EDS also suggests uniform distribution of elements with no phase separation in all the alloy films.

**Table 10.3** Growth conditions for r-$Ge_xSn_{1-x}O_2$ thin films. Reproduced from Takane et al., Phys. Rev. Materials 6, 084,604 (2022). Copyright 2022 The American Physical Society

| Ge precursor | Bis [2-carboxyethylgermanium (IV)] sesquioxide: $C_6H_{10}Ge_2O_7$ |
|---|---|
| Sn precursor | Tin (II) chloride dihydrate ($SnCl_2 \cdot 2H_2O$) |
| Solvent | $H_2O + HCl$ |
| Substrate | r-$TiO_2$ (001) |
| Carrier gas (flow rate) | $O_2$ (3.0 L/min) |
| Dilution gas (flow rate) | $O_2$ (0.5 L/min) |
| Growth temperature | 725 °C |

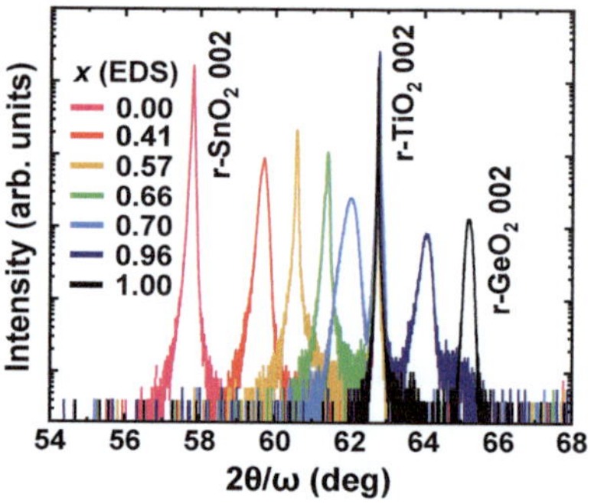


**Fig. 10.8** XRD 2θ/ω scan profiles of the r-$Ge_xSn_{1-x}O_2$ films on (001) r-$TiO_2$ substrates. Reproduced from Takane et al., Phys. Rev. Materials 6, 084,604 (2022). Copyright 2022 The American Physical Society

In the analysis of r-$GeO_2$, it was found that the $GeO_2$ film fluctuated in its thickness and crystallinity in plane. Thus, in order to analyze phase separation in the r-$Ge_xSn_{1-x}O_2$ with $x = 1.00$ and the crystallographic orientation of r-$Ge_xSn_{1-x}O_2$ alloy thin films, electron backscatter diffraction (EBSD) measurements were conducted. In the measurements, the coordinates of the equipment were set as follows: the $z$ axis of the equipment is parallel to the equivalent < 110 > axes of the $TiO_2$ substrate axes, and the $x$ and $y$ of the equipment are parallel to the [001] axis of the $TiO_2$ substrate. Figure 10.9 exhibits the EBSD inverse pole figure (IPF) maps along the $x$, $y$, and $z$ axes. The observed areas are shown by the SEM images on the left-hand side. For the bottom-left area in the film with $x = 1.00$, the EBSD IPF map along the $z$ axis and those along the $x$ and $y$ axes correspond well to (001) and (110) orientations, respectively, of the rutile structure, indicating that both the out-of-plane and in-plane orientations of the area in the film with $x = 1.00$ and r-$TiO_2$ substrate are consistent with each other. However, the EBSD IPF maps along all the axes show almost no orientation of the rutile-type structure in the upper-right area in the film with $x = 1.00$. This result suggests that r-$Ge_xSn_{1-x}O_2$ with $x = 1.00$ (r-$GeO_2$) shows fluctuation in its crystalline quality. The black points in the well-crystallized region are attributed to multiple scattering of diffraction electrons by its rough surface. On the contrary, for the film with $x = 0.96$, the same results as the bottom-left area in the film with $x = 1.00$ are observed and the same out-of-plane and in-plane orientations

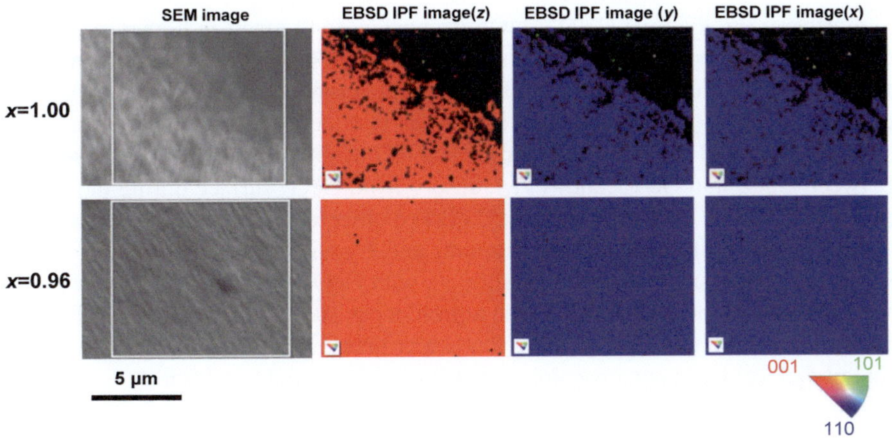


**Fig. 10.9** Plane view energy disperse x-ray spectroscopy (EBSD) mapping of the r-$Ge_xSn_{1-x}O_2$ films with $x = 0.96$ and 1.00. Reproduced from Takane et al., Phys. Rev. Materials 6, 084,604 (2022). Copyright 2022 The American Physical Society

between the film and substrate were confirmed in the entire area in its film. The same results were obtained for $x = 0.00$, 0.41, 0.57, 0.66, and 0.70.

## 10.5 Characteristics of Crystallization of r-$GeO_2$ by the Mist CVD

In a viewpoint of the MBE growth of r-$GeO_2$, Chae et al. reported that because of the high vapor pressure of volatile GeO and the existence of deeply metastable phases for $GeO_2$, in order to avoid the amorphous (glass)-phase, fabrication of crystallized r-$GeO_2$ thin film requires the conditions that balance epitaxial strain, surface adatom kinetic, adsorption/desorption rate, and chemical composition [35, 36]. It is reasonable, to some extent, to consider that these balances also required in other growth techniques including the mist CVD. Regarding the r-$GeO_2$ thin films grown by the mist CVD shown in Fig. 10.5, the growth condition window is also small like that of the MBE [35], suggesting such balances collapsed at $T_g < 675$ °C and > 800 °C, avoiding crystallization of $GeO_2$.

In particular, the crystal structure of fabricated thin films is considered to be more strongly affected by and more likely inherit that of underlying crystal (substrate) in the mist CVD compared to the method under vacuum, probably because the growth occurs under atmospheric pressure, and chemical species in the reaction area possess relatively lower kinetic energy on the surface [42, 43]. This also implies that the balances of surface adatom kinetic can play significant roles for crystallization of $GeO_2$ film on r-$TiO_2$ substrate by the mist CVD.

Furthermore, the mist CVD growth is considered to result in higher oxygen (or ozone) partial pressure in the reaction area compared with that in other growth techniques under vacuum (Fig. 10.10). This is because growth process is performed under conditions open to the atmosphere as well as in aqueous vapor, which incessantly supplies oxygen. Here, it is considered that GeO, which is generated by the reaction between $GeO_2$ and Ge vacancy ($V_O$), desorb from $GeO_2$ surface during the growth processes of r-$GeO_2$ by both the mist CVD and other techniques, because it is grown at high temperature or/and under vacuum and contains more or less $V_O$ [44–46]. Moreover, GeO(g) and $GeO_2$(g) equilibrate with $GeO_2$(s) as Eqs. (10.1) and (10.2), where GeO(g), $GeO_2$(g), and $GeO_2$(s) represent gas-state GeO, $GeO_2$, and solid-state $GeO_2$, respectively [47], indicating that vapor pressure of $GeO_2$(g) is constant and that of GeO(g) has a negative slope against oxygen partial pressure:

$$GeO_2(s) = GeO(g) + \tfrac{1}{2}O_2(g), \tag{10.1}$$

$$GeO_2(s) = GeO_2(g). \tag{10.2}$$

Thus, the desorption rate in the mist CVD growth is considered to be relatively lower as shown in Fig. 10.10a and b, which is considered to result in higher growth rate of r-$GeO_2$ by the mist CVD ($\geq$ 1 μm) than that by MBE (10 nm/h) [35]. These also suggest that the balances of adsorption/desorption rate can play significant roles for crystallization of volatile $GeO_2$.

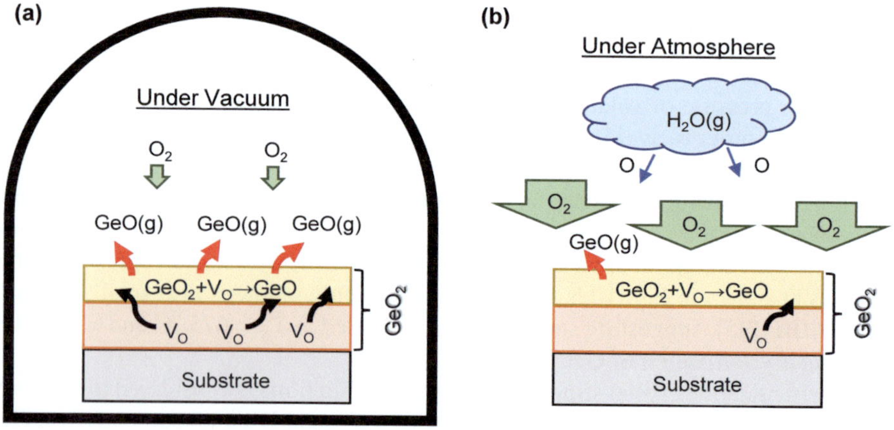


**Fig. 10.10** Schematics of GeO desorption and related phenomena in **a** growth methods under vacuum and **b** mist CVD. The red, black, and blue arrows indicate desorption of GeO from the $GeO_2$ surface, diffusion of $V_O$ in $GeO_2$, and supply of oxygen from $H_2O$ in the source solution. Reproduced from Takane et al., Appl. Phys. Lett. 119, 062,104 [40] with the permission of AIP Publishing

Finally, we discuss other approach for crystallization of materials hard to be crystallized and stabilization of its crystallinity. As previously shown in Fig. 10.9, although r-$Ge_xSn_{1-x}O_2$ film with $x = 1.00$ (r-$GeO_2$) fluctuated its crystallinity in plane, r-$Ge_xSn_{1-x}O_2$ film with $x = 0.96$ well crystallized in the entire area with the same out-of-plane and in-plane orientations between the film and substrate and without any rotational domain. These results suggest that replacing Ge site with a small amount of Sn makes the crystallized r-$GeO_2$ relatively stable. The reason of the stabilization is due to that r-$SnO_2$ have relatively stable in terms of lower vapor pressure of SnO than that of GeO [48] and no other predominant crystal structures except for the rutile-type structure [49]. Such doping or allying of other elements may be one of key approaches for crystallization and stabilization of materials hard to be crystallized.

## 10.6 Conclusions

In this chapter, we showed our previous work about fabrication of r-$GeO_2$ by the mist CVD and discuss the synthesis of materials hard to be crystallized, like metastable or non-equilibrium-phased oxides by the mist CVD method.

We described the experimental results of growth of r-$GeO_2$ and r-$Ge_xSn_{1-x}O_2$ thin films by the mist CVD, then moved on to discussions about the characteristics of crystallization of r-$GeO_2$ by the mist CVD. For the growth of r-$GeO_2$, it is particularly necessary to prepare the growth conditions that can balance epitaxial strain, surface adatom kinetic, adsorption/desorption rate, and chemical composition on the substrate, which results from the high vapor pressure of GeO and the existence of deeply metastable polymorphous for $GeO_2$. Although the growth condition window is narrow, we achieved crystallized r-$GeO_2$ thin films by the mist CVD. For crystallization of r-$GeO_2$ thin films with relatively high growth rates, we consider that the balances of surface adatom kinetic and adsorption/desorption rate, given by the characteristics of the mist CVD, play key roles. Finally, we showed stabilization of crystallinity of r-$GeO_2$ with the addition of a small amount of Sn, suggesting usefulness of doping or allying of other elements for crystallization and stabilization of materials hard to be crystallized.

## References

1. T. Kawaharamura, H. Nishinaka, K. Kametani, Y. Masuda, M. Tanigaki, S. Fujita, Fabrication and properties of ZnO thin films prepared by fine channel mist mehtod. J. Soc. Mater. Sci., Jpn. **55**(2), 153–158 (2006). https://doi.org/10.2472/jsms.55.153
2. Y. Kamada, T. Kawaharamura, H. Nishinaka, S. Fujita, Linear-source ultrasonic spray chemical vapor deposition method for fabrication of ZnMgO films and ultraviolet photodetectors. Jpn. J. Appl. Phys. **45**(32), L857–L859 (2006). https://doi.org/10.1143/JJAP.45.L857

3. T. Kawaharamura, H. Nishinaka, S. Fujita, Growth of crystalline zinc oxide thin films by fine-channel-mist chemical vapor deposition. Jpn. J. Appl. Phys. **47**(6), 4669–4675 (2008). https://doi.org/10.1143/JJAP.47.4669
4. H. Nishinaka, Y. Kamada, N. Kameyama, S. Fujita, Epitaxial ZnO thin films on a -plane sapphire substrates grown by ultrasonic spray-assisted mist chemical vapor deposition. Jpn. J. Appl. Phys. **48**(12), 121103 (2009). https://doi.org/10.1143/JJAP.48.121103
5. T. Okuno, T. Oshima, S.D. Lee, S. Fujita, Growth of $SnO_2$ crystalline thin films by mist chemical vapour deposition method. Phys. Status Solidi C **8**(2), 540–542 (2011). https://doi.org/10.1002/pssc.201000619
6. S.D. Lee, K. Kaneko, S. Fujita, Homoepitaxial growth of beta gallium oxide films by mist chemical vapor deposition. Jpn. J. Appl. Phys., **55**(12), 1202B8 (2016). https://doi.org/10.7567/JJAP.55.1202B8
7. D. Shinohara, S. Fujita, Heteroepitaxy of corundum-structured $\alpha$-$Ga_2O_3$ thin films on $\alpha$-$Al_2O_3$ substrates by ultrasonic mist chemical vapor deposition. Jpn. J. Appl. Phys. **47**(9R), 7311 (2008). https://doi.org/10.1143/JJAP.47.7311
8. N. Suzuki, K. Kaneko, S. Fujita, Growth of corundum-structured In2O3 thin films on sapphire substrates with $Fe_2O_3$ buffer layers. J. Cryst. Growth **364**, 30–33 (2013). https://doi.org/10.1016/j.jcrysgro.2012.11.065
9. N. Suzuki, K. Kaneko, S. Fujita, Growth of corundum-structured $(In_xGa_{1-x})_2O_3$ alloy thin films on sapphire substrates with buffer layers. J. Cryst. Growth **401**, 670–672 (2014). https://doi.org/10.1016/j.jcrysgro.2014.02.051
10. S. Kan, S. Takemoto, K. Kaneko, I. Takahashi, M. Sugimoto, T. Shinohe, S. Fujita, Electrical properties of $\alpha$-$Ir_2O_3$/$\alpha$-$Ga_2O_3$ pn heterojunction diode and band alignment of the heterostructure. Appl. Phys. Lett. **113**(21), 212104 (2018). https://doi.org/10.1063/1.5054054
11. K. Kaneko, Y. Masuda, S.I. Kan, I. Takahashi, Y. Kato, T. Shinohe, S. Fujita, Ultra-wide bandgap corundum-structured p-type $\alpha$-$(Ir,Ga)_2O_3$ alloys for $\alpha$-$Ga_2O_3$ electronics. Appl. Phys. Lett., **118**(10). https://doi.org/10.1063/5.0027297
12. T. Kawaharamura, Physics on development of open-air atmospheric pressure thin film fabrication technique using mist droplets: Control of precursor flow. Jpn. J. Appl. Phys., **53**(5S1), (2014). https://doi.org/10.7567/JJAP.53.05FF08
13. K. Uno, Y. Yamasaki, I. Tanaka, Growth mechanism of zinc oxide and zinc sulfide films by mist chemical vapor deposition. Appl. Phys. Express **10**(1), 015502 (2017). https://doi.org/10.7567/APEX.10.015502
14. P. Rutthongjan, M. Nishi, L. Liu, S. Sato, Y. Okada, G.T. Dang, T. Kawaharamura, Growth mechanism of zinc oxide thin film by mist chemical vapor deposition via the modulation of $[H_2O]/[Zn]$ ratios. Appl. Phys. Express **12**(6), 065505 (2019). https://doi.org/10.7567/1882-0786/ab2134
15. K. Uno, M. Ohta, I. Tanaka, Growth mechanism of $\alpha$-$Ga_2O_3$ on a sapphire substrate by mist chemical vapor deposition using acetylacetonated gallium source solutions. Appl. Phys. Lett., **117**(5), (2020). https://doi.org/10.1063/5.0014056
16. J.W. Fleming, Dispersion in $GeO_2$-$SiO_2$ glasses. Appl. Opt. **23**(24), 4486 (1984). https://doi.org/10.1364/ao.23.004486
17. V. Mizrahi, P.J. Lemaire, T. Erdogan, W.A. Reed, D.J. Digiovanni, R.M. Atkins, Ultraviolet laser fabrication of ultrastrong optical fiber gratings and of germania-doped channel waveguides. Appl. Phys. Lett. **63**(13), 1727–1729 (1993). https://doi.org/10.1063/1.110696
18. S. Sakaguchi, S. Todoroki, Optical properties of $GeO_2$ glass and optical fibers. Appl. Opt. **36**(27), 6809–6814 (1997). https://doi.org/10.1364/ao.36.006809
19. A. Jha, B. Richards, G. Jose, T. Teddy-Fernandez, P. Joshi, X. Jiang, J. Lousteau, Rare-earth ion doped $TeO_2$ and $GeO_2$ glasses as laser materials. Prog. Mater Sci. **57**(8), 1426–1491 (2012). https://doi.org/10.1016/j.pmatsci.2012.04.003
20. A. Toriumi, T. Nishimura, Germanium CMOS potential from material and process perspectives: Be more positive about germanium. Jpn. J. Appl. Phys. **57**(1), 010101 (2018). https://doi.org/10.7567/JJAP.57.010101

21. M. Kobayashi, G. Thareja, M. Ishibashi, Y. Sun, P. Griffin, J. McVittie, P. Pianetta, K. Saraswat, Y. Nishi, Radical oxidation of germanium for interface gate dielectric $GeO_2$ formation in metal-insulator-semiconductor gate stack. J. Appl. Phys. **106**(10), 104117 (2009). https://doi.org/10.1063/1.3259407
22. W.H. Baur, A.A. Khan, Rutile-type compounds. IV. $SiO_2$, $GeO_2$ and a comparison with other rutile-type structures. Acta Crystallogr. Sect. B Struct. Crystallogr. Cryst. Chem., **27**(11), 2133–2139 (1971). https://doi.org/10.1107/S0567740871005466
23. G.S. Smith, P.B. Isaacs, The crystal structure of quartz-like GeO2. Acta Crystallogr. A **17**(7), 842–846 (1964). https://doi.org/10.1107/S0365110X64002262
24. M. Stapelbroek, B.D. Evans, Exciton structure in the u.v.-absorption edge of tetragonal $GeO_2$. Solid State Commun., **25**(11), 959–962 (1978). https://doi.org/10.1016/0038-1098(78)90311-3
25. Chae, S., Lee, J., Mengle, K. A., Heron, J. T., & Kioupakis, E. (2019). Rutile $GeO_2$: An ultrawide-band-gap semiconductor with ambipolar doping. Appl Phys Lett, **114**(10), (2019). https://doi.org/10.1063/1.5088370
26. K.A. Mengle, S. Chae, E. Kioupakis, Quasiparticle band structure and optical properties of rutile $GeO_2$, an ultra-wide-band-gap semiconductor. J. Appl. Phys. **126**(8), 085703 (2019). https://doi.org/10.1063/1.5111318
27. C.A. Niedermeier, K. Ide, T. Katase, H. Hosono, T. Kamiya, Shallow valence band of rutile $GeO_2$ and p-type doping. J. Phys. Chem. C **124**(27), 25721–25728 (2020). https://doi.org/10.1021/acs.jpcc.0c07757
28. K. Bushick, K.A. Mengle, S. Chae, E. Kioupakis, Electron and hole mobility of rutile $GeO_2$ from first principles: An ultrawide-bandgap semiconductor for power electronics. Appl. Phys. Lett. **117**(18), 182104 (2020). https://doi.org/10.1063/5.0033284
29. A.W. Laubengayer, D.S. Morton, Germanium. Xxxix. The polymorphism of germanium dioxide 1. J. Am. Chem. Soc., **54**(6), 2303–2320 (1932). https://doi.org/10.1021/ja01345a019
30. S. Chae, K.A. Mengle, R. Lu, A. Olvera, N. Sanders, J. Lee, P.F.P. Poudeu, J.T. Heron, E. Kioupakis, Thermal conductivity of rutile germanium dioxide. Appl. Phys. Lett., **117**(10), (2020). https://doi.org/10.1063/5.0011358
31. J.W. Goodrum, Top-seeded flux growth of tetragonal $GeO_2$. J. Cryst. Growth **7**(2), 254–256 (1970). https://doi.org/10.1016/0022-0248(70)90020-5
32. D.E. Swets, Crystal growth of tetragonal germanium dioxide. J. Cryst. Growth **8**(3), 311–313 (1971). https://doi.org/10.1016/0022-0248(71)90081-9
33. V. Agafonov, D. Michel, M. Perez, Jorba, M. Fedoroff, Chemical and crystallographic characterization of crystals grown by chemical vapour transport in the $Fe_2O_3-GeO_2$ system. Mater. Res. Bull. **19**(2), 233–239 (1984). https://doi.org/10.1016/0025-5408(84)90095-3
34. H. Takane, Y. Ota, T. Wakamatsu, T. Araki, K. Tanaka, K. Kaneko, Band-gap engineering of rutile-structured $SnO_2$-$GeO_2$-$SiO_2$ alloy system. Phys. Rev. Mater. **6**(8), 084604 (2022). https://doi.org/10.1103/PhysRevMaterials.6.084604
35. S. Chae, H. Paik, N.M. Vu, E. Kioupakis, J.T. Heron, Epitaxial stabilization of rutile germanium oxide thin film by molecular beam epitaxy. Appl. Phys. Lett. **117**(7), 072105 (2020). https://doi.org/10.1063/5.0018031
36. S. Chae, K.A. Mengle, K. Bushick, J. Lee, N. Sanders, Z. Deng, Z. Mi, P.F.P. Poudeu, H. Paik, J.T. Heron, E. Kioupakis, Toward the predictive discovery of ambipolarly dopable ultra-wide-band-gap semiconductors: The case of rutile GeO2. Appl. Phys. Lett. **118**(26), 260501 (2021). https://doi.org/10.1063/5.0056674
37. C.A. Angell, Formation of glasses from liquids and biopolymers. Science **267**(5206), 1924–1935 (1995). https://doi.org/10.1126/science.267.5206.1924
38. G. Deng, K. Saito, T. Tanaka, M. Arita, Q. Guo, Pulsed laser deposition growth of ultra-wide bandgap $GeO_2$ film and its optical properties. Appl. Phys. Lett. **119**(18), 182101 (2021). https://doi.org/10.1063/5.0071918
39. G. Deng, Y. Huang, Z. Chen, K. Saito, T. Tanaka, M. Arita, Q. Guo, Heteroepitaxy of (1 0 0)-oriented rutile $GeO_2$ film on c-plane sapphire by pulsed laser deposition. Mater. Lett. **326**, 132945 (2022). https://doi.org/10.1016/j.matlet.2022.132945

40. H. Takane, K. Kaneko, Establishment of a growth route of crystallized rutile $GeO_2$ thin film (≧ 1 μm/h) and its structural properties. Appl. Phys. Lett. **119**(6), 062104 (2021). https://doi.org/10.1063/5.0060785
41. H. Takane, K. Kaneko, Erratum: "Establishment of a growth route of crystallized rutile $GeO_2$ thin film (≧1 μm/h) and its structural properties." Appl. Phys. Lett. **120**(9), 099903 (2022). https://doi.org/10.1063/5.0088164
42. H. Takane, K. Kaneko, Y. Ota, S. Fujita, Initial nucleation scheme of $Ga_2O_3$ on (0001) sapphire by mist CVD for the growth of α-phase. Jpn. J. Appl. Phys., **60**(5), 055501 (2021). https://doi.org/10.35848/1347-4065/abf47a
43. K. Kaneko, Synthesis of metastable or non-equilibrium-phased oxides by the mist CVD method. J Soc Mater Sci, Jpn **70**(5), 369–373 (2021). https://doi.org/10.2472/jsms.70.369
44. S.K. Wang, K. Kita, C.H. Lee, T. Tabata, T. Nishimura, K. Nagashio, A. Toriumi, Desorption kinetics of GeO from GeO2/Ge structure. J. Appl. Phys. **108**(5), 054104 (2010). https://doi.org/10.1063/1.3475990
45. S.K. Wang, K. Kita, T. Nishimura, K. Nagashio, A. Toriumi, Isotope tracing study of GeO desorption mechanism from $GeO_2$/Ge stack using $^{73}$Ge and $^{18}$O. Jpn. J. Appl. Phys., **50**(4S), 04DA01 (2011). https://doi.org/10.1143/JJAP.50.04DA01
46. S.K. Wang, K. Kita, T. Nishimura, K. Nagashio, A. Toriumi, Kinetic effects of O-vacancy generated by $GeO_2$/Ge interfacial reaction. Jpn. J. Appl. Phys., **50**(4S), 10PE04 (2011). https://doi.org/10.1143/JJAP.50.10PE04
47. C.H. Lee, T. Tabata, T. Nishimura, K. Nagashio, K. Kita, A. Toriumi, Ge/$GeO_2$ Interface control with high-pressure oxidation for improving electrical characteristics. Appl. Phys. Express **2**(7), 071404 (2009). https://doi.org/10.1143/APEX.2.071404
48. R.H. Lamoreaux, D.L. Hildenbrand, L. Brewer, High-temperature vaporization behavior of oxides II. oxides of Be, Mg, Ca, Sr, Ba, B, Al, Ga, In, Tl, Si, Ge, Sn, Pb, Zn, Cd, and Hg. J. Phys. Chem. Ref. Data, **16**(3), 419–443 (1987). https://doi.org/10.1063/1.555799
49. J. Haines, J.M. Léger, X-ray diffraction study of the phase transitions and structural evolution of tin dioxide at high pressure: Relationships between structure types and implications for other rutile-type dioxides. Phys. Rev. B **55**(17), 11144–11154 (1997). https://doi.org/10.1103/PhysRevB.55.11144

# Chapter 11
# Crystalline Microstructure and Versatile Resistive Switching Property in Rutile $TiO_{2-x}$ Four-Terminal Memristors

**Tetsuya Tohei and Akira Sakai**

**Abstract** This chapter presents the basic principles of novel four-terminal memristive devices and proof of concept results obtained in the memristors fabricated on rutile $TiO_{2-x}$ single crystals. Systematic electronic and microstructural characterization show that the behavior of dopant (oxygen vacancy) redistribution and resultant resistive switching properties strongly depend on crystal orientation reflecting crystalline anisotropy or crystal habits of titanium oxides. These experiments prove that the choice of crystal orientation is critical to achieve good reversibility of electronic properties in single crystalline memristors. Feasibility of imparting versatile synaptic functionalities by controlling two dimensional distribution of dopants among multi-terminals configuration is discussed.

## 11.1 Introduction

Recent progress of artificial intelligence technology has created greater expectations for developing brain-inspired computing hardware. To achieve such novel computing architecture, an electrical element that mimics synaptic behaviors of neuromorphic networks (i.e., synaptic plasticity) in the brain is required. A passive element called memristor, whose resistance varies non-volatilely according to the amount of injected charge, has attracted much attention as a candidate for such an element [1–4]. To date variety of artificial synapse implementation has been attempted by using memristive devices. Memristors are typically composed of a metal–insulator-metal (MIM) structure with two electrode terminals. Metal oxides are widely used as an insulator or an active layer of memristors [5]. While various kinds of mechanisms have been proposed to explain resistive switching (RS) phenomena, one of commonly accepted mechanisms rely on dopant redistribution induced by the application of external voltages [6, 7]. Emulation of neuromorphic properties by using memristive devices as artificial synapses has been extensively studied [2–4, 8]. In such operation, variation

T. Tohei (✉) · A. Sakai
Graduate School of Engineering Science, Osaka University, Toyonaka 560-8531, Osaka, Japan
e-mail: tohei@ee.es.osaka-u.ac.jp

 
F. Ishikawa et al. (eds.), *Engineering Crystal Habit*, Topics in Applied Physics 152,
https://doi.org/10.1007/978-981-96-0266-7_11

of synapse strength or synaptic weights in neural systems is mimicked by non-volatile resistive change of devices which occurs under external stimulus of applied voltages. Furthermore, in the biological synapses, there is an intriguing function called hetero-synaptic plasticity, *i.e.*, modulation of synaptic plasticity mediated by multiple connections of synapses rather than homosynaptic interactions [9]. Hetero-synaptic plasticity is considered to be responsible for learning functionalities such as associative learning or other behavioral learning [10, 11]. So far hetero-synaptic properties have been attempted to be implemented by conventional two-terminal memristors in rather complex ways [8, 12]. However, multi-terminal configurations seem to be more essential and natural solution for implementing the versatile hetero-synaptic functions [13–18].

Recently we have developed a memristive devices with a novel four-terminal configuration using rutile $TiO_{2-x}$ single-crystals [13, 14, 19, 20]. Reduced $TiO_2$ ($TiO_{2-x}$) is a prototypic memristive material where mobile oxygen vacancies act as donor dopants contributing to resistive switching [3, 21, 22]. Our four-terminal memristive device is unique in that the resistance is determined by two-dimensional distribution of oxygen vacancies among multiple electrodes. The distribution of oxygen vacancies can be controlled versatilely by the application of external voltages to the four terminal electrodes. Since oxygen vacancies have a positive charge, their distribution are modulated through drift motion caused by applying electric fields. This working principle imparts the possibilities of emulating higher-order brain and neuromorphic functions. In this chapter, we first introduce the basic concept of the unique $TiO_{2-x}$ based four terminal devices. We employ single crystals of rutile $TiO_{2-x}$ to avoid the structural inhomogeneity and electronic scattering due to grain boundaries and to obtain better controllability of dopant distribution and resultant resistive change properties. From a close investigation of crystal microstructure and electronic states examined by high resolution electron microscopy, we argue that the choice of crystal orientation is critical for obtaining better controllability of dopant redistribution. Then we demonstrate wide variety of hetero-synaptic properties including hetero-synaptic plasticity, spike timing dependent plasticity (STDP), and associative learning (Pavlovian conditioning) in a single multi-terminal memristive passive element.

## 11.2 Crystal Structure, Device Structure and Electrical Measurements

### 11.2.1 Crystal Structure of Rutile $TiO_2$ and Device Geometries

We used single crystals of rutile $TiO_2$ as a working material of memristive devices. We employed two major processes to produce reduced $TiO_{2-x}$ single-crystal memristors: thermal reduction of bulk single crystals and homo-epitaxial growth of reduced $TiO_{2-x}$ thin films by pulsed laser deposition (PLD) method [13, 14, 19, 20].

In the former process, a rutile type $TiO_2$ single-crystal substrate was annealed at reduced condition (typically 700 °C under a vacuum of $1.5 \times 10^{-6}$ Pa) to introduce oxygen vacancies into the crystal. Oxygen vacancies were generated according to the following reduction reaction:

$$2n\mathrm{TiO_2} \Leftrightarrow 2n\mathrm{Ti}_n\mathrm{O}_{2n-1} + \mathrm{O_2} + 2\mathrm{V_o^{++}} + 4\mathrm{e}^-$$

where $V_o^{++}$ represents a positively charged oxygen vacancy as a donor and $Ti_nO_{2n-1}$ is referred to as the Magneli phase (typically $n = 4$ or 5), which has higher electrical conductivity than stoichiometric $TiO_2$ [23]. In the latter process, epitaxial thin films of $TiO_{2-x}$ were deposited on rutile $TiO_2$ single crystal substrate in reduced condition. By tuning growth condition of chamber atmosphere and substrate temperature, we could obtained reduced $TiO_{2-x}$ epitaxial thin films with conductivity suitable for memristive operations.

Figure 11.1 shows crystal structure of rutile $TiO_2$. The crystal are composed of edge shearing $TiO_6$ octahedra. The rutile crystal structure shows in-plane anisotropy in (100) surface whereas it is isotropic within (001) plane. Such difference in the anisotropy or isotropy depending on crystalline orientation should bring an important effects on the electronic or microstructural properties in $TiO_2$ based memristive devices. We prepared four-terminals planar device using two kinds of rutile $TiO_2$ crystal orientations: (100) and (001) plane surfaces. We examined surface orientation dependence of resistive change properties and electric field induced microstructural changes upon RS processes.

To create four-terminal planar devices Pt electrodes were deposited on the surface of the reduced $TiO_{2-x}$ substrates. Then, electrical measurements were performed

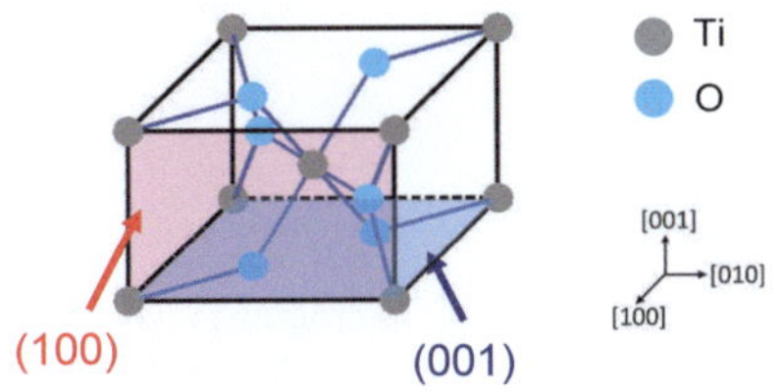


**Fig. 11.1** Schematic of crystal structure of rutile $TiO_2$. (100) and (001) surfaces were employed for fabrication of planar memristive devices

using a semiconductor parameter analyzer under a vacuum condition (base pressure less than $10^{-3}$ Pa).

### 11.2.2 Basic Operation of Four-Terminals Memristive Devices

As mentioned earlier, rutile $TiO_2$ shows structural anisotropy depending on the crystal surface orientations. We investigated difference in the resistive switching behavior and microstructural change for rutile $TiO_{2-x}$ (100) and (001) planar devices [20]. Figure 11.2 shows optical micrographs of the four-terminal memristive devices fabricated on (100) and (001) substrates. The terminals are labeled T1 to T4, with the orthogonal T2 and T4 terminals being used to modify the oxygen vacancy distribution in the region between the T1 and T3 terminals.

The basic protocol for measurements of resistive switching properties consisted of sequential application of write voltages and non-disturbing read voltages. First, the current flow between T1 (T2) and T3 (T4), $I_{1-3}$ ($I_{2-4}$), was measured at a reading voltage (typically 1 V) to probe the initial state. Then, a constant write voltage $V_C$ was applied to T2 and T4 simultaneously for a certain duration time of $T_C$, with T1 and T3 being grounded. Subsequently, $I_{1-3}$ ($I_{2-4}$) was measured again at an applied voltage of 1 V to probe the resistance state of the device. This sequence was repeated with changing the values of $V_C$ and $T_C$ with stepwise. Figure 11.3a shows a typical voltage application protocol for the resistive switching measurement of the four-terminal device. By employing such voltage application schemes, proof-of-concept demonstration of $V_C$-driven control of the two dimensional oxygen vacancy distribution was performed.

Figure 11.3b and c respectively show results for the variations of $I_{1-3}$ and $I_{2-4}$ in the (100) device as a function of the cumulative time $T_C$ for application of the write voltage $V_C$ according to the sequence of Fig. 11.3a. At the initial stage, values

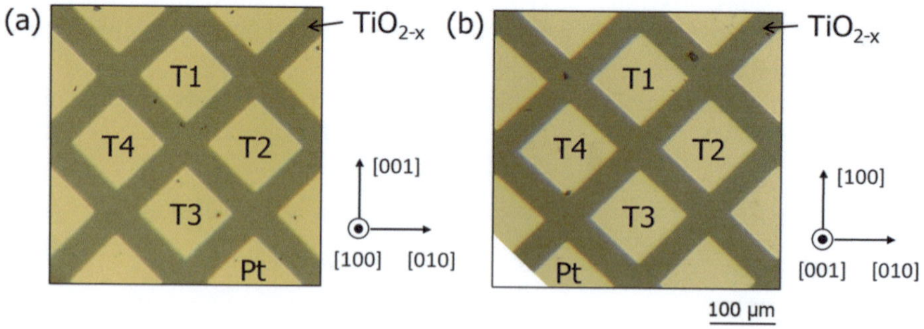


**Fig. 11.2** Optical micrographs showing four-terminal planar device structures fabricated on reduced rutile $TiO_{2-x}$ **a** (100) and **b** (001) surfaces. Reproduced from Ref. [20] with the permission from Springer Nature under Creative Commons CC BY license

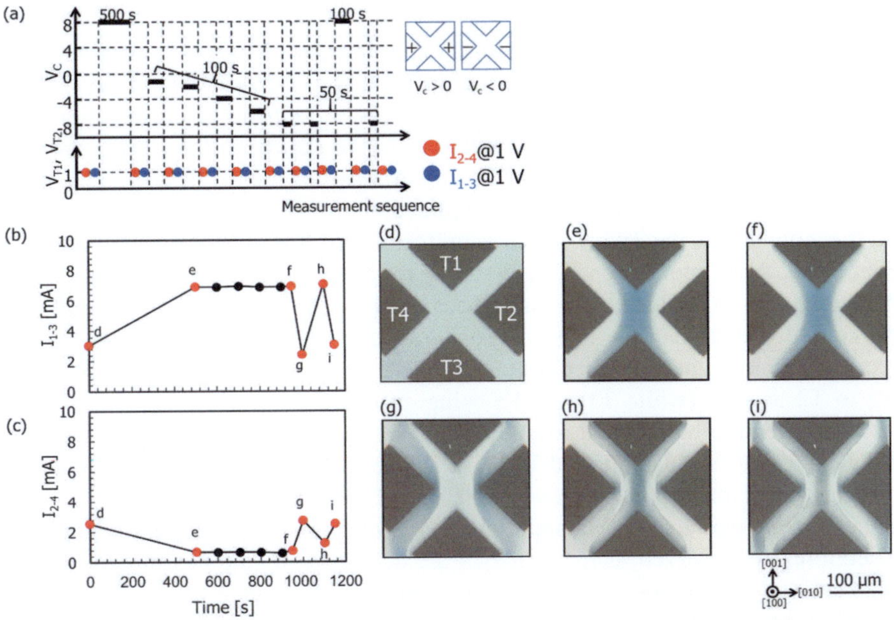


**Fig. 11.3** Voltage application sequences and measurement results for RS properties of the four-terminal devices. **a** A voltage sequence to demonstrate the controllability of the oxygen vacancy distribution in the device. Variations of **b** $I_{1-3}$ and **c** $I_{2-4}$ at 1 V as function of cumulative time $T_C$ for application of voltage $V_C$. **d–i** Optical micrographs showing internal structure of the (100) device on measurement stages of $I_{1-3}$ and $I_{2-4}$ at cumulative times of $T_C$ = **d** 0, **e** 500, **f** 950, **g** 1000, **h** 1100, and **i** 1150 s. Reproduced from Ref. [20] with the permission from Springer Nature under Creative Commons CC BY license

of $I_{1-3}$ and $I_{2-4}$ were almost the same, as expected from the device symmetry and uniform distribution of oxygen vacancy. After application of $V_C$ (8 V for 500 s), we observe that $I_{1-3}$ increased while $I_{2-4}$ decreased. This means that the resistance in the region between T1 (T2) and T3 (T4) was changed due to the application of $V_C$. After that, we switched the polarity of $V_C$ and continued to apply with stepwise increased negative voltage from –1 V to –8 V until 950 s, though we observed both the current values remained constant (Fig. 11.3a–c). A significant decrease (increase) of $I_{1-3}$ ($I_{2-4}$) occurred at cumulative time of 1000 s after application of an additional –8 V for 50 s. Here the resultant $I_{1-3}$ ($I_{2-4}$) was abruptly decreased into slightly lower (higher) value than the initial state. Subsequent application of $V_C$ at + 8 and –8 V causes switching into high and low (low and high) values of $I_{1-3}$ ($I_{2-4}$), respectively. Thus, reversible change of the resistance in the region between diagonally opposing electrodes in the device was proved by the present experiments.

The alternative current variation upon the $V_C$ application to T2 and T4 described above can be correlated to the oxygen vacancy distribution probed by optical observation of the internal structure of the device during the operation. It is known that generation of oxygen vacancies in $TiO_2$ produces trivalent titanium ions ($Ti^{3+}$) which act as color centers. Concentration of color centers upon voltage application is manifested as electro-coloring phenomena of crystals. In reduced $TiO_{2-x}$ based memristors positively charged oxygen vacancies are drifted and redistributed upon the application of external electric field. Change in the distribution (local concentration) of $Ti^{3+}$ ions and oxygen vacancies are reflected in the change in the electro-coloring regions of working $TiO_{2-x}$ crystals [24, 25]. A series of optical micrographs of the device recorded during the above $V_C$ application process is depict in Fig. 11.3d–i. At the initial stage, the $TiO_{2-x}$ area was uniformly colored in faint blue, as shown in Fig. 11.3d. After application of $V_C = +8$ V for 500 s, a dark blue area appeared to bridge T1 and T3 while the other areas became colorless (Fig. 11.3e). This change in the color distribution of the $TiO_{2-x}$ crystal is attributed to the drift motion of charged oxygen vacancy under applied voltage. On the application of positive voltage of $V_C$ to T2 and T4 (while T1 and T3 being grounded), positively-charged oxygen vacancies tend to be repelled from T2 and T4 by drift motion caused by the electric field and accumulate around T1 and T3. The dark blue colored region indicates a higher concentration of oxygen vacancies and should be more conductive, while the colorless region should contain less oxygen vacancies and be less conductive. Therefore, the observed increase of $I_{1-3}$ are accounted by the distribution of dark blue colored region where T1 and T3 are connected through a conductive zone of high concentration oxygen vacancies. Remarkably, this bridge structure between T1 and T3 was then maintained until a time of 950 s (Fig. 11.3f), which is consistent with almost constant $I_{1-3}$ and $I_{2-4}$ regardless of application of inverted polarity voltages. Then, we observed an abrupt change in the optical contrast after the cumulative voltage application until 1000 s. The contrast looks completely inverted; a colorless area bridged T1 and T3, while the other areas were colored dark blue (Fig. 11.3g). These features are again consistent with the observed $I_{1-3}$ decrease; T1 and T3 are connected through a resistive region with a low oxygen vacancy concentration. After subsequent applications of $V_C$ at $+8$ V and –8 V, as shown in Fig. 11.3h and i, the contrast was inverted repeatedly, displaying features of dark blue and faint blue T1-T3 bridges, which correspond to the measured switching into high and low $I_{1-3}$ values, respectively. As the resistive switching was repeated several times the contrast of colored region become complicated and the degree of contrast inversion is reduced. Influence of the formation of irreversible regions to the reversibility of RS property is discussed in the following section. These results demonstrate the interplay between the resistance change and the modulation of the oxygen vacancy distribution which is manifested as internal color change in $TiO_{2-x}$ based four terminals planar device.

### *11.2.3 Resistive Switching Reversibility in (100) and (001) Devices*

Next, we compare reversibility of RS property in the four-terminal device fabricated on (100) and (001) planes of rutile $TiO_{2-x}$. Resistive switching operations were performed using the voltage sequence shown in Fig. 11.4a. In this operation the $V_C$ value was varied to generate a ramp-up and ramp-down waveform. The resistance value $R_{1\text{-}3}$ ($R_{2\text{-}4}$) between terminals T1 (T2) and T3 (T4) was measured by applying a read voltage $V_1$ ($V_2$) to T1 (T2). Then, write voltages $V_C$ (namely $V_{2,4}$) were simultaneously applied to T2 and T4 for a certain period $t$ with T1 and T3 being grounded. This cycle was repeated with ramping the voltage $V_{2,4}$ up and down over the range between the maximum voltage $V_M$ and the minimum voltage $V_m$ in 1 V steps. Figure 11.4b and c shows representative results of resistance change between T1 and T3 (T2 and T4), $R_{1-3}$ ($R_{2-4}$) as a function of the applied $V_C$ in the $TiO_{2\text{-x}}$ (100) device. At the first cycle, we observed that resistance value $R_{1-3}$ ($R_{2-4}$) is gradually decreased (increased) with increasing $V_C$, so that the low resistance state (LRS) is achieved between T1 and T3. At the same time, $R_{2-4}$ shows gradual increase with increasing $V_C$ and reach to high resistance states (HRS) between T2 and T4. The LRS (HRS) is stable while decreasing $V_C$ but a sharp increase (decrease) in $R_{1-3}$ ($R_{2-4}$), a so-called RESET (SET) of the device, occurs at a negative threshold voltage of –7 V, so that the HRS (LRS) is achieved. We observe that when the $R_{1-3}$ is decreased the $R_{2-4}$ is increased, and vice versa when the polarity of $V_C$ is inverted. This indicates that the variations of $R_{1-3}$ and $R_{2-4}$ are complementary to each other. The HRS (LRS) is then stable again while increasing $V_C$. On repeating ramping of voltage Vc with positive and negative polarity, we observe that $R_{1-3}$ and $R_{2-4}$ show cyclic variation in similar manner to the first cycle. However, the maximum values of $R_{1-3}$ and $R_{2-4}$ in each cycle are observed to decrease with the cycle, and this trend is much more obvious for $R_{1-3}$ than for $R_{2-4}$. These results suggest limited reversibility of RS in the (100) devices.

In contrast, we have found that much stable RS reversibility is obtained for the device fabricated on the (001) plane. Typical results for the $R_{1-3}$ and $R_{2-4}$ change as a function of the applied $V_C$ in the (001) device are shown in Fig. 11.4d and e. We see that the variation of $R_{1-3}$ ($R_{2-4}$) shows a repetitive clockwise (counterclockwise) rotation hysteresis on the repeated $V_C$ application cycles. The threshold voltages for SET and RESET operations are well defined (+4 V and 4 V, respectively), and the maximum and minimum resistances did not change significantly after the third cycle. These characteristics clearly demonstrate that the (001) substrate device has advantage of repetition stability of RS properties against the (100) device.

From the cyclic RS operations mentioned above, we found that the (100) device shows limited reversibility of RS on cyclic $V_C$ application, whereas the (001) device shows good reversibility of the resistive switching properties. These difference in the RS property can be understand by the different behaviors in oxygen vacancy redistribution in each devices. Figure 11.5 presents optical micrographs of (100) and (001) devices at each voltage application stage in Fig. 11.4 showing elector-coloring that reflect oxygen vacancy distribution. In the (100) device, a characteristic oxygen vacancy distribution change, that is, inversion of bridging structure between T1 and T3 is observed at the stages of $V_C = +8$ V and –8 V in the first cycle (Fig. 11.5a and b, which is similar to contrasts shown in Fig. 11.3). Then, the internal structures of the device subjected to the second cycle showed distinct features (Fig. 11.5c and d). We observe that strongly colored regions are locally formed within the bridging

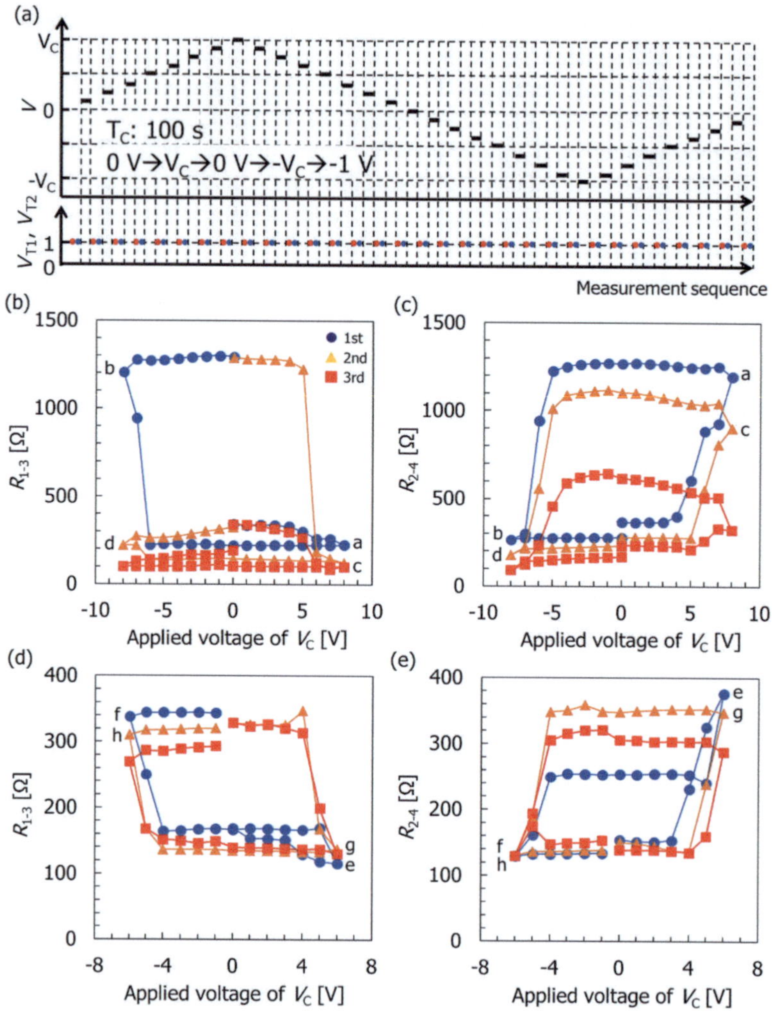


**Fig. 11.4** **a** Voltage application sequence to measure $R_{1\text{-}3}$ ($R_{2\text{-}4}$)—$V_c$ characteristics of $TiO_{2-x}$ four-terminal devices. **b–e** Variations of resistance between T1 (T2) and T3 (T4) as function of gate voltage $V_C$. **b**, **c** (100) device **d**, **e** (001) device. Reproduced from Ref. [20] with the permission from Springer Nature under Creative Commons CC BY license

structure, with additional dark line contrast elongating mainly along the [001] as well as ⟨011⟩ directions. Notably, these strong dark areas or lines still remain in the T1–T3 bridge even after the application of negative $V_C$. (Fig. 11.5d). These presence of irreversible colored regions should be responsible for the detrimental effect on RS reversibility.

As for the (001) device, optical micrographs are shown in Fig. 11.5e–h. Figure 11.5e and f show the inner structures of the device at the stages subjected

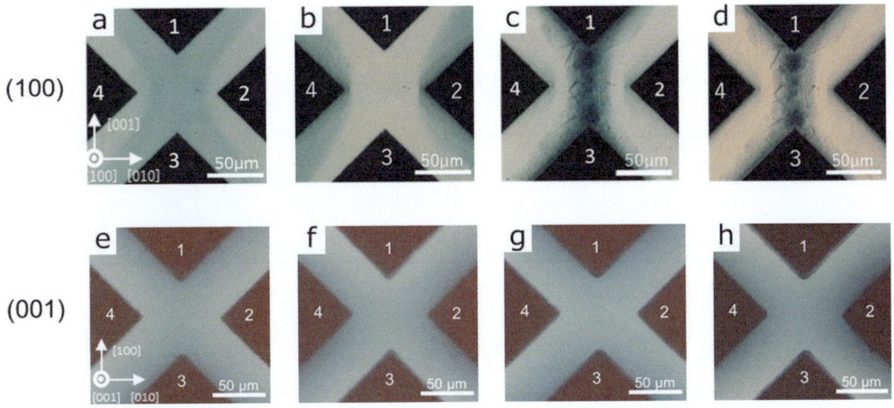


**Fig. 11.5** Optical micrographs showing internal structure of **a–d** (100) plane device and **e–h** (001) plane device. Images were obtained at each voltage application stages in the cyclic $R_{1\text{-}3}$ ($R_{2\text{-}4}$) measurements shown in Fig. 11.4b–e. Reproduced from Ref. [20] with the permission from Springer Nature under Creative Commons CC BY license

to application of $V_C$ at + 6 V and –6 V in the first cycle, respectively. We observe that the dark-blue-colored region bridging T1 and T3 is formed at $V_C = +6$ V (corresponds to the lowest $R_{1-3}$), and this configuration is flipped to bridge between T2 and T4 at $V_C= -6$ V (corresponds to the highest $R_{1-3}$). It is notable that interfaces between the dark blue and colorless areas in (001) device are much broader than the case of (100) device. Remarkably, the flipping of the dark blue bridging structure in the (001) device was found to repeat with good reversibility in the second and further $V_C$ cycling, without producing any irreversible regions (Fig. 11.5g and h). Comparison of the vacancy redistribution behavior between the (100) and (001) device clearly indicates that the drift and diffusion of oxygen vacancies is significantly dependent on the crystal orientation of $TiO_{2\text{-}x}$ and the direction of the electric field applied to the device. The different dopant (oxygen vacancy) redistribution due to crystalline anisotropy or crystal habits in $TiO_{2\text{-}x}$ accounts for reversibility of RS properties in the memristive devices. The absence of distinct crystal structure changes in the (001) device was also confirmed by STEM observations, which will be presented in the latter section.

### *11.2.4 Resistive Switching in (001) $TiO_{2-x}$ Thin-Film Device*

Four-terminal devices based on bulk crystal of reduced $TiO_2$ gave proof of concept demonstration of memristors based on 2D distribution of oxygen vacancies. In order to peruse possibility of further miniaturization and reduction of power consumption, we have fabricated four-terminal devices based on epitaxially-grown $TiO_{2\text{-}x}$ thin-films [14]. In thin film devices electronic active regions are confined to several tens

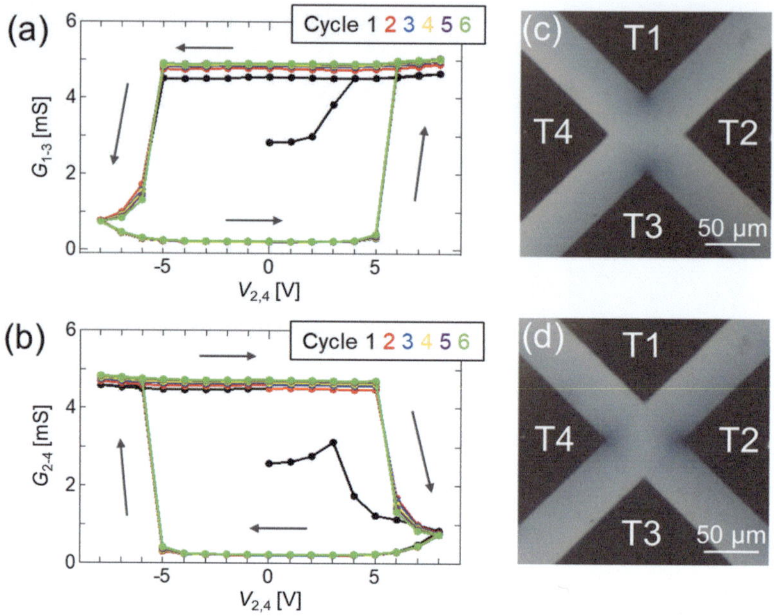


**Fig. 11.6** **a**, **b** Change in the conductance between T1 (T2) and T3 (T4) as function of $V_{2,4}$ in the $TiO_{2-x}$ (001) thin film device. **a** $G_{1-3}$ and **b** $G_{2-4}$. **c**, **d** Optical micrographs showing internal structure of device at applied voltages of **c** $V_{2,4} = 8$ V and **d** $V_{2,4} = -8$ V in cycle 2. Reproduced from Ref. [14] with the permission from American Chemical Society under CC BY-NC-ND 4.0 license: < https://pubs.acs.org/doi/full/10.1021/acsaelm.2c00161 > further permissions related to the material excerpted should be directed to the ACS

of nanometers thickness, and preparing higher resistance $TiO_{2-x}$ thin films and/or down-scaling the device size can help reduce the power consumption. A $TiO_{2-x}$ film of 60–80 nm thickness was deposited on a rutile $TiO_2$ (001) single-crystal substrate using pulsed laser deposition (PLD). The substrate temperature was 500 °C, and the oxygen partial pressure during the growth was held at $10^{-3}$–$10^{-4}$ Pa to control the content of oxygen vacancies in the deposited thin film. Squared Pt electrodes were fabricated in tetragonal alignment on the surface of the $TiO_{2-x}$ layer after deposition.

Figure 11.6a and b shows variations of conductance $G_{1-3}$ and $G_{2-4}$ as a function of applied $V_{2,4}$ ($V_C$), measured for a $TiO_{2-x}$ (001) thin film device with a voltage sequence similar to Fig. 11.4a. We observe that upon applying sweeping $V_{2,4}$ with positive and negative polarities, $G_{1-3}$ and $G_{2-4}$ show characteristic loop behavior with hysteresis between high and low conductance states. When the $G_{1-3}$ increases the $G_{2-4}$ decreases, and vice versa, indicating that the $G_{1-3}$ and $G_{2-4}$ variations are complementary. The unique hysteric characteristic of four-terminal device is more clearly evidenced in the thin film based device. The cyclic operation shows good repeatability of the conductance loops. Optical micrographs at the stages of $V_{2,4} = +8$ V and -8 V clearly show the concentration of oxygen vacancy around T1–T3 and T2–T4, respectively, evidencing the clear flipping of bridging conducting regions (Fig. 11.6c and d). Also the maximum HCS-to-LCS conductance ratio was found to be more than 20 in the thin-film based device. This is a remarkable improvement (by 1000% increase) over the bulk single crystal based devices whose resistance ratio was 2–3 [20]. The resistance ratio in the present devices tend to be smaller than typical values reported in filamentary based memristors where switching process is sharper but stochastic. Our device is based on two dimensional redistribution of dopants and therefore conducting mechanism is bulk like in nature, rather than filamentary mechanism. This can contribute to improve the yield and stability of devices while acquiring moderate on/off ratio. We will present that tuning of multi-valued conductance states (or synaptic weight) is possible in enough resolution in later sections.

## 11.3 Crystal Microstructure and Electronic States by STEM Observations

In order to investigate the microscopic origins of the aforementioned resistive switching properties in the four-terminal memristive devices, electron micrographic observations were performed by using scanning transmission electron microscopy (STEM) [20]. Specimens for STEM observations were extracted using a focused ion beam (FIB) system from different parts of the device samples: as reduced sample (pristine device sample before the voltage application, labeled as AR001), dark blue colored regions in the (001) and (100) plane four-terminal devices after the voltage application (labeled as DB001 and PS100, respectively) in Fig. 11.7). Here we compare the microstructural change in (100) and (001) devices which show a notable difference in dopant redistribution and resistive change behaviors. Figure 11.7 shows cross-sectional high-angle annular dark field (HAADF)-STEM images taken from the electrically active zones (dark blue colored regions) in the (100) and (001) plane four-terminal devices. For the colored sample in (100) device (Fig. 11.7a), several high-contrast streaks extending along the [100] and [010] directions are observed in the cross-section. In HAADF-STEM observations regions containing heavy elements show higher intensity (imaged brighter) than those containing light elements (*Z* contrast method). Observed high-contrast (bright) streaks indicates the formation of new phases with different composition from the matrix in the colored region under the application of voltage. These phases are considered to be formed by the accumulation of oxygen vacancies along specific crystallographic directions in the $TiO_{2-x}$ crystal and form planer structures (PSs). The high resolution STEM image observed in the PS shows that there are stacking disordered regions which are indicative of strong structural perturbation (red arrows in Fig. 11.7c). For the dark-blue-colored region of the (001) device (DB001), on the other hand, the cross-sectional STEM image in Fig. 11.7b shows no significant microstructural change. The electron diffraction pattern and the high-resolution STEM image of the DB001 sample confirmed a single-crystal rutile structure, as shown in Fig. 11.7b and d.

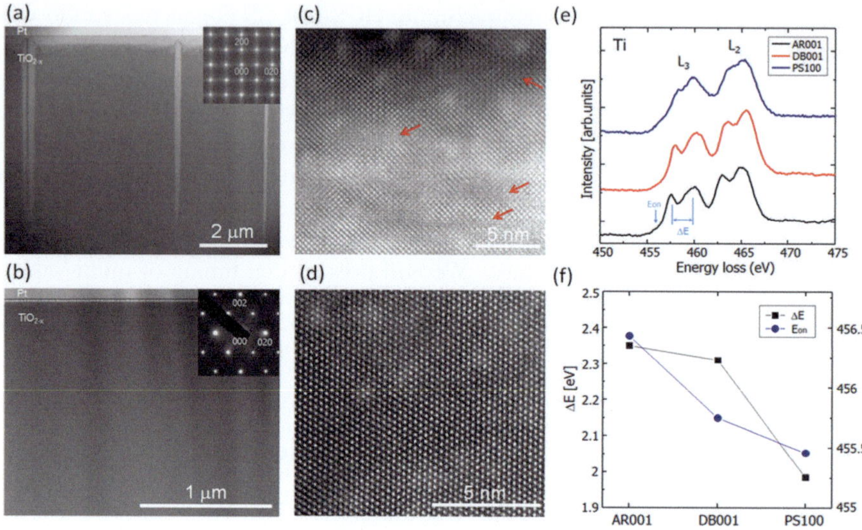


**Fig. 11.7** **a–d** Cross-sectional HAADF-STEM images and electron diffraction patterns (insets) of electrically active zones in $TiO_{2-x}$ four-terminal devices. **a**, **c** (100) device and **b**, **d** (001) device. **e** Ti $L_{2,3}$ edge EELS spectra of active zones in different devices of AR001, PS100, and DB001. **f** Onset energy $E_{on}$, and $e_g$–$t_{2g}$ peak splitting $\Delta E$, of Ti $L_3$ edge. Reproduced from Ref. [20] with the permission from Springer Nature under Creative Commons CC BY license

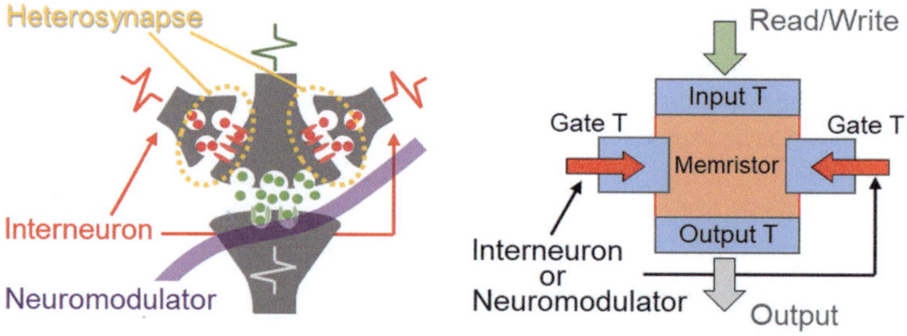


**Fig. 11.8** Comparison of a biological heterosynapse and neuromodulator (left) and a four-terminal memristive device (right). Reproduced from Ref. [14] with the permission from American Chemical Society under CC BY-NC-ND 4.0 license < https://pubs.acs.org/doi/full/10.1021/acsaelm.2c00161 > further permissions related to the material excerpted should be directed to the ACS

These observations prove the absence of clear crystal structure change in working (001) device. This is in clear contrast to the case of the (100) devices where strong modification of the crystal structure was observed.

To analyze the local electronic structure (valence state of Ti element) in the RS regions, electron energy loss spectroscopy (EELS) measurements using STEM were performed on the samples from each devices. We focused on the Ti-$L_{2,3}$ edge EELS to analyze the Ti valence states.

In order to quantify the variations in the EELS spectra, the energy splitting between the $e_g$ and $t_{2g}$ peaks in the Ti-$L_3$ edges ($\Delta E$) and the onset energy of Ti-$L_{2,3}$ edges ($E_{on}$) were extracted from the spectra. The $e_g$ and $t_{2g}$ peak splitting is influenced by the oxygen composition in $TiO_2$ crystals: when octahedrally coordinated oxygen ions around Ti are deficient in $TiO_{2-x}$, $\Delta E$ is reduced. Also $E_{on}$ is known to reflect the composition ratio of $TiO_{2-x}$ [25–27]. The Ti-$L_{2,3}$ EELS spectra acquired at the planar structure (PS) in the (100) device (PS100), dark blue region in (001) device (DB001), and as-reduced (001) $TiO_{2-x}$ (AR001) are shown in Fig. 11.7e. We observed that the spectrum of the PS in the (100) device is clearly changed from that of the as-reduced $TiO_{2-x}$ (AR001). This change in the EELS spectrum is evidenced by values of the onset energy $E_{on}$ and the energy splitting $\Delta E$: both the $E_{on}$ and the $\Delta E$ for the PS are smaller than those for AR001 (Fig. 11.7f), which indicates that the Ti ions in the PS is more strongly reduced than the as-reduced sample. These changes in the atomic and electronic structures are attributed to the accumulation of oxygen vacancies caused by an external voltage. In $TiO_2$ based memristors, shear plane defects or the Magnéli phases with oxygen deficiencies are known to act as highly conductive filaments [28, 29]. Also theoretical calculations based on the density functional theory have predicted that the ordered arrangement of oxygen vacancies along the [001] direction in rutile $TiO_2$ is energetically stable and can form a conductive filament [30]. Thus the PS found in TEM observations possibly constitutes current leakage paths formed in the T1-T3 bridge during application of the positive $V_C$. Some of leakage paths are not decomposed even under the negative $V_C$, so that the resistance of the T1–T3 bridge gradually decreases on the repeated $V_C$ cycles. This mechanism eventually tends to deteriorate the controllability of the RS properties. These irreversible microstructural change should be responsible for the formation of irreversible colored regions and the limited RS reversibility of (100) devices. On the other hand, the Ti-L edge EELS spectrum acquired for the dark blue region in (001) device (DB001) shows a certain change from the as-reduced $TiO_{2-x}$: the onset energy $E_{on}$ and the $L_3$ peak splitting $\Delta E$ have lower values than those for the as-reduced (001) device (AR001), but larger than those for PS in (100) device (PS100) (Fig. 11.7f). This indicates that the active region in (001) device is moderately reduced by application of the voltage, while its atomistic structure does not change significantly. This change in the electronic structure without a distinct change in the crystalline structure should be the origin of the reversible RS characteristics in the (001) device.

To summarize, we observed a remarkable dependence of resistive switching characteristics and atomic and electronic structure changes on the crystal orientations of the device plane in our four-terminal $TiO_{2-x}$ memristors. STEM analysis demonstrated that dark lines observed in optical images are closely related to planer defects in the $TiO_{2-x}$ matrix, which result from a local accumulation of excess oxygen vacancies. As we mentioned earlier, (001) device shows better reversibility of optical color changes and resistive switching properties (Figs. 11.4 and 11.5). This is explained by the electronic structure change without a crystal structural change occurring in (001) devices upon application of voltages. The present result is a paradigm of the essential role of crystalline anisotropy or crystal habits in the correlated electronic and crystalline switching phenomena inherent to the Rutile structure employed in the memristive device experiments.

## 11.4 Versatile Synaptic Functionality in Four-Terminals $TiO_2$ Memristors

### *11.4.1 Heterosynaptic Implementation by Four-Terminals Memristor*

As we presented in above sections, electrical properties of $TiO_{2-x}$ four-terminals memristor are modified by oxygen vacancy distribution, which is optically visualized by electrocoloring phenomena. A strongly (weakly) colored region contains higher (lower) concentration of oxygen vacancies and becomes more (less) conductive. The device can be configured into LRS or HRS by changing the morphology of the colored region; LRS (HRS) correspond to the state where colored region bridges (separates) T1 and T3. Thus, morphological control of the oxygen vacancy distribution enables a resistance change in the device. By changing two dimensional distribution of oxygen vacancies between multiple electrodes, versatile and tunable resistive switching properties can be attained. This can bring a realization of complex synaptic functions performed by heterosynapse in biological brains (Fig. 11.8). We will introduce various hetero-synaptic functions implemented by our four-terminal memristive devices in the following sections.

### *11.4.2 Multi-valued RS Property by Changing Sweep Rate*

First we tried to implement multi valued resistance states in our four-terminal device by applying voltages with varied sweep rate in the *I*-*V* hysteric measurements [13]. RS characteristics were measured by sweeping the voltage $V_1$ applied to T1, with T3 grounded, and by applying $V_{2,4}$ as a gate voltage (Fig. 11.9). To demonstrate the multilevel resistance modulation, consecutive positive and negative voltage sweeps of $V_1$ were conducted with stepwise decreased sweep rates (from 92.7 to 19.6 mV/s). Observed *I*-*V* curves clearly exhibit hysteresis, indicating RS from LRS to HRS for positive sweeps and HRS to LRS for negative sweeps. The current value gradually decreased through repeated $V_1$ sweeping where the sweep rate was decrementally changed. Resistance is defined by a slope of *I*-*V* curve and it gradually increased (decreased) and show multilevel values on the consecutive voltage $V_1$ sweeping processes.

Figure 11.9d and e shows the dependence of the resistance on the voltage sweep count, where sweeping rate was changed depending on the increment of resistance. By combining repetitive sweeps and a gradual reduction in the sweep rate, we

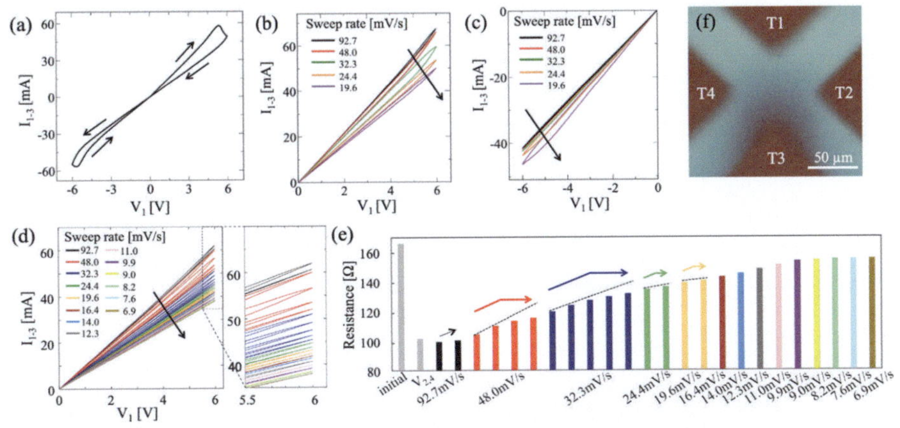


**Fig. 11.9** Electrical Characterization of the device. **a** $I_{1-3}$-$V_1$ characteristics of the device during a voltage sweep (at sweep rate of 19.6 mV/s). **b, c** $I_{1-3}$-$V_1$ characteristics during consecutive positive and negative voltage sweeps, respectively. The $V_1$ voltage sweep rate was 92.7–19.6 mV/s. **d, e** $I_{1-3}$-$V_1$ characteristics and variation of resistance during consecutive positive voltage sweeps (sweep rate changed between 92.7 and 6.9 mV/s). **f** Optical micrograph of the device taken in the last stage of the measurement. Reproduced from Ref. [13] with the permission from Springer Nature under Creative Commons CC BY license

obtained very fine multilevel resistance states. These RS properties are attributable to the drift and redistribution of oxygen vacancies in the device. An optical image observed after the final sweep is shown in Fig. 11.9f. The dark region around T3 and the colorless region around T1 indicate that the oxygen vacancies were repelled from T1 and accumulated around T3 as a result of the consecutive voltage sweeps. These results show that the observed gradual resistance change is attributed to oxygen vacancy distribution and it is proved that multi-resistance states are attainable with the present device.

### *11.4.3 Gate Tunable Heterosynaptic Plasticity by Pulse Drive Mode*

Next we show the gate modulation of the RS characteristics in the four-terminals device for potential use as an artificial synapse with tunable plasticity [13]. Here we apply voltage pulses $V_1$ to T1 (with T3 grounded) to modulate the resistance (or conductance) between T1 and T3. This operation emulates the synaptic plasticity between pre- and post-synapse. The potentiation and depression characteristics of the memristive device are determined using voltage protocols consisting of write, gating, and non-disturbing read pulses as shown in Fig. 11.10. First, $V_{2,4}$ is applied to accumulate oxygen vacancies between T1 and T3. Then, as a depression process, consecutive positive write pulses $V_{write}$ are applied to T1 until the device resistance

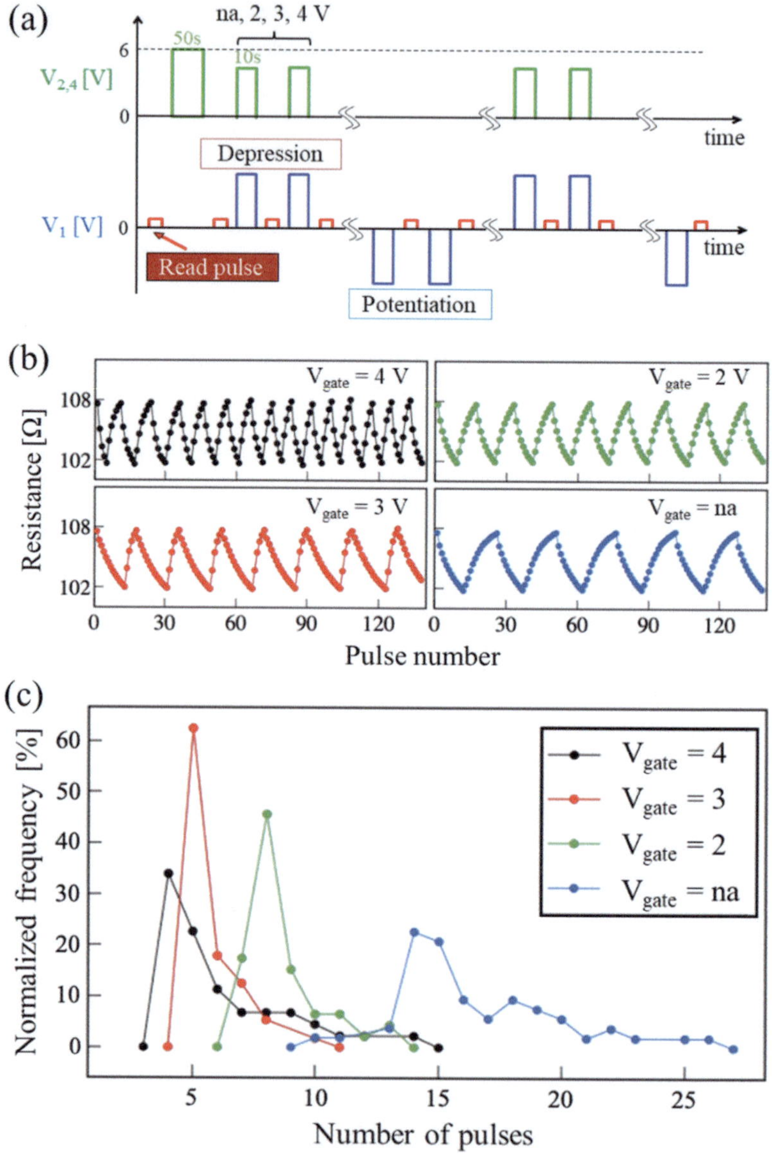


**Fig. 11.10** Gate modulation of the RS characteristics of the four-terminal device. **a** Voltage application protocol employing gate ($V_{2,4}$), write and read ($V_1$) pulses to emulate synaptic potentiation and depression. **b** Potentiation and depression of the device with different gate voltages **c** Gate voltage dependence of the number of pulses required to reach the target resistance (HRS). Reproduced from Ref. [13] with the permission from Springer Nature under Creative Commons CC BY license

reaches a target value of HRS. Subsequently, consecutive negative write pulses are applied to T1 until the resistance reaches a target value of LRS for the potentiation process. In addition, a gate voltage pulse of amplitude $V_{gate}$, which mimic the signal transmission from interneurons or a neuromodulator as illustrated in Fig. 11.8, is applied to T2 and T4 simultaneously with or prior to each write pulse during depression or potentiation processes.

An example of voltage protocol for the gating during depression is shown in Fig. 11.10a. After the initial application of $V_{2,4}$ to concentrate the oxygen vacancy, we applied a series of positive and negative write pulses to T1 until the resistance reached the target value of HRS (LRS) for the depression (potentiation) process. During the depression process, positive gating pulses of amplitude $V_{gate}$ were applied coincident with a positive $V_{write}$ pulses. Figure 11.10b shows results of resistance modulation with gate voltages $V_{gate}$ of 4, 3, 2 V and na (non-applied) in the protocol shown in Fig. 11.10a. We observe that the period of resistance modulation as a function of the pulse number varies systematically with the changed gate voltages. Figure 11.10c shows the gate voltage dependence of the number of pulses required to reach the target HRS during the depression process. It is notable that the required number of pulses changes depending on applied gate voltages: number of pulse tends to decrease with increasing gate voltage. Thus, the depression rate as a synaptic device is tunable by the synchronized application of the gate voltage. This gate effect can be understood by considering the oxygen vacancy distribution. The application of a positive gate voltage makes oxygen vacancies more concentrated in the region between T1 and T3 compared to the case with no gate voltage. Such an oxygen vacancy distribution brings a larger resistance change per write pulse since the rate of resistance change increases with dopant concentration. Thus the target HRS in the depression is reached with fewer write pulses under the condition of higher gate voltages.

The present results demonstrate the feasibility of gate tuning on synaptic properties in four-terminal memristive devices. A pair of gating electrodes (T2 and T4 here) can mimic the role of a neuromodulator and regulate the plasticity of hetero-synaptic systems.

### *11.4.4 Tunable Spike-Timing-Dependent Plasticity (STDP)*

In the previous section we have focused on rather 'static' RS behavior. Here we present an implementation of several dynamic functions such as spike-timing-dependent potentiation (STDP) [14]. STDP is a Hebbian synaptic learning rule that is often demonstrated in biological experiments [31]. We demonstrate that such temporal synaptic functions can be mimicked by tuning voltage application protocols, i.e., switching polarities and (de)synchronization of input voltage pulses, for the electrodes in the four-terminal device.

In the STDP, the synaptic weight or strength of the synaptic connection is modulated with the relative timing of the pre- and post-synaptic spikes. That is, the synaptic weight change $\Delta w$ varies depending on time difference between two spikes $\Delta t$. In canonical asymmetric Hebbian STDP, the connection of synapses is strengthened if the pre-spike precedes the post-spike ($\Delta t > 0$), while the connection is weakened when the post-spike precedes the pre-spike ($\Delta t < 0$). Other kinds of Hebbian STDP such as asymmetric anti-Hebbian and symmetric Hebbian rules have also been confirmed in biological systems [32].

Though implementation of STDP in artificial synapses has been demonstrated in several memristive devices with conventional two-terminals configuration [33, 34], STDP synaptic weight modulation in a hetero-synapse configuration has been rarely reported. We attempted to impart STDP function to four-terminal memristors and demonstrated gate modulation of STDP characteristics. Figure 11.11a and b show the designed spike pulse schemes for pre- and post-spikes and a gate spike employed in electronic STDP experiments. As the pre- and post-spikes, voltage waveforms composed of positive and negative pulses (amplitude of $\pm$ 7 V and width of 25 ms) were used and applied to the device with a time difference $\Delta t$. For the gate modulated operation, the gate spike pulse $V_{\text{gate}}$ that has the same pulse width and timing as the pre-spike but a different amplitude was applied (the upper part of Fig. 11.11b). The total effective voltage pulses of $V_1$ and $V_{\text{gate}}$ that reflect above spike pulse components and $\Delta t$ are shown in the lower parts of Fig. 11.11a and b. The $V_1$ pulse was applied to T1 to mimic pre- and post-spike excitations and $V_{\text{gate}}$ was applied to T2 and T4 to simulate the neuro-modulator effect. Figure 11.11c shows the voltage application sequence, which consists of initializing processes to adjust the conductance, and consecutive voltage spikes with/without simultaneous $V_{\text{gate}}$ application with systematically changing $\Delta t$. The synaptic weight change $\Delta w$ was defined by following formula.

$$\Delta w = \frac{G_{\text{after}} - G_{\text{before}}}{G_{\text{before}}} \times 100\,[\%]. \tag{11.1}$$

where and $G_{\text{before}}$ and $G_{\text{after}}$ represents conductance between T1 and T3 before and after the application of $V_1$ (and $V_{\text{gate}}$) spike pulse.

The observed variations of synaptic weight $\Delta w$ as a function of $\Delta t$ is shown in Fig. 11.11d–k. Figure 11.11d shows the result of $\Delta w$–$\Delta t$ relation for the case with no gate spike voltages. We observe that synaptic weights show potentiation (positive $\Delta w$) for $\Delta t > 0$ and depression (negative $\Delta w$) for $\Delta t < 0$. The result clearly reproduces the behavior of canonical asymmetric Hebbian STDP. The mechanism of this response is understood as follows; for positive $\Delta t$, effective negative voltage pulses are applied to T1 so that the accumulation of oxygen vacancies are enhanced near T1, leading to a conductance increase. On the other hand, for negative $\Delta t$, the effective positive voltage pulses applied to T1 could repel oxygen vacancies, resulting in a conductance decrease. By applying simultaneous gate spikes, the STDP characteristics can be modulated systematically with changing gate voltage amplitudes. With increasing $V_{\text{gate}}$ from 0 to 5 V, $|\Delta w|$ decreases irrespective of the polarity of $\Delta t$. At $V_{\text{gate}} = 5$ V, $\Delta w$ becomes almost zero for whole range of $\Delta t$. And further increasing $V_{\text{gate}}$ to 7 V (Fig. 11.11k) results in asymmetric anti-Hebbian

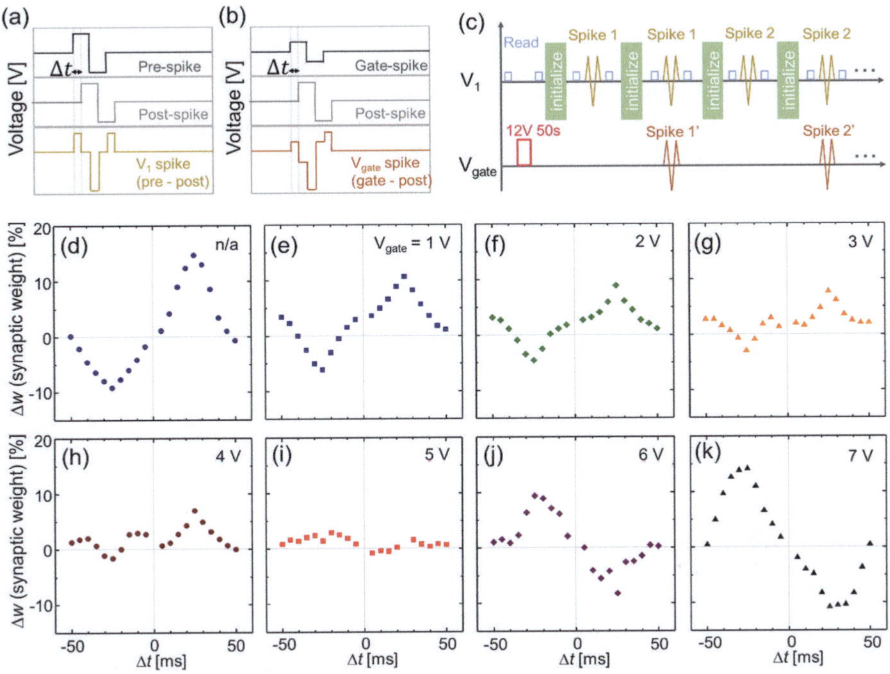


**Fig. 11.11** **a**, **b** Designed spike pulse schemes used for electronic STDP experiments. Pre-spike, post-spike, and total effective voltage applied to T1 are shown in the upper, middle, and lower parts in **a**, respectively. Design of gate voltage spikes applied to T2 and T4 is shown in the lower part in (**b**). **c** Voltage sequence of $V_1$ and $V_{gate}$ pulses. **d–k** The STDP responses ($\Delta w$—$\Delta t$ relation) under various gate voltages $V_{gate}$. Reproduced from Ref. [14] with the permission from American Chemical Society under CC BY-NC-ND 4.0 license < https://pubs.acs.org/doi/full/10.1021/acsaelm.2c00161 > further permissions related to the material excerpted should be directed to the ACS

STDP in which the synaptic weight change occurs in an inverse relationship to spike timing compared to canonical Hebbian STDP: precedence of post-spike to pre-spike ($\Delta t < 0$) results in potentiation and leading of pre-spike ($\Delta t > 0$) brings depression. Since the $V_{gate}$ spike waveform is determined by the amplitude of $V_{gate}$ and $\Delta t$ as shown in Fig. 11.11b, increasing $V_{gate}$ causes an increase in the negative (positive) components of $V_{gate}$ pulse for a positive (negative) $\Delta t$. Such an effective negative (positive) voltage pulse should enhance dispersion (accumulation) of oxygen vacancies between T1 and T3 and results in the decrease (increase) of the device conductance. The gate spike systematically tunes the amount of synaptic weight change in STDP and also control symmetry of the STDP rule from canonical Hebbian and anti-Hebbian, including the form of transient states.

It is notable that the gating operation in the four-terminal memristor can mimic the neuromodulatory control of STDP behaviors. In biological nervous systems, implemantaion of anti-Hebbian STDP as well as canonical Hebbian may prove advantageous for effective association and reinforcement learning [35], which could also enhance the capability of neural networks composed of artificial synapses. In addition, the transient states observed at $V_{gate} = 5$ V upon the canonical Hebbian to anti-Hebbian convresion, can be regarded as 'silent synapses' where neurotransmission does not produce a response in the receiving cell. The maturation of silent synapses via unsilencing and stabilization is known to be one fundamental step in the development of neural networks in the brain [36]. In the context of deep neural network based machine learning, the formation of silent synapses corresponds to 'dropout', that is a technique used for regularization of neural network models: 'dropping' some nodes from the network during training is effective to prevent the model from overfitting [37]. The present device enables modulation of STDP behaviors by gate operation and this allows dynamic control of sensitivity of artificial synapses including silent synapse state. Such operation based on a single passive element is not achieved in conventional synaptic hardware and the present device may provide new opportunities for imparting timing dependent hetero-synaptic functions.

### *11.4.5 Pavlovian Conditioning*

In this section we present an electronic implementation of Pavlovian conditioning [38] as a typical example of associative learning. The original Pavlov's experiment is summarized as follows: as a natural response a dog salivates when it senses the presence of food. Conditioning is performed by ringing a bell on every feeding time of the dog. After the conditioning, the dog is observed to salivate on hearing bell sounds even without any food. On the experience of feeding and bell ringing with temporal contiguity, the dog learns to associate the bell sound (neutral stimulus) with food (unconditioned stimulus). The behavior of Pavlov's dog is summarized in the truth table shown in Table 11.1.

To implement Pavlovian conditioning electronically, we designed electric measurements on the four-terminal memristive device as follows. We assigned applied voltages to T1 and T4 (named $V_1$ and $V_4$, respectively) as the bell sound and food, respectively (Fig. 11.12a). The dog's response is measured as the resultant conductance between T1 and T3, that is $G_{1\text{-}3}$. Firstly we made initialization process to adjust $G_{1\text{–}3}$ to a certain value of low conductance state (LCS) and the conductance $G_{4\text{–}3}$ between T4 and T3 to a high conductance state (HCS) (Fig. 11.12). High $G_{4\text{-}3}$ and low $G_{1\text{-}3}$ represents the dog's natural (unconditioned) reaction to food and bell

**Table 11.1** Truth table of Pavlovian conditioning

| Bell sound | Food | Response (before conditioning) | Response (after conditioning) |
|---|---|---|---|
| 0 | 0 | 0 | 0 |
| 0 | 1 | 1 | 1 |
| 1 | 1 | 1 | 1 |
| 1 | 0 | 0 | 1 |

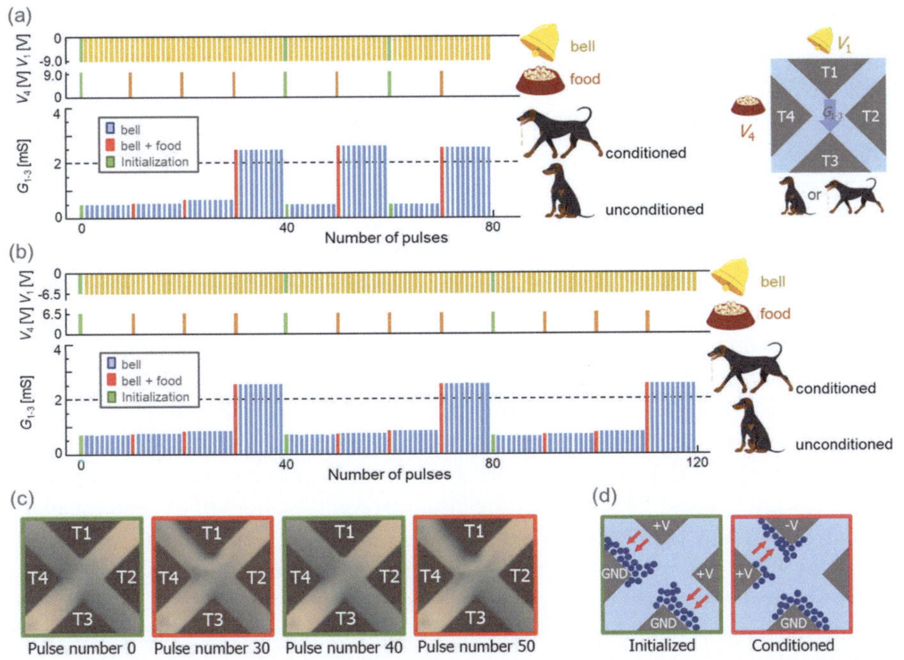


**Fig. 11.12** Voltage protocols for emulated Pavlovian conditioning and resultant conductance $G_{1\text{–}3}$ with voltage amplitudes of **a** $V_1 = –9$ V, $V_4 = 9$ V and **b** $V_1 = –6.5$ V, $V_4 = 6.5$ V. Inset of **a** shows correspondence between the four-terminal device and Pavlovian conditioning; $V_1$: bell sound, $V_4$: food, and $G_{1\text{–}3}$: dog's response. **c** Optical micrographs after the application of pulse numbers 0, 30, 40, and 50 in (**a**). **d** Schematics of oxygen vacancy distribution for initialized and conditioned states. Reproduced from Ref. [14] with the permission from American Chemical Society under CC BY-NC-ND 4.0 license. <https://pubs.acs.org/doi/full/10.1021/acsaelm.2c00161> further permissions related to the material excerpted should be directed to the ACS

sounds, respectively. Then, voltage pulses of $V_1$ were applied repeatedly to simulate the bell sound. At every tenth application of $V_1$, a voltage pulse $V_4$ as a food was applied simultaneously, which emulate the conditioning process.

Figure 11.12a shows a voltage application sequence of $V_1$ and $V_4$ and the resultant conductance values of $G_{1\text{-}3}$. At first application of $V_1$ pulses caused no significant change on $G_{1\text{-}3}$ before the $V_4$ application. After the simultaneous application of $V_1$ and $V_4$, $G_{1\text{-}3}$ was observed to increase slightly. At the third simultaneous application of $V_1$ and $V_4$ at a pulse number of 30, $G_{1\text{–}3}$ increased sharply, and afterwards the high $G_{1\text{-}3}$ was maintained only with the application of $V_1$. This result reproduce the phenomenon of conditioning: the dog associates the bell sound with food and learns to react to the bell sound alone. We also confirmed the process of forced forgetting; $G_{1\text{-}3}$ could be reset to its initial value by the initialization processes (at pulse number of 40 and 60 in Fig. 11.12a) and LCS was retained against the application of $V_1$. The conditioning state was recalled by the subsequent simultaneous application of $V_1$

and $V_4$ at pulse numbers of 50 and 70. We note that only a single simultaneous application of $V_1$ and $V_4$ is enough for the second and third conditioning, whereas three trials are required for the first conditioning process. This conditioning efficiency can be modified by changing the applied voltage. Another set of conditioning operation with different amplitude of $V_1$ and $V_4$ is shown in Fig. 11.12b. In this case, conditioning was achieved after the third simultaneous application of $V_1$ and $V_4$ for every conditioning process after initialization. Thus, the required number of conditioning for learning (conditioning efficiency) is controllable by changing the amplitude of $V_1$ and $V_4$. These results demonstrate that the tunable Pavlovian conditioning scheme can be implemented in four-terminal memristive devices.

To verify the microscopic origin of the Pavlovian conditioning behavior in the four-terminal device, the oxygen vacancy distribution during the conditioning process was observed through electro-coloring phenomena (Fig. 11.12c). At the initialized state (pulse number: 0), dark blue regions were observed mainly around the tips of T3 and T4. During initialization, the electric field drives the oxygen vacancies around T1 and T2 towards T3 and T4. The resulting dilution (condensation) of oxygen vacancies around T1 (T3 and T4) likely leads to low $G_{1\text{-}3}$ (high $G_{4\text{-}3}$) (as schematically shown in Fig. 11.12d). After the first conditioning process (pulse number: 30), the optical image showed a strongly colored region was formed at the tip of T1. During the conditioning operation, simultaneous application of negative $V_1$ and positive $V_4$ cause the effectively large electric field directed from T4 to T1, which makes oxygen vacancies to drift and accumulate around T1, leading to a high $G_{1–3}$ (a schematic shown in the right panel of Fig. 11.12d). A similar optical contrast change was observed after the second initialization (pulse number: 40) and conditioning (pulse number: 50) processes, indicating the good repeatability of two dimensional oxygen vacancy redistribution. In this way, the conductance state of $G_{1\text{-}3}$ can be switched between LCS and HCS by redistribution of oxygen vacancies driven by associated application of voltage stimuli of $V_1$ (bell) and $V_4$ (Food).

The present results directly indicate that two dimensional distribution of oxygen vacancies in the device's active region plays a crucial role in tuning the electrical conductance between multiple electrodes. A rich variety of synaptic connectionism realized in the four-terminal memristor allows the implementation of complex hetero-synaptic function such as associative learning based on single memristive passive device with simple architecture.

## 11.5 Conclusions

In this chapter we presented recent developments in four-terminals memristive devices based on rutile $TiO_2$ single crystals. Reflecting the crystalline anisotropy or crystal habits, devices based on (100) and (001) planes shows a significant difference in dopant (oxygen vacancy) redistribution behaviors and memristive characteristics. Particularly good reversibility of dopant distribution and resistive switching has been achieved in (001) devices. Implementation of versatile synaptic functionalities are proved to be feasible based on simple multiterminal passive elements. In synaptic operations described in this chapter, not only electrical interaction between two orthogonal pairs of electrodes, interaction between neighboring electrodes that brings oblique modification of dopant distribution is exploited as in the emulation of Pavlovian conditioning. This controllable variety of RS property relies on in-plane isotropy of the rutile $TiO_2$ (001) surface employed. By changing shape, number and configuration of electrodes we can expect further extension of RS properties. Furthermore, adapting amorphous materials which are free from crystalline anisotropy as RS layer should be beneficial to obtain uniform and reversible RS properties. Exploration of different device configurations and networking are subjects of interest in future works.

## References


1. L.O. Chua, IEEE T. Circuits Syst. **18**, 507 (1971)
2. D.B. Strukov, G.S. Snider, D.R. Stewart, R.S. Williams, Nature **453**, 80 (2008)
3. M. Prezioso, F.M. Bayat, B.D. Hoskins, G.C. Adam, K.K. Likharev, D.B. Strukov. Nature **521**, 61 (2015)
4. T. Ohno, T. Hasegawa, K. Terabe, J.K. Gimzewski, M. Aono, Nat. Mater. **10**, 591 (2011)
5. A. Sawa, Mater. Today **11**, 28 (2008)
6. Y. Aoki, C. Wiemann, V. Feyer, H.-S. Kim, C. M. Schneider, H.I. Yoo, M. Martin. Nat. Commun. **5**, 3473 (2014)
7. D.B. Strukov, J.L. Borghetti, R.S. Williams, Small **5**, 1058 (2009)
8. M. Ziegler, R. Soni, T. Patelczyk, M. Ignatov, T. Bartsch, P. Meuffels, H. Kohlstedt, Adv. Funct. Mater. **22**, 2744 (2012)
9. R.D. Hawkins, T.W. Abrams, T.J. Carew, E.R. Kandel, Science **219**, 400–405 (1983)
10. V. Castellucci, E.R. Kandel, Science **194**, 1176–1178 (1976)
11. C.H. Bailey, M. Giustetto, Y.-Y. Huang, R.D. Hawkins, E.R. Kandel, Nat. Rev. Neurosci. **1**, 11–20 (2000)
12. X. Yang, Y. Fang, Z. Yu, Z. Wang, T. Zhang, M. Yin, M. Lin, Y. Yang, Y. Cai, R. Huang, Nanoscale **8**, 18897–18904 (2016)
13. Z. Nagata, T. Shimizu, T. Isaka, T. Tohei, N. Ikarashi, Sakai. Sci. Rep. **9**, 10013 (2019)
14. R. Miyake, Z. Nagata, K. Adachi, Y. Hayashi, T. Tohei, A. Sakai, ACS Appl. Electron. Mater. **4**, 2326 (2022)
15. Y. Yang, B. Chen, W.D. Lu, Adv. Mater. **27**, 7720–7727 (2015)
16. V.K. Sangwan, H.-S. Lee, H. Bergeron, I. Balla, M.E. Beck, K.-S. Chen, M.C. Hersam, Nature **554**, 500–504 (2018)
17. Y. Yang, M. Yin, Z. Yu, Z. Wang, T. Zhang, Y. Cai, W.D. Lu, R. Huang, Adv. Electron. Mater. **3**, 1700032 (2017)
18. X. Zhu, D. Li, X. Liang, W.D. Lu, Nat. Mater. **18**, 141–148 (2019)
19. K. Yamaguchi, S. Takeuchi, T. Tohei, N. Ikarashi, A. Sakai, Jpn. J. Appl. Phys. **57**, 06KB02 (2018)
20. S. Takeuchi, T. Shimizu, T. Isaka, T. Tohei, N. Ikarashi, A. Sakai, Sci. Rep. **9**, 2601 (2019)
21. R. Waser, R. Dittmann, G. Staikov, K. Szot, Adv. Mater. **21**, 2632–2663 (2009)
22. K. Szot, M. Rogala, W. Speier, Z. Klusek, A. Besmehn, R. Waser, Nanotechnology **22**, 254001 (2011)
23. R.F. Bartholomew, D.R. Frankl, Phys. Rev. **187**, 828 (1969)

24. H. Miyaoka, G. Mizutani, H. Sano, M. Omote, K. Nakatsuji, F. Komori, Solid State Commun. **123**, 399–404 (2002)
25. A. Moballegh, E.C. Dickey, Acta Mater. **86**, 352–360 (2015)
26. R.D. Leapman, L.A. Grunes, Phys. Rev. Lett. **45**, 398 (1980)
27. E. Stoyanov, F. Langenhorst, G. S. -Neumann. Am. Mineral. **92**, 577 (2007)
28. D. -H, Kwon, K.M. Kim, J.H. Jang, J.M. Jeon, M.H. Lee, G.H. Kim, X.–S. Li, G.–S. Park, B. Lee, S. Han, M. Kim, C.S. Hwang, Nat. Nanotech. **5**, 148 (2010)
29. J.P. Strachan, M.D. Pickett, J.J. Yang, S. Aloni, A.L.D. Kilcoyne, G.M-Ribeiro, R.S. Williams, Adv. Mater. **22**, 3573 (2010)
30. S.-G. Park, B. Magyari-Köpe, Y. Nishi, Phys. Rev. B **82**, 115109 (2010)
31. H. Markram, J. Lübke, M. Frotscher, B. Sakmann, Science **275**, 213–215 (1997)
32. D. Shulz, D. Feldman, Spike timing-dependent plasticity.in *Comprehensive developmental neuroscience: neural circuit development and function in the healthy and diseased brain,* ed. by J.L.R. Rubenstein, P. Rakic (Elsevier, San Diego, CA, 2013) Ch. 9
33. D. Kuzum, R.G. Jeyasingh, B. Lee, H.-S.P. Wong, Nano Lett. **12**, 2179–2186 (2012)
34. Y. Li, Y. Zhong, J. Zhang, L. Xu, Q. Wang, H. Sun, H. Tong, X. Cheng, X. Miao, Sci. Rep. **4**, 4906 (2014)
35. Ł Kuśmierz, T. Isomura, T. Toyoizumi, Curr. Opin. Neurobiol. **46**, 170–177 (2017)
36. G.A. Kerchner, R.A. Nicoll, Nat. Rev. Neurosci. **9**, 813–825 (2008)
37. N. Srivastava, G. Hinton, A. Krizhevsky, I. Sutskever, R.J. Salakhutdinov, Machine Learning Research **15**, 1929–1958 (2014)
38. I. P. Pavlov, Experimental psychology and psychopathology in animals1, in *Psychopathology and psychiatry* (Taylor & Francis Group, New York, 1994) Ch. 2

# Index

**A**
Adjuvant, 46, 66
α-form, 96
Aluminium(III), 45
Ambipolar, 186
Artificial synapse, 199, 213, 216, 218

**B**
Bandgap, 111, 126, 127, 135, 163, 164, 172, 186

**C**
Calcium carbonate, 64, 67, 79, 81, 85
Chaotropic, 69
Colloform, 3
Coronene, 17, 19, 20
Crystal growth, 3–5, 8–10, 14, 17, 18, 24, 29, 38, 39, 51, 69, 89, 94, 103, 109–113, 134
C-terminus, 68

**D**
Diagenesis, 8
Dimers, 49, 50, 54, 60, 61
Dissolution, 51, 52, 60, 61, 79–81, 84, 85, 89, 96, 102
Dopant, 112, 173, 199, 200, 207, 209, 215, 220, 221
Down-conversion, 171, 172
Driving force, 3, 4, 13, 50, 51, 53, 57, 58, 63–65, 67

**E**
EELS spectra, 210, 211
Epitaxial, 109, 111–114, 125, 134, 201
Epitaxy, 10, 143, 144, 164, 187
Exciton, 163, 164, 175, 178
Excitons, 20

**F**
Floc, 60, 61
Fluorescent probe, 79–82, 84, 87
Flux ratio, 154, 156–159, 161, 165
Framboid, 11
Framboidal, 3, 7, 10, 11, 13, 14
FWHM, 175, 178

**G**
GaAs, 110–113, 115–122, 124–126, 129–135, 143–145, 148, 151, 153, 154, 156, 158, 159, 161, 163–165
Gallium arsenide, 143
γ-form, 96, 97
GaNAs shell, 116, 117, 130–132, 135
Gate spike, 216, 217
Gate spike pulse, 216

**H**
HAADF image, 122, 124, 125, 127, 128, 131
Hebbian, 215–218
Heterosynaptic plasticity, 213
Hexa-aqua, 47, 63
High-angle annular dark field, 121, 123, 125, 128, 132, 209


F. Ishikawa et al. (eds.), *Engineering Crystal Habit*, Topics in Applied Physics 152,
https://doi.org/10.1007/978-981-96-0266-7

Homoepitaxy, 187
Hydrolysis, 45, 47, 49, 50, 54, 57, 58, 63, 70
(Hydr)oxides, 17

**I**
Impingement, 151, 153, 155, 156
Ion implantation, 31, 33, 37
Iron(III), 45
Iron-oleate, 67, 68

**K**
Keggin-$Al_{13}$, 54, 55, 65

**L**
Ligand, 48, 50, 52, 58, 61–66, 68

**M**
Magnetotactic, 66, 69
Megacryst, 21, 23–25
Memristive, 199–202, 207, 209, 210, 212, 213, 215, 216, 218, 220
Memristor, 199–201, 204, 207, 211, 212, 216, 217, 220
Metalorganic chemical vapor deposition, 172
Mist CVD, 185–188, 191, 193–195
Morphology, 3, 4, 7, 10, 11, 13, 17, 18, 53, 61, 79, 85, 87, 88, 174

**N**
Nanoparticles, 11, 29, 37–39, 46, 49, 51, 61, 67, 69
Nanophotonic, 134, 172
Nanowire, 109, 115, 117, 118, 120, 122, 126, 128, 133, 143, 165, 171, 172
Neuromorphic, 199, 200
Nucleation, 3, 4, 10, 11, 14, 24, 45–48, 50–54, 57, 58, 60–62, 64–70, 94–99, 101–103, 143, 144, 146, 148, 149, 153, 159

**O**
Olated, 50, 56
Oligomeric, 47, 50, 55, 60
Orthoclase, 22
Oxo-bridged, 61
Oxolated, 49, 51, 53, 57, 58
Oxygen vacancy, 199, 201–205, 207, 208, 212, 213, 215, 219, 220
(Oxy)(hydr)oxides, 45, 47, 53, 54, 60, 62, 63, 66, 69

**P**
pH, 47–53, 55–58, 60, 63, 64, 66, 67, 79–82, 84, 85, 87, 88
Plagioclase, 21, 22, 24, 25, 31, 33
PL intensity, 175, 176, 178
Plutonic, 21
Polytypism, 127, 128, 130–132, 135
Post-synaptic, 215
Prenucleation, 50, 52, 66, 69, 70
Protonation, 58
Pulse number, 96, 215, 219, 220
Pyrite, 3–11, 13, 14

**Q**
Quantum cutting, 171

**R**
Rare-earth-doped, 173, 174
Regolith, 30–39
Resistive change memory, 199
R-$GeO_2$, 185–195
R-$Ge_xSn_{1-x}O_2$ alloy, 185, 186, 191, 192
RHEED pattern, 145–148, 150, 158–160, 165
Rocking curve, 190

**S**
Self-assisted GaAs, 143–145, 148, 151, 153, 154, 156
Self-assisted GaAs Nanowire, 165
Solar wind implantation, 31, 33, 35, 37, 39
Solidus, 191
Space weathering, 29–31, 34, 37–39
Spike pulse, 216, 217
Sputtering, 35, 37, 172, 173, 179
Sub-euhedral, 10

**T**
Texture, 3, 4, 7, 9, 14, 38, 109, 118, 134
Theranostic, 46
Titanium oxides, 31, 199
Transmission electron microscopy, 30, 33, 51, 68, 116, 117, 123, 125, 128, 132, 133, 161, 190, 209

**V**
Visualization, 79–81, 84, 87, 88

**W**
Water structure, 52, 53
Whisker, 8, 29, 36, 38–40, 109, 112–114
Wurtzite, 25, 113, 128, 129, 132, 133, 143

**Z**
Zinc-blende, 143
ZnO, 171–181

MIX
Papier aus verantwortungsvollen Quellen
Paper from responsible sources
FSC® C105338

If you have any concerns about our products,
you can contact us on
**ProductSafety@springernature.com**

In case Publisher is established outside the EU,
the EU authorized representative is:
**Springer Nature Customer Service Center GmbH**
**Europaplatz 3, 69115 Heidelberg, Germany**

Printed by Libri Plureos GmbH
in Hamburg, Germany